Lecture Notes in Computer Science 5170

Commenced Publication in 1973
Founding and Former Series Editors:
Gerhard Goos, Juris Hartmanis, and Jan van Leeuwen

Otmane Ait Mohamed (Ed.)

Theorem Proving in Higher Order Logics

21st International Conference, TPHOLs 2008
Montreal, Canada, August 18-21, 2008
Proceedings

 Springer

Volume Editor

Otmane Ait Mohamed
Concordia University
Department of Electrical and Computer Engineering
1455 de Maisonneuve Blvd. W. Montreal
Quebec Canada H3G 1M8
E-mail: ait@encs.concordia.ca

Library of Congress Control Number: 2008931582

CR Subject Classification (1998): F.4.1, I.2.3, F.3.1, D.2.4, B.6.3

LNCS Sublibrary: SL 1 – Theoretical Computer Science and General Issues

ISSN   0302-9743
ISBN-10   3-540-71065-5 Springer Berlin Heidelberg New York
ISBN-13   978-3-540-71065-3 Springer Berlin Heidelberg New York

Springer is a part of Springer Science+Business Media

springer.com

© Springer-Verlag Berlin Heidelberg 2008
Printed in Germany

Typesetting: Camera-ready by author, data conversion by Scientific Publishing Services, Chennai, India
Printed on acid-free paper   SPIN: 12446179   06/3180   5 4 3 2 1 0

Preface

This volume constitutes the proceedings of the 21st International Conference on Theorem Proving in Higher Order Logics (TPHOLs 2008), which was held during August 18–21, 2008 in Montreal, Canada. TPHOLs covers all aspects of theorem proving in higher order logics as well as related topics in theorem proving and verification.

There were 40 papers submitted to TPHOLs 2008 in the full research category, each of which was refereed by at least four reviewers selected by the Program Committee. Of these submissions, 17 research papers and 1 proof pearl were accepted for presentation at the conference and publication in this volume. In keeping with longstanding tradition, TPHOLs 2008 also offered a venue for the presentation of emerging trends, where researchers invited discussion by means of a brief introductory talk and then discussed their work at a poster session. A supplementary proceedings volume was published as a 2008 technical report of Concordia University.

The organizers are grateful to Michael Gordon and Steven Miller for agreeing to give invited talks at TPHOLs 2008. As part of the celebration of the 20 years of TPHOLs, TPHOLs 2008 invited tool developers and expert users to give special tool presentations of the most representative theorem provers in higher order logics. The following speakers kindly accepted our invitation and we are grateful to them: Yves Bertot (Coq), Matt Kaufmann (ACL2), Sam Owre (PVS), Konrad Slind (HOL), and Makarius Wenzel (Isabelle).

The TPHOLs conference traditionally changes continents each year to maximize the chances that researchers around the world can attend. TPHOLs started in 1998 in the University of Cambridge as an informal users' meeting for the HOL system. Since 1993, the proceedings of TPHOLs have been published in the Springer *Lecture Notes in Computer Science* series:

1993 (Canada)	Vol. 780	2001 (UK)	Vol. 2152
1994 (Malta)	Vol. 859	2002 (USA)	Vol. 2410
1995 (USA)	Vol. 971	2003 (Italy)	Vol. 2758
1996 (Finland)	Vol. 1125	2004 (USA)	Vol. 3223
1197 (USA)	Vol. 1275	2005 (UK)	Vol. 3603
1998 (Australia)	Vol. 1479	2006 (USA)	Vol. 4130
1999 (France)	Vol. 1690	2007 (Germany)	Vol. 4732
2000 (USA)	Vol. 1869	2008 (Canada)	Vol. 5170

We would like to thank our local organizers at Concordia University for their help in many aspects of planning and running TPHOLs.

Finally, we thank our sponsors: Intel Corporation, Concordia University, the National Institute of Aerospace, and the Regroupement Strategique en Microsystèmes du Québec, for their support.

May 2008   Otmane Ait Mohamed
                                                                César Muñoz
                                                               Sofiène Tahar

Organization

Conference Chair

Sofiène Tahar (Concordia)

Program Chairs

Otmane Ait Mohamed (Concordia)   César Muñoz (NIA)

Program Committee

Mark Aagaard (Waterloo)
Hasan Amjad (Cambridge)
Yves Bertot (INRIA)
Jens Brandt (Kaiserslautern)
Thierry Coquand (Chalmers)
Jean-Christophe Filliâtre (CNRS)
Ganesh Gopalakrishnan (Utah)
Mike Gordon (Cambridge)
Hanne Gottliebsen (Queen Mary)
Jim Grundy (Intel)
Elsa Gunter (Urbana-Champaign)
John Harrison (Intel)
Jason Hickey (Caltech)

Peter Homeier (US DoD)
Joe Hurd (Galois)
Paul Jackson (Edinburgh)
Thomas Kropf (Tübingen and Bosch)
John Matthews (Galois)
Tobias Nipkow (München)
Sam Owre (SRI)
Christine Paulin-Mohring (Paris Sud)
Lawrence Paulson (Cambridge)
Klaus Schneider (Kaiserslautern)
Konrad Slind (Utah)
Matthew Wilding (Rockwell Collins)
Burkhart Wolff (ETH Zürich)

External Reviewers

Behzad Akbarpour
Brian Aydemir
Yves Bertot
Pierre Casteran
Lucas Dixon
Catherine Dubois
Bruno Dutertre
Amjad Gawanmeh
Georges Gonthier
Alexey Gotsman
David Greve
David Hardin

Rebekah Leslie
Guodong Li
Ulf Norell
Nicolas Oury
Grant Passmore
Raymond Richards
Tarek Sadani
Susmit Sarkar
Norbert Schirmer
Murali Talupur
Andrew Tolmach
Thomas Tuerk

Nicolas Julien   Christian Urban
Florent Kirchner   Tjark Weber
Vladimir Komendantsky   Makarius Wenzel
Alexander Krauss

Table of Contents

Invited Papers

Tutorials

Regular Papers

Proof Pearls

Twenty Years of Theorem Proving for HOLs
Past, Present and Future

Mike Gordon

The University of Cambridge Computer Laboratory
William Gates Building
15 JJ Thomson Avenue, Cambridge CB3 0FD, United Kingdom
Mike.Gordon@cl.cam.ac.uk,
http://wwww.cl.cam.ac.uk/~mjcg

1 Theorem Proving for HOLs?

There are two kinds of theorem provers for higher order logics: fully automatic (e.g. TPS and Leo) and user guided (e.g. HOL4, HOL Light, ProofPower, Isabelle/HOL, Coq, Nuprl and PVS). All the user guided systems, except PVS, are based on the LCF "fully expansive" approach invented by Robin Milner. PVS evolved from a different tradition that doesn't expand everything down to primitive inferences.

The emphasis here is on user guided proof assistants, but future developments in automatic higher order proof methods are likely to be incorporated into these as the automatic methods available today are mainly propositional or first order.

2 From 1988 to 2008

Twenty years ago Nuprl was already going strong and HOL88 had just been released. By the mid 1990s Coq, Isabelle/HOL, HOL Light, ProofPower and PVS had arrived and mechanised proof in higher order logic was an active area, already with many impressive applications. The following subsections are comments on some of the major developments up to the present.

2.1 Automation

Early LCF-style proof assistants forced users to construct proofs by tedious low-level steps such as explicit quantifier instantiation and Modus Ponens. It was amazing that so much managed to get done this way. PVS introduced a more automated proof development framework based on a few primitive powerful methods. Theorem proving with PVS was much faster and easier than with old style LCF systems, but soon LCF-style provers began to catch up using derived rules that automated low level reasoning. Ideas were imported from the automatic theorem proving community, such as resolution for first order reasoning and decision procedures for specific theories like linear arithmetic. These methods were programmed as "tactics" in ML. It had been thought that

O. Ait Mohamed, C. Muñoz, and S. Tahar (Eds.): TPHOLs 2008, LNCS 5170, pp. 1–5, 2008.

one might never make fully-expansive automatic proof tools efficient enough, but the art of tactic programming advanced very rapidly and remarkably good performance was achieved. One can never have enough automation, and adding more automatic proof methods will always be an active topic of research.

Although fully-expansive theorem proving made amazing advances, there are cases where it is hard to achieve adequate performance, so linkups to external solvers have been explored, e.g. for model checking, SAT and some decision procedures. Sometimes the solvers are regarded as oracles and their results trusted and sometimes they are used to find proofs or counterexamples, which are then replayed inside the proof assistant to retain the security of full expansiveness. The efficiency/trust trade-offs of linking to external solvers is still an active area.

2.2   User Interfaces

Beginners approaching the early proof assistants derived from LCF were usually not impressed by their command line user interfaces, especially if they were used to graphical programming environments. This was regarded as an embarrassment and several substantial efforts to create better user interfaces were undertaken in the 1990s (I can think of three such projects at Cambridge alone). The resulting GUIs never caught on: although they looked impressive in demos, the need to manage multiple windows and access tools via menus got in the way for experts. PVS introduced a middle ground between a raw command line and a full windows style GUI. The PVS interface, built in emacs, was neither clunkily primitive or got in the way of experts. This approach caught on and now the majority of proof assistants support an emacs-based front-end. Both Coq and Isabelle/HOL have interfaces based on Proof General, a generic front-end developed at Edinburgh and implemented in emacs.

The distinction between procedural and declarative proof styles emerged during the 1990s, partly through the discovery by the HOL community of the amazing Mizar proof checker. Attempts were made to duplicate the naturalness of Mizar's proof specification language within an LCF-style tactic framework. A lasting trace of this is the Isar interface to Isabelle/HOL, though fragments of the original "Mizar mode" experiments in HOL Light have propagated to other HOL systems. It seems that a declarative style is well suited to textbook mathematics (which is where Mizar excelled). It is less clear how suited declarative proof is to verification applications that need tightly-coupled combinations of algorithmic and user-guided deduction. Some pure mathematics developments (Four color and Jordan Curve theorems) have been done using the procedural front ends of Coq and HOL Light, whilst others (sub-proofs of the Kepler conjecture) have been done using the declarative Isar front end to Isabelle/HOL. Declarative proof has had a clear influence, but it hasn't pushed aside the earlier procedural proof-as-programming approaches.

2.3   Functional and Logic Programming

Higher order logic is based around the typed lambda-calculus and so its terms can be viewed as functional programs. Proof assistants, especially Coq, have developed a role as functional programming environments for creating 'certified' programs.

Functions in higher order logic are total, so functional programming inside proof assistants can only create terminating programs, and thus proof of termination is always necessary (though is often accomplished automatically). Partial functions can be considered as single-valued relations, and all the main proof assistants support inductive definitions of relations. In Isabelle/HOL these can then be efficiently executed using ideas from logic programming. Programs in logic can either be executed 'securely' by deduction or translated to external languages like ML or Prolog when higher execution performance is needed. Some proof assistants have imported functional software engineering ideas. For example, Isabelle/HOL and PVS have versions of QuickCheck for debugging by testing.

Computation can be a very important part of proof. This has been especially developed by Coq where there are impressive applications involving combinations of proof and calculation linked by reflection principles (e.g. primality testing using elliptic curves).

2.4   Theorem Prover as an Implementation Platform

Higher order logic can encode the semantics of many of the specification languages used in formal methods (e.g. Z, LCF, mu-calculus). Encodings of such languages have been constructed and a methodology has evolved (e.g. "deep" and "shallow" embedding). The LCF-style proof assistants provide an ML-based programming environment for implementing tools like verification condition generators. However, the assurance provided by semantic embedding comes at a price: the considerable effort needed to formalise and then mechanise the semantics.

Industrial verification tools, both for hardware and software, are usually implemented by directly coding up algorithms in a raw programming language. Tools built using semantic embedding implemented inside proof assistants largely remain as academic prototypes (an exception is ProofPower/Z, which provides support for the Z specification language via semantic embedding in HOL and has been successful in industrial applications).

2.5   Versions of Higher Order Logic

HOL4, HOL Light and ProofPower all use Church's simple type theory extended with the Hindley-Milner decidable polymorphic type discipline. Isabelle/HOL enhances this with Haskell-style type-classes to manage overloading in a systematic manner and locales for structuring theory developments. Certain kinds of specification and reasoning about functional programs are hard to express using Hindley-Milner polymorphism and there are substantial implementation experiments in progress (HOL2P and HOL-Omega) to extend the Church simple type system to increase expressive power whilst retaining tractability of type checking.

At the other end of the type simplicity spectrum lies Coq. This uses a non-classical constructive logic based around propositions-as-types, so that type checking is a core part of theorem proving rather than being a kind of decidable static analysis as in the HOL systems.

PVS adopts a middle ground: it has a classical logic, but with dependent subtypes that need theorem proving for checking type correctness conditions.

It seems, however, that this diversity is less significant than one might think. For example, the authors of a recent Coq paper say: "Using decidable types and relying heavily on rewriting for our proofs gives a 'classical' flavour to our development that is more familiar to what can be found in provers like Isabelle or Hol than what is usually done in Coq".

2.6 Impressive Proofs

The progress in complexity of what can be proved today compared with the state-of-the-art in 1988 is quite stunning. Not only have enormously complex pure mathematical theorems been proved (e.g. Four Color Theorem, Jordan Curve Theorem, large developments in measure theory and multivariate real analysis) but there have been very substantial applications to industrial examples (e.g. floating point verifications, processor correctness proofs, compiler verifications, formal analysis of aircraft flight safety rules). Although some of these proofs took years, it has become clear that there is no fundamental problem in proving pretty much anything: higher order logic appears capable of representing both textbook mathematics and industrial scale formal models.

3   Future

Three things are particularly striking about the evolution of theorem proving for higher order logics over the last 20 years:

1. differences between the various logics and proof methodologies are growing less significant;
2. increases in automation are spectacular both through new kinds of automation and via links to external tools;
3. the mechanisation of pure mathematics has made stunning progress and is not only driving proof method developments but also finding applications.

These will be continuing trends and point to directions for the future. It is impossible to have any confidence in predicting what will actually happen, but below are some thoughts on two particular challenges.

3.1   Library of Formalised Mathematics

It seems clear that cultural and historical pressures will prevent convergence to a single theorem prover or logic. In the 1990s Larry Paulson and I had a funded project called something like "Combining HOL and Isabelle", but the systems were not combined and there are still two proof assistants being developed at Cambridge! The reason why the HOL and Isabelle developments didn't merge was, I think, not primarily technical but due to the hassle and cultural adjustment needed in switching tools when there was no really compelling reason to do so. Each tool development is a little community and merging communities is hard. Besides not wanting to give up ones cosy sub-culture, there is hard work involved in switching that isn't on the critical path to immediate goals. A current illustration is the ARM processor development in HOL4. We would like to port this to other logics, particularly Isabelle/HOL and ACL2, but this will take months, maybe more, and is hard to get off the ground as the work involved is not at all alluring! It is not just that it is hard to merge tool communities, it is not even clear that it is desirable. There is a certain amount of friendly competition between different systems, and the process of reimplementing ideas from system A into system B can lead to new ideas and improvements.

It would be wonderful for everyone to continue to use their favourite tools, but somehow to be able to contribute to building a library of formalised mathematics, both pure and applied. How can this dream be reconciled with the reality of different HOL tool communities? This is surely a grand challenge problem whose solution would benefit everyone. A possible solution: generate proof libraries in some standard set theory?

3.2   Substantial Implementation Experiments Directly in Logic

Higher order logic is very expressive: its terms support functional and logic programming, and maybe even imperative programming via monads. Proof assistants can be used as implementation platforms, e.g. for implementing compilers (the CompCert compiler from Clight to PowerPC and the Utah compiler from HOL to ARM) and hardware synthesis (lots of work over the years). One can even run compiled code inside a theorem prover: ACL2 has demonstrated that formal processor models can be executed with performance not that much less than that of C simulators. Thus, in principle, complete systems developments could be done inside a proof assistant, coding in higher order logic and then compiling using certified execution of certified algorithms.

A challenge for the future is to go beyond academic scale examples and build a significant system this way. The CLI Stack project in ACL2 was an early project along these lines. Can higher order logic technology be the basis for a Stack 2.0 project?

4   Conclusion

Theorem proving for HOLs has made incredible progress in the last twenty years and is now mature, powerful and poised for a golden age!

Will This Be Formal?

Steven P. Miller*

Advanced Technology Center, Rockwell Collins,
400 Collins Rd NE, Cedar Rapids, Iowa 52498
spmiller@rockwellcollins.com

Abstract. While adding formal methods to traditional software development processes can provide very high levels of assurance and reduce costs by finding errors earlier in the development cycle, there are at least four criteria that should be considered before introducing formal methods into a project. This paper describes five successful examples of the use of formal methods in the development of high integrity systems and discusses how each project satisfied these criteria.

Keywords: Formal methods, model checking, theorem proving, avionics.

1 Introduction

Adding formal methods to traditional software development processes can provide very high levels of assurance and reduce costs by finding errors earlier in the development cycle. However, to be successful there are at least four criteria that should be considered before introducing formal methods into a project. This paper describes five successful examples of the use of formal methods in industry and discusses how each of the five projects satisfied these criteria.

2 Examples of the Successful Use of Formal Methods

This section describes five successful examples of the use of formal methods in the development of high integrity systems. In three projects model checking was used to verify the functional correctness of Simulink® models. In two projects theorem proving was used to verify security properties of microcode and source code.

2.1 FCS 5000 Flight Control System

One of the first applications of model checking at Rockwell Collins was to the mode logic of the FCS 5000 Flight Control System [1]. The FCS 5000 is a family of Flight Control Systems for use in business and regional jet aircraft. The mode logic determines

* This work was supported in part by the NASA Langley Research Center under contract NCC-01001 of the Aviation Safety Program (AvSP), the Air Force Research Lab under contract FA8650-05-C-3564 of the Certification Technologies for Advanced Flight Control Systems program (CerTA FCS) and the Air Force Research Lab under the EISTS program Delivery Order 4.

O. Ait Mohamed, C. Muñoz, and S. Tahar (Eds.): TPHOLs 2008, LNCS 5170, pp. 6–11, 2008.
© Springer-Verlag Berlin Heidelberg 2008

which lateral and vertical flight modes are armed and active at any time. While inherently complex, the mode logic consists almost entirely of Boolean and enumerated types and is written in Simulink. The mode logic analyzed consisted of five inter-related mode transition diagrams with a total of 36 modes, 172 events, and 488 transitions.

Desired properties of the mode logic were formally verified using the NuSMV model checker. To accomplish this, the Simulink models were automatically translated into NuSMV using a translation framework developed by Rockwell Collins and the University of Minnesota. This same translation framework also optimized the models for efficient analysis by the NuSMV BDD-based model checker.

Analysis of an early specification of the mode logic found 26 errors, seventeen of which were found by the model checker. Of these 17 errors, 13 were classified by the FCS 5000 engineers as being possible to miss by traditional verification techniques such as testing and inspections. One was classified as being unlikely to be found by traditional verification techniques.

2.2 ADGS-2100 Adaptive Display and Guidance System

One of the most complete examples of model checking at Rockwell Collins was the analysis of the Window Manager logic in the ADGS-2100 Adaptive Display and Guidance System [2]. The ADGS-2100 is a Rockwell Collins product that provides the display management software for next-generation commercial aircraft. The Window Manager (WM) is a component of the ADGS-2100 that ensures that data from different applications is routed to the correct display panel, even in the event of physical failure of one or more components.

Like the FCS 5000 mode logic, the WM is specified in Simulink and was verified by translating it into NuSMV and applying the NuSMV model checker. While the WM contains only Booleans and enumerated types, it is still quite complex. It is divided into five main components that contain a total of 16,117 primitive Simulink blocks that are grouped into 4,295 instances of Simulink subsystems. The reachable state space of the five components ranges from 9.8×10^9 to 1.5×10^{37} states.

Ultimately, 593 properties about the WM were developed and checked, and 98 errors were found and corrected in early versions of the model. As with the FCS 5000 mode logic, this verification was done early in the design process while the design was still changing. While the verification was initially performed by formal methods experts, by the end of the project, the WM developers themselves were doing virtually all the model checking.

2.3 Lockheed Martin Operational Flight Program

The Air Force Research Labs (AFRL) sponsored Rockwell Collins to apply model checking to the Operational Flight Program (OFP) of an Unmanned Aerial Vehicle developed by Lockheed Martin Aerospace as part of the CerTA FCS project [3]. The OFP is an adaptive flight control system that modifies its behavior in response to flight conditions. Phase I of the project concentrated on applying model checking to portions of the OFP, specifically the Redundancy Management (RM) logic, which were well suited to analysis with the NuSMV model checker. While relatively small (the RM logic consisted of three components containing a total of 169 primitive

Simulink blocks organized into 23 subsystems, with reachable state spaces ranging from 2.1×10^4 to 6.0×10^{13} states), they were replicated once for each of the ten control surfaces on the aircraft, making them a significant portion of the total OFP logic.

The verification of the RM logic took approximately 130 hours, with about half of that time spent preparing the models to be verified, correcting the errors found, and running the verification on the corrected models. A total of 62 properties were checked and 12 errors were found and corrected.

In Phase II of this project, the translator framework was extended so that an SMT-solver model checker could be used to verify portions of the numerically intensive inner loop control components in the OFP model. This phase is just being completed and will be reported on at a later date.

2.4 AAMP7G Intrinsic Partitioning

The AAMP7G is microprocessor developed by Rockwell Collins for use in its products. The AAMP7G provides high code density, low power consumption, long life cycle, and is screened for the full military temperature range. In addition, the AAMP7G includes a micro-coded separation kernel that provides MILS capability by ensuring the separation of data at different security classification levels.

To formally verify the AAMP7G intrinsic partitioning mechanism, it was first necessary to develop a formal description of what "data separation" means. This definition, now referred to as the GWV theorem, was specified as a formal property in the language of the ACL2 theorem prover [4]. To prove that the GWV theorem was satisfied by the AAMP7G, the microcode implementing the security kernel was modeled in ACL2 and the GWV theorem proven using the ACL2 theorem prover. To ensure that the ACL2 model of the microcode truly specified the behavior of the microcode on the AAMP7G, it was subjected to a painstaking code-to-spec review overseen by the National Security Agency (NSA) [5].

In May of 2005, the AAMP7G was certified as meeting the EAL-7 requirements of the Common Criteria as "… capable of simultaneously processing unclassified through Top Secret Codeword information".

2.5 Greenhills Integrity-178B Real Time Operating System

The Greenhills Integrity-178B Real Time Operating System implements an ARINC-653 compliant APEX interface that has been certified to DO-178B Level A. It also includes a security kernel written in C that ensures the separation of data at different security classification levels.

To formally verify the Integrity-178B security kernel, the GWV specification of data separation developed for the AAMP7G was generalized to the GWVr2 theorem in order to describe the more dynamic scheduling managed by the OS [6]. As with the AAMP7G, the separation kernel was modeled in ACL2 language and the GWVr2 theorem was proven using the ACL2 theorem prover. To ensure that the ACL2 model of the C code truly specified the behavior of the Integrity-178B security kernel, it was also subjected to a painstaking code-to-spec review overseen by the NIAP/NSA.

Formal verification of the Integrity-178B security kernel satisfied the U.S. Government Protection Profile for Separation Kernels in Environments Requiring High

Robustness and the Common Criteria v2.3 EAL7 ADV requirements. Final certification of the Integrity-178B is now pending completion of NSA penetration testing.

3   Requirements for the Successful Use of Formal Methods

This section identifies four criteria for the successful use of formal verification on a problem and discusses how the examples described earlier satisfy these criteria.

3.1   Is the Problem Important?

While the cost of formal verification has been decreasing with the introduction of more powerful computers and analysis tools, it is still unusual for it to be accepted as an alternative to traditional verification techniques such as reviews and testing. To provide value, formal methods have to either satisfy a need for assurance greater than that provided by traditional means or have to reduce overall development costs by finding errors earlier in the life cycle. In either case, the problem being addressed should be inherently important.

This is the case for each of the examples cited earlier. The ADGS 2100 Window Manager is part of a DO-178B Level A system that provides critical functionality on Air Transport class aircraft. While the FCS 5000 Mode Logic is part of a DO-178B Level C system, errors in its implementation are highly visible to pilots of the aircraft, making the elimination of such errors very desirable. The Lockheed Martin Redundancy Management Logic implements important fault tolerance mechanisms essential for the correct operation of the UAV. Both the intrinsic partitioning mechanism of the AAMP7 and the Green Hills Integrity-178B Real-Time OS needed to provide separation of security domains and to satisfy the Common Criteria.

3.2   Are High Fidelity Models Available for Analysis?

Unlike testing, which verifies the actual implementation of a system, formal verification can only be applied to models of a system such as its design or code. While formal verification will typically find many errors that testing will miss, the availability of high fidelity models is critical for formal verification to be successful.

In each of the examples described earlier, high fidelity models were readily available or could be created at an acceptable cost. For the FCS 5000 Mode Logic, the ADGS 2100 Window Manager, and the Lockheed Martin OFP, unambiguous, machine-readable Simulink models had been created by the designers and used to generate source code. These models were automatically translated into high fidelity models for verification using the NuSMV and PROVER model-checkers.

In contrast, formal models of the AAMP7G microcode and the Greenhills Integrity-178B security kernel were created by hand and verified through painstaking code-to-spec reviews that were directly overseen by the NSA. While creation of these models was expensive, the cost was acceptable in order to achieve the extremely high levels of assurance required for certification.

3.3 Can the Properties of Interest be Stated Formally?

Even if a problem is inherently important, distilling the critical requirements into mathematical relationships that can be formally verified is not always possible. For the FCS 5000 Mode Logic, the ADGS 2100 Window Manager, and the Lockheed Martin OFP, the main challenge was identifying all the properties to be verified. Usually, this was done by formalizing the existing requirements and through discussions with the developers. Frequently, this process uncovered undocumented assumptions and missing requirements that had to be resolved through discussion.

In contrast, the challenge in formalizing the separation requirements for the AAMP7G intrinsic partitioning and the Greenhills Integrity-178B security kernel was developing a precise statement of the abstract concept of separation. While this was ultimately stated as a single ACL2 theorem (the GWV theorem for the AAMP7G and the GWVr2 theorem for the Integrity-178B security kernel), the process took several months of discussion and refinement.

3.4 Are the Right Analysis Tools Available?

The final consideration is whether the right analysis tools are available for the problem. To be effective, the formal verification tools must be able to verify the properties of interest, produce results sufficiently quickly, produce results at acceptable cost, and only require expertise that their users can be expected to know or acquire. Typically, the capability of the analysis tools will play as large a role in selecting which problems will be formally verified as will the inherent need for high assurance.

For the FCS 5000 Mode Logic, the ADGS 2100 Window Manager, and the first phase of the Lockheed Martin OFP, the models all had state spaces of less than 10^{50} reachable states, making them well suited for verification with BDD-based model checkers. Even so, the success of all these projects depended on the ability to automatically generate high fidelity models that were optimized for analysis. This was especially true for the ADG-2100 Window Manager where the design models were being revised daily. If the analysis had taken more than a few hours, the project would not have been a success. In the second phase of the Lockheed Martin OFP analysis, the numerically intensive nature of the problem required the use of the Prover model checker. While successful, the greater expertise required to use an SMT-solver instead of a BDD-based model checker poses real challenges to the transfer of this technology to production developments.

Verification of the security separation provided by the AAMP7G and the Greenhills Integrity-178B security kernel was performed using the ACL2 theorem prover. These efforts took several months and significant expertise to complete. Even so, the project was highly successful due to the inherent importance of the problem, the stability of the AAMP7G microcode and Integrity-178B source code, and the availability of experts to formulate the formal properties and complete the proofs.

4 Future Directions

This paper has briefly described five industrial applications of formal methods and identified some of the main reasons those projects were successful. While the availability of the right tools played a key role in each example, there are still many directions for research that could make formal methods even more useful. An obvious need is to extend the domain of models for which model checking is feasible to include numerically intensive models with transcendental functions. While SMT-solvers hold great promise, there may be difficulty in getting practicing engineers to use them on a routine basis. Industrial users could also use help in determining when they have an optimal set of properties. Also valuable would be a sound basis for determining what testing can be replaced by analysis. Finding ways to compose the verification of subsystems to verify entire systems will be essential to cope with the increasing size of digital systems. Techniques for modeling and verifying asynchronous systems using message passing is another area of need.

References

1. Miller, S., Anderson, E., Wagner, L., Whalen, M., Heimdahl, M.: Formal Verification of Flight Critical Software. In: AIAA Guidance, Navigation and Control Conference and Exhibit, AIAA-2005-6431, American Institute of Aeronautics and Astronautics (2005)
2. Whalen, M., Innis, J., Miller, S., Wagner, L.: ADGS-2100 Adaptive Display & Guidance System Window Manager Analysis, CR-2006-213952, NASA (2006)
3. Whalen, M., Cofer, D., Miller, S., Krogh, B., Storm, W.: Integration of Formal Analysis into a Model-Based Software Development Process. In: 12th International Workshop on Formal Methods for Industrial Critical Systems (FMICS 2007), Berlin, Germany (2007)
4. Greve, D., Wilding, M., Vanfleet, W.M.: A Separation Kernel Formal Security Policy. In: Fourth International Workshop on the ACL2 Prover and Its Applications (ACL2-2003) (2003)
5. Greve, D., Richards, R., Wilding, M.: A Summary of Intrinsic Partitioning Verification. In: Fifth International Workshop on the ACL2 Prover and Its Applications (ACL2-2004) (2004)
6. Greve, D., Wilding, M., Richards, R., Vanfleet, W.M.: Formalizing Security Policies for Dynamic and Distributed Systems. In: Systems and Software Technology Conference (SSTC 2005), Utah State University (2005)

A Short Presentation of Coq

Yves Bertot

INRIA Sophia Antipolis Méditerranée

1 Introduction

The Coq proof assistant has been developed at INRIA, Ecole Normale Supérieure de Lyon, and University of Paris South for more than twenty years [6]. Its theoretical foundation is known as the "Calculus of Inductive Constructions" [4,5]. Versions of the system were distributed regularly from 1989 (version 4.10). The current revision is 8.1 and a revision 8.2 is about to come out. This 8th generation was started in 2004, at the time when a radical change in syntax was enforced and a textbook [2] was published. A more complete historical overview, provided by G. Huet and C. Paulin-Mohring, is available in the book foreword.

The calculus of Inductive constructions is a variant of typed lambda-calculus based on dependent types. Theorems are directly represented by terms of the lambda-calculus, in the same language that is also used to describe formulas and programs. Having all elements of the logic at the same level makes it possible to mix computation and theorem proving in productive ways.

2 The Gallina Specification Language

2.1 Types and Formulas

In the Coq programming language, types can express very precise specifications, like "a function that takes as input an even number and produces as output a natural number that is the half of the input". With a simple type as in Haskell or ML, the type can only be described as "a function that takes as input a number and produces as output a number".

To describe the information that the output satisfies a given relation with the input, we need to add a new notation to the typing language. For instance, if the relation "`half x y`" means "x is the half of y", then we need to have a typing notation to express that the function input will be named y. In Coq, this is given by the "`forall`" notation. Another notation makes it possible to write "{x : nat | half x y}" to describe a type where all elements are pairs, where the first component is a natural number, and the second component is a proof that this number is the half of y. Thus, if we also assume that the predicate `even y` means "y is even", we can write the specification for the function we described in the previous paragraph as

```
forall y: nat, even y -> {x : nat | half x y}
```

O. Ait Mohamed, C. Muñoz, and S. Tahar (Eds.): TPHOLs 2008, LNCS 5170, pp. 12–16, 2008.

This actually is the type of a function that takes two arguments as inputs: the first argument is a natural number, the second argument is a proof that this number is even. The result of this function is a pair where the first component is a number and the second component is a proof that this number is the half of the input. The dependency between the first input, the proof on this input and the output is apparent in the reuse of the name y at various places in the specification. Thus, we can mix concrete data (like y and x) and more immaterial proofs about this data. Theorems are just fully specified functions like the one above: they are distinguished only by the fact that they return a proof.

The distinction between functions as programs and functions as theorems is blurred, so that the function type that we wrote explicitly above can also be read as "for every y, if y is even, then there exists a x so that `half x y` holds." When a function type is read as a logical formula, the "`forall`" keyword can really be read as a universal quantification and the arrow "`->`" can really be read as an implication. This correspondance is known as the *Curry-Howard* isomorphism. In its pure form, the logical setting is restricted with respect to the logical setting of HOL or Isabelle/HOL: some facts that would be taken for granted do not hold. Important examples are the excluded middle (any proposition is either true or false) and extensionality (two functions that coincide on every input are equal). The advantage of working in this fragment of logic is that any proof of existence always contains a constructive process. This is used for the *extraction* tool, which makes it possible to derive programs in functional programming languages from formal proofs [12,11]. However, users can also require a package with the missing axioms to get full classical logic.

2.2 Inductive Types

Users can define new types using the *inductive type* capability. An inductive type is given by providing simultaneously a type or a type family and canonical ways to construct elements of this type. For instance, the type of natural number is not primitive in the Coq logic. It is rather defined as an inductive type `nat` with two constructors.

```
Inductive nat : Set := O | S : nat -> nat.
```

The type that is actually defined is the minimal solution of an "abstract equation" between types. This is reflected by an associated inductive principle named `nat_ind` that is generated automatically from the definition.

The same syntax can be used to define type families, i.e., types that are indexed over another type. For instance, the notion of even number can be described as the following inductive type:

```
Inductive even : nat -> Prop :=
  e0 : even 0
| e2 : forall n, even n -> even (S (S n)).
```

This definition defines `even` as the minimal property that is satisfied for 0 and gets inherited from n to n+2. It is interesting that the type `even 1` cannot contain an element, and this is used to express that `even 1` is logically false.

Defining propositions as inductive types is very powerful and actually all logical connectives apart from universal quantification and implication are described
using inductive types. This also includes equality, which thus does not benefit
from a specific treatement.

Coq also provides facilities to define *coinductive types*. These types are not
minimal solutions to type equations but maximal solutions. In practice, coinductive types can be used to define potentially infinite datatypes, like infinite
lists and recursive functions for these datatypes. Computation with these infinite
values is supported through lazy evaluation.

2.3   Expressions, Functions, and Proofs

Functions in Coq are written in a style that is very similar to the style that is
available in functional programming languages. Anonymous functions are written "fun x => e" and function application is written by simple juxtaposition,
parentheses being added only to disambiguate applications. Numbers receive a
specific treatement, with a notion of scope to help work with natural numbers,
integers, or real numbers, and the usual infix notations for elementary operations. To compute on inductive types, the language provides a pattern-matching
construct that is very close to the one found in Ocaml. The following example
shows a use of Coq's syntax to describe a simple recursive function.

```
Fixpoint div2 (n:nat) : nat :=
  match n with S (S p) => S (div2 p) | _ => 0 end.
```

However, pattern-matching constructs may sometimes be *dependently typed*: each
branch returns a value in a different type, which depends on the pattern being
recognized. This dependent pattern-matching is very powerful and makes it possible to show that some cases cannot occur.

Thus, the Coq system provides a programming language that makes it possible to perform most kinds of side-effect-free functional programming. Several
kinds of recursive definitions are supported: the first kind is *structural recursion*,
the second is *well-founded recursion*. Structurally recursive functions have the
characteristic that they can be executed directly inside Coq. A special attention was devoted to the efficiency of in-system execution, so that executing a
function inside Coq often runs with the same speed as in a "byte-code" execution using Ocaml, which means only one order of magnitude slower than executing native code. Coq also provides an *extraction* feature, where algorithms
from the calculus of constructions are translated to programs in conventional
functional programming languages. Extracted programs can then be compiled
efficiently.

A recent extension to Coq makes it possible to define simultaneously a function and various tools to reason about this function [1]. This command, known
as the Function command also attempts to smoothe the difference between
structurally recursive functions and *well-founded* recursive functions. Discussions
on user mailing-lists indicate that this approach makes the system easier for
newcomers.

Coq is now used as the main development tool for the the CompCert compiler, a certified compiler for a large subset of the C language [10]. In the long run, this effort should lead to a completely formally verified production chain from coq specifications to binary code.

3   Goal Directed Proof

Proofs in Coq can be constructed directly, by composing functions that represent theorems. However, it is often much easier to perform *goal-directed proof* where a statement is first proposed to the system and then decomposed in simpler statement using tactics, following the LCF tradition. A language with basic tactics is provided, with a variety of composition operators: sequence, alternation, etc. Users can define new tactics using a specific language known as `Ltac`. They can also use directly the programming language provided in the calculus of constructions to support their own proof procedures.

A collection of automatic proof procedures are provided for a variety of proof problems: equalities between polynomial or fractional formulas (`ring` or `field`), inequations between linear formulas (`omega` for integers and `fourier` for real numbers), automatic proof in first-order logic, incomplete resolution of systems of inequation between polynomial formulas using sums-of-squares [3], as suggested by [9], etc.

4   The Reflection Feature

An important feature of the Coq system is that the algorithms that are described using the calculus of constructions can also be run directly in the system to solve some proof problems. In a typical application, a data-type is defined to describe abstractly the formulas manipulated in some class of proof problem, a function is described to compute on this data-type, and a theorem is proved about this function (for instance, if the function's target type is a boolean value, one asserts that when this boolean result is `true`, then some property holds for the input). The new theorem is then used directly to reason on arbitrary instances of the proof problem. The proof process involves a computation of the function. This feature relies on efficient in-system computation of structurally recursive functions. The tactic `ring` (to solve equations between polynomial formulas) is a good example of a proof procedure relying on this reflective approach.

The solution of the four-colour theorem by G. Gonthier [7] makes an intensive use of this approach. In that proof, two levels of reflection are used. On the one hand, a large procedure is defined to cover the collection of cases that need to be verified mechanically. On the other hand, the type of boolean values is used directly to represent logical truth value in a systematic way, so many small steps of logical reasoning can also be handled using functional evaluation. This relies on an extension of the Coq system known as *small-scale reflection* [8].

5   Sharing Proofs and Results

The Coq user community shares questions and announcements through a mailing list, known as `coq-club` and a library of user contributions is maintained as a companion to the coq distributions. Pointers are available at the address `coq.inria.fr`.

References

1. Barthe, G., Forest, J., Pichardie, D., Rusu, V.: Defining and reasoning about recursive functions: a practical tool for the coq proof assistant. In: Hagiya, M., Wadler, P. (eds.) FLOPS 2006. LNCS, vol. 3945, Springer, Heidelberg (2006)
2. Bertot, Y., Castéran, P.: Interactive theorem proving and program development, Coq'art: the calculus of inductive constructions. Texts in Theoretical Computer Science: an EATCS series. Springer, Heidelberg (2004)
3. Besson, F.: Fast reflexive arithmetic tactics: the linear case and beyond. In: Altenkirch, T., McBride, C. (eds.) TYPES 2006. LNCS, vol. 4502, pp. 48–62. Springer, Heidelberg (2007)
4. Coquand, T., Huet, G.: The Calculus of Constructions. Information and Computation 76 (1988)
5. Coquand, T., Paulin-Mohring, C.: Inductively defined types. In: Martin-Löf, P., Mints, G. (eds.) COLOG 1988. LNCS, vol. 417. Springer, Heidelberg (1990)
6. Dowek, G., Felty, A., Herbelin, H., Huet, G., Murthy, C., Parent, C., Paulin-Mohring, C., Werner, B.: The Coq Proof Assistant User's Guide. INRIA, Version 5.8 (May 1993)
7. Gonthier, G.: A computer-checked proof of the four colour theorem (2004), `http://research.microsoft.com/~gonthier/4colproof.pdf`
8. Gonthier, G., Mahboubi, A.: A small scale reflection extension for the coq system. Technical Report 6455, Centre Commun INRIA Microsoft fresearch (2008), `http://hal.inria.fr/intria/00258384`
9. Harrison, J.: Verifying nonlinear real formulas via sums of squares. In: Schneider, K., Brandt, J. (eds.) TPHOLs 2007. LNCS, vol. 4732, pp. 102–118. Springer, Heidelberg (2007)
10. Leroy, X.: Formal certification of a compiler back-end, or programmin a compiler with a proof assistant. In: Principles of Programming Languages (POPL 2006). ACM Press, New York (2006)
11. Letouzey, P.: A new extraction for Coq. In: Geuvers, H., Wiedijk, F. (eds.) TYPES 2002. LNCS, vol. 2646. Springer, Heidelberg (2003)
12. Paulin-Mohring, C., Werner, B.: Synthesis of ML programs in the system Coq. Journal of Symbolic Computation 15, 607–640 (1993)

An ACL2 Tutorial

Matt Kaufmann and J Strother Moore

Department of Computer Sciences, University of Texas at Austin,
Taylor Hall 2.124, Austin, Texas 78712
{kaufmann,moore}@cs.utexas.edu

Abstract. We describe a tutorial that demonstrates the use of the ACL2 theorem prover. We have three goals: to enable a motivated reader to start on a path towards effective use of ACL2; to provide ideas for other interactive theorem prover projects; and to elicit feedback on how we might incorporate features of other proof tools into ACL2.

1 Introduction

The name "ACL2" [14] stands for "A Computational Logic for Applicative Common Lisp" and refers to a functional programming language, a formal logic, and a mechanized proof system. It was developed by the authors of this paper, with early contributions by Bob Boyer, and is the latest in the line of "Boyer-Moore" theorem provers [2,3] starting with the 1973 Edinburgh Pure Lisp Prover' [1].

The ACL2 logic and programming language are first-order and admit total recursive function definitions, and are based on a non-trivial purely functional subset of the Common Lisp [20] programming language. Thus, ACL2 can be built on many Lisp platforms. We have extended this subset in some important ways, in no small part because ACL2 is written primarily in itself! Extensions include additional primitives; a *program mode* that avoids proof obligations; a *state* with applicative semantics supporting file I/O and global variables; and applicative property lists and arrays with efficient under-the-hood implementations.

This extended abstract describes a one-hour tutorial, not presented here, but accessible from the "demos" link on the ACL2 home page [14]. Our ambitious goal is to create effective ACL2 users. Of course, no such system can be absorbed deeply in just one hour, so we point to useful documentation and references. Our focus on demos and references suits a second and probably more important goal of this talk, given that most in our audience will already be committed to their favorite theorem prover: To provide a sense of ACL2 interaction that could provide ideas for other interactive theorem prover projects. Conversely, we hope that this introduction to ACL2 will stimulate suggestions for how to improve ACL2 by incorporating ideas from other proof tools.

2 About ACL2

We invite the reader to browse the ACL2 web pages starting with the home page [14], to find: papers on applications and on foundations; tutorials and demos; documentation; mailing list pages; and other useful information.

O. Ait Mohamed, C. Muñoz, and S. Tahar (Eds.): TPHOLs 2008, LNCS 5170, pp. 17–21, 2008.
© Springer-Verlag Berlin Heidelberg 2008

The remaining sections summarize a few ACL2 demos and ACL2 features that they illustrate. In those demos we will refer to sections of the online hypertext user's manual [16] with the notation "see <u>documentation</u>". This note skips the logical foundations [12,13], focusing instead on the use of ACL2. We conclude this section with a few words about how the ACL2 proof engine attempts an individual proof and how ACL2 supports interactive proof development.

Figure 1 shows the ACL2 proof engine as an orchestrated collection of automated tactics, including a simplifier that incorporates conditional congruence-based rewriting as well as decision procedures.

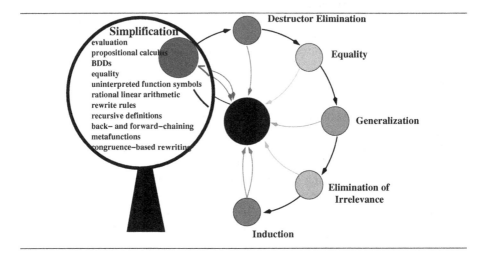

Fig. 1. The ACL2 Waterfall, highlighting the Simplifier

Proof attempts often fail at first! Figure 2 illustrates user interaction with ACL2. The user submits a definition or theorem, which ACL2 attempts to prove using definitions and rules stored in the *logical <u>world</u>*, a database that includes rules stored from definitions and theorems. If the attempt succeeds, then ACL2 makes a corresponding extension to its logical world. Otherwise, ACL2 provides output that can suggest lemmas to prove, and it also offers a variety of other proof debugging tools [15]. Ultimately, a completed proof may be checked by *certifying* the resulting <u>books</u>: input files of <u>events</u>, in particular <u>definitions</u> and proved theorems. Books can be developed independently and combined into libraries of rules that are valuable for automating future proof attempts.

3   Demo: Basics of Interaction with ACL2

This demo develops a proof that for a recursively-defined notion of permutation, the reverse of a list is a permutation of the list. We illustrate how to use ACL2

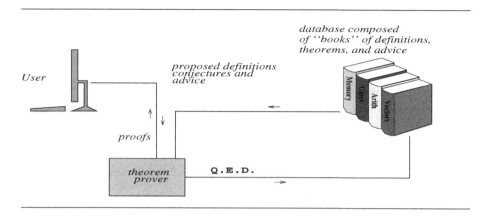

Fig. 2. Basic ACL2 Architecture

interactively, specifically highlighting several aspects of using ACL2. We use a shell running under Emacs, but an Eclipse-based interface is available [7].

- Top-down proof development (see the-method [11]), using simplification checkpoints to debug proof failures
- Helpful proof automation, in particular conditional rewriting with respect to equivalence and congruence relations and a loop-stopper heuristic
- Library development and local scopes
- The proof-checker goal manager

4 Demo: A JVM Model

Next, we demonstrate the so-called M5 model of the Java Virtual Machine (JVM). In the process we illustrate several aspects of ACL2:

- Library re-use via include-book
- Namespace support through Lisp *packages*
- Efficient evaluation supporting simulation of models
- Lisp *macros*, providing user-defined syntax

5 Demo: Proof Debugging and Theory Management

This short demo shows how ACL2 can help to identify rules that slow it down.

- Management of theories (disabling and enabling rules)
- Helpful suggestions from the tool
- Accumulated-persistence for statistics on rule use
- DMR (Dynamic Monitoring of the Rewrite stack)

6   Concluding Remarks

Below is a partial list of useful ACL2 features not mentioned above, many of them requested by users. For more about ACL2, see the home page [14].

- a primitive for user-installed executable counterparts [9]
- proof control [10], including user-installed metatheoretic simplifiers, user-supplied syntactic conditions for rewrite rules, and dynamically computed hints
- traditional tactics (macro-commands) for the proof-checker
- partially-defined functions (see encapsulate) and, mimicking second-order logic, *functional instantiation* [5,13]
- a defpun utility [18] for defining non-terminating tail-recursive functions, built on top of ACL2 with macros
- capabilities for system-level control, such as user-defined tables, state with applicative semantics, and an extended macro capability (make-event) useful in defining templates for creating events
- guards, which may be viewed as a general, semantic replacement for types
- many modes and switches (see switches-parameters-and-modes)
- hooks to external tools [17], built on a *trust tag* mechanism (defttag) [6]
- experimental extensions others have initiated, to support:
  - Real numbers (through non-standard analysis) [8]
  - Hash cons, function memoization, and applicative hash tables [4]
  - Parallel evaluation [19]
- more debugging tools, e.g. to trace or to inspect the rewrite stack
- diverse tools for querying the logical database (see history)
- quantification via Skolemization (see defun-sk)

Acknowledgements. This material is based upon work supported by DARPA and the National Science Foundation (NSF) under Grant No. CNS-0429591 and also NSF grant EIA-0303609. We also thank Sandip Ray for helpful feedback on a preliminary version of this paper.

References

1. Boyer, R.S., Moore, J S.: Proving theorems about pure lisp functions. JACM 22(1), 129–144 (1975)
2. Boyer, R.S., Moore, J S.: A Computational Logic. Academic Press, NY (1979)
3. Boyer, R.S., Moore, J S.: A Computational Logic Handbook. Academic Press, London (1997)
4. Boyer, R.S., Hunt Jr., W.A.: Function memoization and unique object representation for ACL2 functions. In: Proceedings of the Sixth International Workshop on the ACL2 Theorem Prover and Its Applications, pp. 81–89. ACM, New York (2006)

5. Boyer, R.S., Goldschlag, D.M., Kaufmann, M., Moore, J S.: Functional instantiation in first-order logic. In: Lifschitz, V. (ed.) Artificial Intelligence and Mathematical Theory of Computation: Papers in Honor of John McCarthy, pp. 7–26. Academic Press, London (1991)
6. Dillinger, P., Kaufmann, M., Manolios, P.: Hacking and extending ACL2. In: ACL2 Workshop 2007, Austin, Texas (November 2007),
   http://www.cs.uwyo.edu/~ruben/acl2-07/
7. Dillinger, P., Manolios, P., Moore, J S., Vroon, D.: ACL2s: The ACL2 Sedan. Theoretical Computer Science 174(2), 3–18 (2006)
8. Gamboa, R., Kaufmann, M.: Non-Standard Analysis in ACL2. Journal of Automated Reasoning 27(4), 323–351 (2001)
9. Greve, D.A., Kaufmann, M., Manolios, P., Moore, J S., Ray, S., Ruiz-Reina, J.L., Sumners, R., Vroon, D., Wilding, M.: Efficient execution in an automated reasoning environment. Journal of Functional Programming 18(1), January 3–18, 2008; Tech. Rpt. TR-06-59, Dept. of Computer Sciences, Univ. of Texas, Austin, http://www.cs.utexas.edu/ftp/pub/techreports/tr06-59.pdf
10. Hunt Jr., W.A., Kaufmann, M., Krug, R., Moore, J S., Smith, E.: Meta reasoning in ACL2. In: Hurd, J., Melham, T. (eds.) TPHOLs 2005. LNCS, vol. 3603, pp. 163–178. Springer, Heidelberg (2005)
11. Kaufmann, M.: Modular proof: The fundamental theorem of calculus. In: Kaufmann, M., Manolios, P., Moore, J S. (eds.) Computer-Aided Reasoning: ACL2 Case Studies, Boston, MA., pp. 75–92. Kluwer Academic Publishers, Dordrecht (2000)
12. Kaufmann, M., Moore, J S.: A precise description of the ACL2 logic. In: Dept. of Computer Sciences, University of Texas, Austin (1997), http://www.cs.utexas.edu/users/moore/publications/km97a.ps.gz
13. Kaufmann, M., Moore, J S.: Structured Theory Development for a Mechanized Logic. Journal of Automated Reasoning 26(2), 161–203 (2001)
14. Kaufmann, M., Moore, J S.: The ACL2 home page. In: Dept. of Computer Sciences, University of Texas, Austin (2008), http://www.cs.utexas.edu/users/moore/acl2/
15. Kaufmann, M., Moore, J S.: Proof Search Debugging Tools in ACL2. In: Boulton, R., Hurd, J., Slind, K. (eds.) Tools and Techniques for Verification of System Infrastructure, A Festschrift in honour of Prof. Michael J. C. Gordon FRS, Royal Society, London (March 2008), http://www.ttvsi.org/
16. Kaufmann, M., Moore, J S.: The ACL2 User's Manual. In: Dept. of Computer Sciences, University of Texas, Austin (2008), http://www.cs.utexas.edu/users/moore/acl2/#User's-Manual
17. Kaufmann, M., Moore, J S., Ray, S., Reeber, E.: Integrating external deduction tools with ACL2. In: Benzmueller, C., Fischer, B., Sutcliffe, G. (eds.) Proceedings of the 6th International Workshop on Implementation of Logics (IWIL 2006). CEUR Workshop Proceedings, vol. 212, pp. 7–26 (2006); The Journal of Applied Logic (to appear)
18. Manolios, P., Moore, J S.: Partial functions in ACL2. Journal of Automated Reasoning 31(2), 107–127 (2003)
19. Rager, D.L.: Adding parallelism capabilities to ACL2. In: Proceedings of the Sixth International Workshop on the ACL2 Theorem Prover and its applications, pp. 90–94. ACM Press, New York (2006)
20. Steele Jr., G.L.: Common Lisp The Language, 2nd edn. Digital Press (1990)

A Brief Overview of PVS

Sam Owre and Natarajan Shankar*

Computer Science Laboratory,
SRI International

Abstract. PVS is now 15 years old, and has been extensively used in research, industry, and teaching. The system is very expressive, with unique features such as predicate subtypes, recursive and corecursive datatypes, inductive and coinductive definitions, judgements, conversions, tables, and theory interpretations. The prover supports a combination of decision procedures, automatic simplification, rewriting, ground evaluation, random test case generation, induction, model checking, predicate abstraction, MONA, BDDs, and user-defined proof strategies. In this paper we give a very brief overview of the features of PVS, some illustrative examples, and a summary of the libraries and PVS applications.

1 Introduction

PVS is a verification system [17], combining language expressiveness with automated tools. The language is based on higher-order logic, and is strongly typed. The language includes types and terms such as: numbers, records, tuples, functions, quantifiers, and recursive definitions. Full predicate subtype is supported, which makes typechecking undecidable. For example, division is defined such that the second argument is nonzero, where nonzero is defined:

```
nonzero_real: TYPE = {r: real | r /= 0}
```

Note that this means PVS is total; partiality is only supported via subtyping. Dependent types for records, tuples, and function types is also supported. Here is a record type (introduced with `[# #]`) representing a finite sequence, where the `seq` is an array with domain depending on the `length`.

```
finseq: TYPE = [# length: nat, seq: [below[length] -> T] #]
```

Beyond this, the PVS language has structural subtypes (i.e., a record that adds new fields to a given record), dependent types for record, tuple, and functions, recursive and corecursive datatypes, inductive and coinductive definitions, theory interpretations, and theories as parameters, conversions, and judgements that provide control over the generation of proof obligations. Specifications are

---

* This material is based on work performed at SRI and supported by the National Science Foundation under Grants No. CCR-ITR-0326540 and CCR-ITR-0325808.

O. Ait Mohamed, C. Muñoz, and S. Tahar (Eds.): TPHOLs 2008, LNCS 5170, pp. 22–27, 2008.

given as collections of parameterized theories, which consist of declarations and formulas, and are organized by means of importings.

The PVS prover is interactive, but with a large amount of automation built in. It is closely integrated with the typechecker, and features a combination of decision procedures, BDDs, automatic simplification, rewriting, and induction. There are also rules for ground evaluation, random test case generation [11], model checking, predicate abstraction, and MONA. The prover may be extended with user-defined proof strategies.

PVS has been used as a platform for integration. It has a rich API, making it relatively easy to add new proof rules and integrate with other systems. Examples of this include the model checker, Duration Calculus [19], MONA [12], Maple [1], Ag [14], and Yices. The system is normally used through a customized Emacs interface, though it is possible to run it standalone (PVSio does this). PVS is open source, and is available at http://pvs.csl.sri.com. PVS is a part of SRI's Formal Methods Program [6].

In the following sections we will describe a basic example, give a brief description of some more advanced examples, describe some of the available libraries, and finally describe some of the applications.

2   PVS Examples

Ordered Insertion Ordered binary trees are a fundamental data structure used to represent more abstract data structures such as sets, multisets, and associative arrays. The basic idea is that the values in the nodes are totally ordered, and the values of the nodes on the left are less than the current node, which in turn is less than those on the right. The first step is to define a binary tree. In PVS, this can be specified using a datatype.

```
binary_tree[T: TYPE]: DATATYPE BEGIN
  leaf: leaf?
  node(val: T, left, right: binary_tree): node?
END binary_tree
```

A binary_tree is constructed from the leaf and node *constructors*; the node constructor takes three arguments, and has *accessors* val, left, and right, and *recognizers* leaf? and node?. This is all parameterized by the type T. Here is an example creating a binary_tree where the type T is int:

When this is typechecked by PVS, theories are created that include many axioms such as extensionality and induction, and various mapping and reduction combinators [17].

The theory of ordered binary trees is parameterized with a type and a total ordering, an ordered? predicate is defined, and an insert operation specified. The main theorem states that if a tree is ordered, then the tree obtained after inserting an element is also ordered. This is naturally an inductive argument. In the actual specification available at http://pvs.csl.sri.com/examples/datatypes/datatypes.dmp a helper lemma is used to make the induction easier. The proof then is quite simple; both lemmas use `induct-and-rewrite!`, including the helper lemma as a rewrite rule in the proof of the main lemma.

Inductive and Coinductive Definitions. PVS provides a mechanism for defining inductive and coinductive definitions. A simple example of an inductive definition is the transitive closure of an arbitrary relation R.

```
TC(R)(x, y): INDUCTIVE bool =
  R(x, y) OR (EXISTS z: R(x, z) AND TC(R)(z, y))
```

This is simply a *least fixedpoint* with respect to a given domain of elements and a set of rules, which is well-defined if the rules are *monotonic*, by the well known Knaster-Tarski theorem. Under these conditions the greatest fixedpoint also exists and corresponds to *coinductive* definitions. Inductive and coinductive definitions have induction principles, and both must satisfy additional constraints to guarantee that they are well defined.

Corecursive Datatypes. The ordered binary tree example is based on an inductively defined *datatype*, which is defined inductively and describes an *algebra*, as described by Jacobs and Rutten [7]. It is also possible to define *codatatypes*, corresponding to *coalgebras*. The simplest codatatype is the stream:

```
stream[T: TYPE]: CODATATYPE BEGIN
  cons(first:T, rest:stream): cons? END stream
```

This describes an infinite stream. Instead of induction, properties of coalgebras are generally proved using bisimulation, which is automatically generated for codatatypes, as induction is for datatypes.

```
colist[T: TYPE]: CODATATYPE BEGIN
  null: null?
  cons(first: T, rest: colist): cons?
  END colist
```

The colist example abovce looks like the `list` datatype, but includes both finite and infinite lists. This is often useful in specifications; for example, to describe a machine that may run forever, or may halt after some steps. Without coalgebras, the usual approaches are to model halting as stuttering, or to model it as a union type—both of which have various drawbacks.

Theory Interpretations. PVS has support for interpreting one theory in terms of another. This allows uninterpreted types and constants to be interpreted, and the axioms of the uninterpreted theory are translated into proof obligations, thus guaranteeing soundness. Theory interpretations are primarily used for refinement and to prove consistency for an axiomatically defined theory. PVS theory interpretations are described in the theory interpretations report [17].

Model Checking and Predicate Abstraction. PVS includes an integrated model checker that is based on the μ-calculus.[1]. To use the model checker, a finite transition system must be defined with an initial predicate and a transition relation. The PVS prelude provides CTL temporal operators, as well as several definitions of fairness. Examples making use of the model checker may be found in the report [13], which also describes the PVS table construct in some detail.

Model checking requires finite (and usually small) domains, but real systems are generally not finite. Thus in order to apply model checking to a system one must first map it to a finite abstraction. One powerful technique for doing this semi-automatically is *predicate abstraction* [16], which has been integrated into the PVS theorem prover as the `abstract` rule.

Ground Evaluation and PVSio. PVS is primarily a specification language, but it is possible to recognize an executable subset, and hence to actually run PVS. This is efficiently done with the ground evaluator, described in [18,5].

César Muñoz has extended the ground evaluator with *PVSio* [10], which includes a predefined library of imperative programming language features such as side effects, unbounded loops, input/output operations, floating point arithmetic, exception handling, pretty printing, and parsing. The PVSio library is implemented via semantic attachments. PVSio is now a part of the PVS distribution.

3   PVS Libraries and Applications

PVS has an extensive set of libraries available. To begin with, there is the prelude—a preloaded set of theories defining many concepts, ranging from booleans through relations, functions, sets, numbers, lists, CTL, bit-vectors, and equivalence classes (see the prelude report [17] for details). The PVS distribution includes extensions of the basic finite sets and bit-vector theories given in the prelude.

NASA Langley has been working with PVS for many years, and has developed extensive libraries, available at http://shemesh.larc.nasa.gov/fm/fm-pvs.html. This includes libraries for algebra, analysis, calculus, complex numbers, graphs/digraphs, number theory, orders, series, trigonometric functions, and vectors. They have also contributed the Manip and Field packages, which make it easier to do numeric reasoning.

PVS has been used for applications in teaching, research, and industry. Formal specification and verification are inherently difficult, so the focus tends to be on applications with high cost of failure, or critical to life or national security. Thus most applications are to requirements [2,8], hardware [9], safety-critical applications [3], and security [15]

4   Conclusions and Future Work

PVS contains several features that we have omitted from this brief introduction. PVS is still under active maintenance and development. We have many

---

[1] See Chapter 7 of *Model Checking* [4] for an introduction to the μ-calculus.

features we hope to add in the future, including polymorphism, reflection, proof generation, faster rewriting and simplification, a declarative proof mode, counterexamples, proof search and other target languages for the ground evaluator.

References

1. Adams, A., Dunstan, M., Gottliebsen, H., Kelsey, T., Martin, U., Owre, S.: Computer algebra meets automated theorem proving: Integrating Maple and PVS. In: Boulton, R.J., Jackson, P.B. (eds.) TPHOLs 2001. LNCS, vol. 2152, pp. 27–42. Springer, Heidelberg (2001)
2. Archer, M.: TAME: Using PVS strategies for special-purpose theorem proving. Annals of Mathematics and Artificial Intelligence 29(1–4), 139–181 (2000)
3. Carreño, V., Muñoz, C.: Aircraft trajectory modeling and alerting algorithm verification. In: Aagaard, M.D., Harrison, J. (eds.) TPHOLs 2000. LNCS, vol. 1869, pp. 90–105. Springer, Heidelberg (2000)
4. Clarke, E.M., Grumberg, O., Peled, D.A.: Model Checking. MIT Press, Cambridge (1999)
5. Crow, J., Owre, S., Rushby, J., Shankar, N., Stringer-Calvert, D.: Evaluating, testing, and animating PVS specifications. Technical report, Computer Science Laboratory, SRI International, Menlo Park, CA (March 2001), http://www.csl.sri.com/users/rushby/abstracts/attachments
6. Formal Methods Program. Formal methods roadmap: PVS, ICS, and SAL. Technical Report SRI-CSL-03-05, Computer Science Laboratory, SRI International, Menlo Park, CA (October 2003), http://fm.csl.sri.com/doc/roadmap03
7. Jacobs, B., Rutten, J.: A tutorial on (co)algebras and (co)induction. EATCS Bulletin 62, 222–259 (1997)
8. Kim, T., Stringer-Calvert, D., Cha, S.: Formal verification of functional properties of an SCR-style software requirements specification using PVS. In: Katoen, J.-P., Stevens, P. (eds.) ETAPS 2002 and TACAS 2002. LNCS, vol. 2280, pp. 205–220. Springer, Heidelberg (2002)
9. Miller, S.P., Srivas, M.: Formal verification of the AAMP5 microprocessor: A case study in the industrial use of formal methods. In: WIFT 1995: Workshop on Industrial-Strength Formal Specification Techniques, Boca Raton, FL, pp. 2–16. IEEE Computer Society, Los Alamitos (1995)
10. Muñoz, C.: Rapid Prototyping in PVS. National Institute of Aerospace, Hampton, VA (2003), http://research.nianet.org/~munoz/PVSio/
11. Owre, S.: Random testing in PVS. In: Workshop on Automated Formal Methods (AFM), Seattle, WA (August 2006), http://fm.csl.sri.com/AFM06/papers/5-Owre.pdf
12. Owre, S., Rueß, H.: Integrating WS1S with PVS. In: Emerson, E.A., Sistla, A.P. (eds.) CAV 2000. LNCS, vol. 1855, pp. 548–551. Springer, Heidelberg (2000)
13. Owre, S., Rushby, J., Shankar, N.: Analyzing tabular and state-transition specifications in PVS. Technical Report SRI-CSL-95-12, (1995); also published as NASA Contractor Report 201729, http://www.csl.sri.com/csl-95-12.html
14. Pombo, C.L., Owre, S., Shankar, N.: A semantic embedding of the Ag dynamic logic in PVS. Technical Report SRI-CSL-02-04, Computer Science Laboratory, SRI International, Menlo Park, CA (October 2004)
15. Rushby, J.: A separation kernel formal security policy in PVS. Technical note, Computer Science Laboratory, SRI International, Menlo Park, CA (March 2004)

16. Saïdi, H., Graf, S.: Construction of abstract state graphs with PVS. In: Grumberg, O. (ed.) CAV 1997. LNCS, vol. 1254, pp. 72–83. Springer, Heidelberg (1997)
17. Shankar, N., Owre, S., Rushby, J.M., Stringer-Calvert, D.W.J.: PVS System Guide, PVS Language Reference, PVS Prover Guide, PVS Prelude Library, Abstract Datatypes in PVS, and Theory Interpretations in PVS. Computer Science Laboratory, SRI International, Menlo Park, CA (1999), http://pvs.csl.sri.com/documentation.shtml
18. Shankar, N.: Static analysis for safe destructive updates in a functional language. In: Pettorossi, A. (ed.) LOPSTR 2001. LNCS, vol. 2372, pp. 1–24. Springer, Heidelberg (2002), ftp://ftp.csl.sri.com/pub/users/shankar/lopstr01.pdf
19. Skakkebæk, J.U., Shankar, N.: A Duration Calculus proof checker: Using PVS as a semantic framework. Technical Report SRI-CSL-93-10, Computer Science Laboratory, SRI International, Menlo Park, CA (December 1993)

A Brief Overview of HOL4

Konrad Slind[1] and Michael Norrish[2]

[1] School of Computing, University of Utah
slind@cs.utah.edu
[2] National ICT Australia
Michael.Norrish@nicta.com.au

Abstract. The HOL4 proof assistant supports specification and proof in classical higher order logic. It is the latest in a long line of similar systems. In this short overview, we give an outline of the HOL4 system and how it may be applied in formal verification.

1 Introduction

HOL4 is an ML-based environment for constructing proofs in higher order logic. It provides a hierarchy of logical theories which serves as a rich specification library for verification. It also provides a collection of interactive and fully automatic proof tools which can be used in further theory building or in the provision of bespoke verification procedures.

Implementation history. The original HOL system was created in the mid-1980's when Mike Gordon at Cambridge University performed surgery on the Edinburgh LCF system, installing a version of Church's higher order logic as the object language of the system. The metalanguage in which the logic was encoded was Edinburgh ML, itself implemented on top of Lisp. An enhanced version of HOL, called HOL88 [6], was publically released (in 1988), after several years of further development. HOL90 (released in 1990) was a port of HOL88 to SML by Slind at the University of Calgary. The Lisp substrate was abandoned, and some of the more recondite underlying LCF technology was trimmed away or reimplemented. HOL90 ran on Poly/ML and SML/NJ. HOL98 (released in 1998) was a new design, emphasizing separate compilation of theories and proof procedures [12], thus allowing HOL to be ported to MoscowML.

HOL4 is the latest version of HOL, featuring a number of novelties compared to its predecessors. HOL4 continues to be implemented in SML; it currently runs atop Poly/ML and MoscowML. HOL4 is also the supported version of the system for the international HOL community [11].

Project management. The HOL project[1] is open source, managed using the facilities of SourceForge, and currently has about 25 developers, not all of whom are active. In general, control of user contributions is relaxed; anyone who wishes to make a contribution to the system may do so, provided they are willing to

---

[1] Located at http://hol.sourceforge.net

O. Ait Mohamed, C. Muñoz, and S. Tahar (Eds.): TPHOLs 2008, LNCS 5170, pp. 28–32, 2008.
© Springer-Verlag Berlin Heidelberg 2008

provide support. However, modifications to the kernel are scrutinized closely by the project managers (the present authors) before being accepted.

2   Technical Features

We now summarize some notable aspects of HOL4.

2.1   Logic

The logic implemented by HOL4 is essentially Church's Simple Type Theory [3], with polymorphic type variables. The logic implemented by HOL systems, including ProofPower and HOL-Light, has been unchanged since the release of HOL88. An extremely important aspect of the HOL logic, not mentioned by Church, is primitive definition principles for consistently introducing new types and new constants.

An ongoing theme in HOL systems has been adherence to the derivation judgement of the logic: all theorems have to be obtained by performing proofs in higher order logic. However, in some cases, it is practical to allow external proof tools to be treated as oracles delivering HOL theorems *sans* proof. Such theorems are tagged in such a way that the provenance of subsequent theorems can be ascertained.

2.2   Kernels

As is common with HOL designs, the kernel implementation of the logic is kept intentionally small. Only a few simple axioms and rules of inference are encapsulated in the abstract type of theorems implemented in the logic kernel. Currently in HOL4 we maintain two kernels, one based on an explicit substitution calculus, and one based on a standard name-carrying lambda calculus. The desired kernel implementation may be chosen at build time. Informal testing indicates that each kernel outperforms the other on some important classes of input, but that neither outperforms the other in general.

2.3   Derived Rules and Definition Principles

Given such a simple basis, serious verification efforts would be impossible were it not for the fact that ML is a programmable metalanguage for the proof system. Derived inference rules and high-level definition principles are built by programming: such complex logical steps are reduced to a sequence of kernel inferences. For example, some of the current high-level definition principles for types are those supporting the introduction of quotient types and ML-style datatypes. Datatypes can be mutually and nested recursive and may also use record notation. At the term level, support is provided for defining inductively specified predicates and relations; mutual recursion and infinitary premises are allowed. Total recursive functions specified as recursion equations, possibly using ML-style pattern matching, are defined by a package that mechanizes the wellfounded recursion theorem. Mutual and nested recursions are supported. Simple termination proofs have been automated; however, more serious termination proofs have of course to be performed interactively.

2.4   Proof Tools

The view of proof in HOL4 is that the user guides the proof at a high level, leaving subsidiary proofs to automated reasoners. Towards this, we provide a database of type-indexed theorems (case analysis, induction, *etc*) which supports user control of decisive proof steps. In combination with a few 'declarative proof' constructs, this allows many proofs to be conducted at a high level.

HOL4 has a healthy suite of automated reasoners. All produce HOL proofs. Propositional logic formulas can be sent off to external SAT tools and the resulting resolution-style proofs are backtranslated into HOL proofs. For formulas involving $\mathbb{N}$, $\mathbb{Z}$, or $\mathbb{R}$, decision procedures for linear arithmetic may be used. A decision procedure for n-bit words has recently been released. For formulas falling (roughly) into first order logic, a robust implementation of ordered resolution has become very popular.

Probably the most commonly used proof tool is simplification. We provide a call-by-value evaluation mechanism which reduces ground, and some symbolic, terms to normal form [1]. A more general (and more heavily used) tool, the simplifier, provides conditional and contextual ordered rewriting, using matching for higher order patterns. The simplifier may be extended with arbitrary context-aware decision procedures.

For experienced users, most simple proofs can be accomplished via a small amount of interactive guidance (specifying induction or case-analysis, for example) followed by application of the simplifier and first order proof search.

2.5   Theories and Libraries

The system provides a wide collection of theories on which to base further verifications: booleans, pairs, sums, options, numbers ($\mathbb{N}$, $\mathbb{Z}$, $\mathbb{Q}$, $\mathbb{R}$, fixed point, floating point, n-bit words), lists, lazy lists, character strings, partial orders, monad instances, predicate sets, multisets, finite maps, polynomials, probability, abstract algebra, elliptic curves, lambda calculus, program logics (Hoare logic, separation logic), machine models (ARM, PPC, and IA32), temporal logics (ω-automata, CTL, μ-calculus, PSL) and so on. All theories have been built up definitionally.

HOL4 also has an informal notion of a *library*, which is a collection of theories, APIs, and proof procedures supporting a particular domain. For example, the library for $\mathbb{N}$ provides theories formalizing Peano Arithmetic and extensions (numerals, gcd, and simple number theory), a decision procedure, simplification sets for arithmetic expressions, and an extensive collection of syntactic procedures for manipulating arithmetic terms. Loading a library extends the logical context with the types, constants, definitions, and theorems of the comprised theories; it also automatically extends general proof tools, such as the simplifier and the evaluator, with library-specific contributions.

Both theories and libraries are persistent: this is achieved by representing them as separately compiled ML structures. A 'make'-like dependency maintenance tool is used to automatically rebuild formalizations involving disparate collections of HOL4 libraries and theories, as well as ML or external source code in other programming languages.

2.6   External Interfaces

There is a variety of ways for a logic implementation to interface with external tools. On the input side, as we have mentioned, purported theorems coming from external tools need to be accompanied with enough information to reconstruct a HOL proof of the theorem. An example of this is the interface with SAT solvers which can supply proof objects (we currently favour minisat).

Another approach is illustrated by the integration of a BDD library into HOL. This has been used to support the formalization and application of model-checking algorithms for temporal logic. Since HOL theorems are eventually derived from operations on BDDs representing HOL terms, the oracle mechanism mentioned earlier is used to tag such theorems as having been constructed extra-logically.

On the output side, HOL formalizations confining themselves to the 'functional programming' subset of HOL may be exported to ML. This gives a pathway from formalizations to executables. The generated code is exported as separately compilable ML source with no dependencies on the HOL4 implementation. Thus, the theory hierarchy of HOL4 is paralleled by a hierarchy of ML modules containing exported definitions of datatypes and computable functions formalized in HOL. We support the substitution of more efficient versions of such modules; for example, the GMP library used in the mlton compiler may be used instead of the relatively slow code generated from our theory of numerals.

Finally, higher order logic can be used as a metalogic in which to formalize another logic; such has been done for ACL2 [4,5]. HOL4 is used to show that ACL2 is sound. This allows a two-way connection between the two systems in which a HOL formalization may be translated to the HOL theory of ACL2, this formalization is then transported to the ACL2 system and processed in some way (*e.g.*, reduced using the powerful ACL2 evaluation engine) and then the result is transported back to HOL4 and backtranslated to the original HOL theory.

3   Current Projects

Network specification and validation. Peter Sewell and colleagues have used HOL4 to give the first detailed formal specifications of commonly used network infrastructure (UDP, TCP) [2]. This work has heavily used the tools available in HOL4 for operational semantics. They also implemented an inference rule which tested the conformance of real-world traces with their semantics.

Programming language semantics. As an application of the HOL4 backend of the Ott tool [14], Scott Owens has formalized the operational semantics of a large subset of OCaml and proved type soundness [13]. The formalization heavily relied upon the definition packages for datatypes, inductive relations, and recursive functions. Most of the proofs proceeded by rule induction, case analysis, simplification, and first order proof search with user-selected lemmas. In recent work, Norrish has formalized the semantics of C++ [10].

Machine models. An extremely detailed formalization of the ARM due to Anthony Fox sits at the center of much current work in HOL4 focusing on the verification of low-level software. The development is based on a proof that a micro-architecture implements the ARM instruction set architecture. In turn, the ISA has been extended with so-called 'Thumb' instructions (which support compact code) and co-processor instructions. On top of the ISA semantics, Myreen has built a separation logic for the ARM and provided proof automation [8]. *Compiling from logic; decompiling to logic.* It is possible to compile a 'functional programming subset' of the HOL logic to hardware [15] and also to ARM code [7]. This supports high-level correctness proofs of low-level implementations. As well, one can map in the other direction and *decompile* machine code to HOL functions with equivalent semantics [9].

References

1. Barras, B.: Proving and computing in HOL. In: Aagaard, M.D., Harrison, J. (eds.) TPHOLs 2000. LNCS, vol. 1869, pp. 17–37. Springer, Heidelberg (2000)
2. Bishop, S., Fairbairn, M., Norrish, M., Sewell, P., Smith, M., Wansbrough, K.: Rigorous specification and conformance testing techniques for network protocols, as applied to TCP, UDP, and Sockets. In: Proceedings of SIGCOMM. ACM Press, New York (2005)
3. Church, A.: A formulation of the Simple Theory of Types. Journal of Symbolic Logic 5, 56–68 (1940)
4. Gordon, M.J.C., Hunt, W.A., Kaufmann, M., Reynolds, J.: An embedding of the ACL2 logic in HOL. In: Proceedings of ACL2 2006, ACM International Conference Proceeding Series, vol. 205, pp. 40–46. ACM Press, New York (2006)
5. Gordon, M.J.C., Reynolds, J., Hunt, W.A., Kaufmann, M.: An integration of HOL and ACL2. In: Proceedings of FMCAD 2006, pp. 153–160. IEEE Computer Society, Los Alamitos (2006)
6. Gordon, M., Melham, T.: Introduction to HOL, a theorem proving environment for higher order logic. Cambridge University Press, Cambridge (1993)
7. Li, G., Slind, K.: Compilation as rewriting in higher order logic. In: Pfenning, F. (ed.) CADE 2007. LNCS (LNAI), vol. 4603. Springer, Heidelberg (2007)
8. Myreen, M., Gordon, M.: Hoare logic for realistically modelled machine code. In: Grumberg, O., Huth, M. (eds.) TACAS 2007. LNCS, vol. 4424. Springer, Heidelberg (2007)
9. Myreen, M., Slind, K., Gordon, M.: Machine-code verification for multiple architectures: An application of decompilation into logic. In: FMCAD 2008 (submitted 2008)
10. Norrish, M.: A formal semantics for C++. In: Informal proceedings of TTVSI 2008 (2008)
11. Norrish, M., Slind, K.: HOL-4 manuals (1998-2008),
    http://hol.sourceforge.net/
12. Norrish, M., Slind, K.: A thread of HOL development. Computer Journal 45(1), 37–45 (2002)
13. Owens, S.: A sound semantics for OCaml-Light. In: Proceedings of ESOP 2008. LNCS, vol. 4960. Springer, Heidelberg (2008)
14. Sewell, P., Nardelli, F., Owens, S., Peskine, G., Ridge, T., Sarkar, S., Strnisa, R.: Ott: Effective tool support for the working semanticist. In: Proceedings of ICFP 2007. ACM Press, New York (2007)
15. Slind, K., Owens, S., Iyoda, J., Gordon, M.: Proof producing synthesis of arithmetic and cryptographic hardware. Formal Aspects of Computing 19(3), 343–362 (2007)

The Isabelle Framework

Makarius Wenzel[1], Lawrence C. Paulson[2], and Tobias Nipkow[1]

[1] Technische Universität München, Institut für Informatik
[2] University of Cambridge, Computer Laboratory

1 Overview

Isabelle, which is available from `http://isabelle.in.tum.de`, is a generic framework for interactive theorem proving. The *Isabelle/Pure* meta-logic allows the formalization of the syntax and inference rules of a broad range of object-logics following the general idea of natural deduction [32, 33]. The logical core is implemented according to the well-known "LCF approach" of secure inferences as abstract datatype constructors in ML [16]; explicit proof terms are also available [8]. *Isabelle/Isar* provides sophisticated extra-logical infrastructure supporting structured proofs and specifications, including concepts for modular theory development. *Isabelle/HOL* is a large application within the generic framework, with plenty of logic-specific add-on tools and a large theory library. Other notable object-logics are *Isabelle/ZF* (Zermelo-Fraenkel set-theory, see [34, 36]) and *Isabelle/HOLCF* [26] (Scott's domain theory within HOL). Users can build further formal-methods tools on top, e.g. see [53].

Beginners are advised to start working with Isabelle/HOL; see the tutorial volume [30], and the companion tutorial [28] covering structured proofs. A general impression of Isabelle/HOL and ZF compared to other systems like Coq, PVS, Mizar etc. is given in [52]. The Proof General Emacs interface [3] is still the de-facto standard for interaction with Isabelle. The Isabelle document preparation system enables one to generate high-quality PDF-LaTeX documents from the original theory sources, with full checking of the formal content.

The *Archive of Formal Proofs* `http://afp.sf.net` collects proof libraries, examples, and larger scientific developments, mechanically checked with Isabelle. AFP is organized like a journal everybody can contribute to. Submitting formal theories there helps to maintain applications in the longer term, synchronized with the ongoing development of Isabelle itself.

2 Specification Mechanisms

Isabelle/Pure is a minimal version of higher-order logic; object-logics are specified by stating their characteristic rules as new axioms. Any later additions in application theories are usually restricted to *definitional specifications*, and the desired properties are being proven explicitly. Working directly from primitive definitions can be tedious, and higher-level specification mechanisms have emerged over the years, implemented as derived concepts within the existing background logic. This includes (co)inductive sets [35], inductive datatypes [11], and recursive functions [42, 23].

O. Ait Mohamed, C. Muñoz, and S. Tahar (Eds.): TPHOLs 2008, LNCS 5170, pp. 33–38, 2008.

3   Structured Proofs

The Isar proof language [49] continues the natural deduction principles of Isabelle/Pure, working with human readable proof texts instead of primitive inferences. *"Isar"* abbreviates *"Intelligible semi-automated reasoning"*; the language is also related to Mizar, but many underlying principles are quite different [54].

The Isabelle/Isar design also follows the generic framework idea [51]. Starting with a small selection of common principles of natural deduction, various advanced concepts are defined as derived elements (e.g. for calculational reasoning [7] and complex induction proofs [50]). The demands for structured proof composition have also influenced the way of writing definitions and statements, using extra language elements corresponding to Isar proofs, instead of going through the object-logic again [12].

4   Modular Theory Development

Isabelle theories are organized as a graph, with monotonic operations to extend and to merge logical environments. Smaller units of context elements are managed by separate mechanisms for modular theory development, notably *axiomatic type-classes* [27, 48, 40] and *locales* [21, 5, 6]. More recent work integrates type-classes and locales [18], joining the simplicity of classes with the flexibility of locales.

The generic *local theory* concept [19] integrates user-defined module mechanisms smoothly into the Isabelle/Isar framework. The Isabelle distribution already incorporates locales, classes, and class instantiation contexts into the local theory infrastructure. Other approaches to modular theories like AWE [13] could be integrated as well.

Internally, all aspects of locality in Isabelle are centered around the notions of *proof context* and *morphism* — to transfer entities from one context into another. This covers primitive types / terms / theorems of Isabelle/Pure, and any extra-logical context data defined in Isabelle/Isar. This idea of "local everything" allows us to implement tools within an abstract theory and apply them in concrete application contexts later on. One example is an implementation [15] of algebraic methods on abstract rings that can be used for concrete rings.

5   Reasoning Tools

Isabelle has traditionally supported a fair amount of automated reasoning tools. The basic framework is centered around higher-order unification. The Simplifier supports higher-order rewriting, with plug-in interfaces for extra simplification procedures written in ML. The Classical Reasoner [37] and Classical Tableau Prover [38] have been recently complemented by the Metis prover due to Joe Hurd. Various arithmetic proof procedures are available as well. Sledgehammer [41] uses external automated provers (E, Vampire, SPASS) as untrusted search tools to find the necessary lemmas for a particular goal; the actual proof is then performed internally with Metis.

6   Counterexample Search

Because much of the time one (unwittingly) tries to prove non-theorems, Isabelle/HOL offers two facilities to find counterexamples: *Quickcheck* [10] tries randomized instantiation of the free variables and is restricted to executable formulae (see §7). *Refute* [47] searches for small finite countermodels by translating (unrestricted) HOL formulae into propositional logic and hitting them with a SAT solver.

7   Code Generation

Executable HOL theories, including recursive functions and inductive definitions, can be translated into various functional languages, notably SML, OCaml, Haskell [9, 17]. Efficient imperative code can be generated from functions written in monadic style [14]. Results of ML-level computations can be re-imported as theorems ("reflection") to allow efficient computations in proofs. These code generators are restricted to evaluation of closed terms. Efficient evaluation of terms with free variables is supported by a compiled implementation of "normalization by evaluation" [1].

8   Major Applications

In the past 20 years, Isabelle has been used by numerous researchers and students of computer-science and mathematics world wide. Below we summarize some representative large-scale applications.

Pure Mathematics. Here the largest applications are: a) The verification by Bauer, Nipkow [29] and Obua [31] of two of the algorithmic parts of Hales' proof of the *Kepler Conjecture* (What is the densest arrangement of spheres in space?). This is part of Hales' *Flyspeck* project, the complete verification of his proof. b) Avigad's verification of the *Prime Number Theorem* [4] (about the distribution of primes). c) Paulson's proof [39] of the relative consistency of the axiom of choice in ZF, formalized in Isabelle/ZF.

Systems verification. The *Verisoft* project http://www.verisoft.de formalized a whole computer system from the hardware up to an operating system kernel [2] and a compiler for C-dialect [24]. The *L4.verified* project [20, 43] verifies the L4 operating system microkernel, relating an abstract specification, a Haskell model, and the C code.

Programming languages. A large amount of work has gone into formalizations of a sequential Java-like language *Jinja* [22], including bytecode-verification, virtual machine and compiler. Jinja has become the basis for further extensions like multithreading [25] and multiple inheritance [46]. *Isabelle/HOL-Nominal* [44] extends Isabelle/HOL with a unique infrastructure for defining and reasoning about languages with bound variables. Many case studies have been carried out, for example about the meta theory of LF [45].

References

[1] Aehlig, K., Haftmann, F., Nipkow, T.: A compiled implementation of normalization by evaluation. In: Theorem Proving in Higher Order Logics (TPHOLs 2008). LNCS. Springer, Heidelberg (2008)

[2] Alkassar, E., Schirmer, N., Starostin, A.: Formal pervasive verification of a paging mechanism. In: Ramakrishnan, C.R., Rehof, J. (eds.) Tools and Algorithms for the Construction and Analysis of Systems (TACAS 2008). LNCS, vol. 4963, pp. 109–123. Springer, Heidelberg (2008)

[3] Aspinall, D.: Proof General: A generic tool for proof development. In: European Joint Conferences on Theory and Practice of Software (ETAPS) (2000)
[4] Avigad, J., Donnelly, K., Gray, D., Raff, P.: A formally verified proof of the prime number theorem. ACM Trans. Comput. Logic 9(1:2), 1–23 (2007)
[5] Ballarin, C.: Locales and locale expressions in Isabelle/Isar. In: Berardi, S., Coppo, M., Damiani, F. (eds.) TYPES 2003. LNCS, vol. 3085. Springer, Heidelberg (2004)
[6] Ballarin, C.: Interpretation of locales in Isabelle: Theories and proof contexts. In: Borwein, J.M., Farmer, W.M. (eds.) MKM 2006. LNCS (LNAI), vol. 4108. Springer, Heidelberg (2006)
[7] Bauer, G., Wenzel, M.: Calculational reasoning revisited — an Isabelle/Isar experience. In: Boulton, R.J., Jackson, P.B. (eds.) TPHOLs 2001. LNCS, vol. 2152. Springer, Heidelberg (2001)
[8] Berghofer, S., Nipkow, T.: Proof terms for simply typed higher order logic. In: Aagaard, M.D., Harrison, J. (eds.) TPHOLs 2000. LNCS, vol. 1869. Springer, Heidelberg (2000)
[9] Berghofer, S., Nipkow, T.: Executing higher order logic. In: Callaghan, P., Luo, Z., McKinna, J., Pollack, R. (eds.) TYPES 2000. LNCS, vol. 2277, pp. 24–40. Springer, Heidelberg (2002)
[10] Berghofer, S., Nipkow, T.: Random testing in Isabelle/HOL. In: Cuellar, J., Liu, Z. (eds.) Software Engineering and Formal Methods (SEFM 2004), pp. 230–239. IEEE Computer Society Press, Los Alamitos (2004)
[11] Berghofer, S., Wenzel, M.: Inductive datatypes in HOL — lessons learned in Formal-Logic Engineering. In: Bertot, Y., Dowek, G., Hirschowitz, A., Paulin, C., Théry, L. (eds.) TPHOLs 1999. LNCS, vol. 1690. Springer, Heidelberg (1999)
[12] Berghofer, S., Wenzel, M.: Logic-free reasoning in Isabelle/Isar. In: Mathematical Knowledge Management (MKM 2008), LNCS (LNAI). Springer, Heidelberg (2008)
[13] Bortin, M., Broch Johnsen, E., Lüth, C.: Structured formal development in Isabelle. Nordic Journal of Computing 13 (2006)
[14] Bulwahn, L., Krauss, A., Haftmann, F., Erkök, L., Matthews, J.: Imperative functional programming in Isabelle/HOL. In: Theorem Proving in Higher Order Logics (TPHOLs 2008). LNCS. Springer, Heidelberg (2008)
[15] Chaieb, A., Wenzel, M.: Context aware calculation and deduction — ring equalities via Gröbner Bases in Isabelle. In: Kauers, M., et al. (eds.) MKM/CALCULEMUS 2007. LNCS (LNAI), vol. 4573. Springer, Heidelberg (2007)
[16] Gordon, M.J.C., Milner, R., Wadsworth, C.P.: Edinburgh LCF. LNCS, vol. 78. Springer, Heidelberg (1979)
[17] Haftmann, F., Nipkow, T.: A code generator framework for Isabelle/HOL. In: K. Schneider, J. Brandt (eds.) Theorem Proving in Higher Order Logics: Emerging Trends Proceedings. Dept. Comp. Sci., U. Kaiserslautern (2007)
[18] Haftmann, F., Wenzel, M.: Constructive type classes in Isabelle. In: Altenkirch, T., McBride, C. (eds.) TYPES 2006. LNCS, vol. 4502. Springer, Heidelberg (2007)
[19] Haftmann, F., Wenzel, M.: Local theory specifications in Isabelle/Isar (2008), http://www.in.tum.de/~wenzelm/papers/local-theory.pdf
[20] Heiser, G., Elphinstone, K., Kuz, I., Klein, G., Petters, S.M.: Towards trustworthy computing systems: taking microkernels to the next level. SIGOPS Operating Systems Review 41(4), 3–11 (2007)
[21] Kammüller, F., Wenzel, M., Paulson, L.C.: Locales: A sectioning concept for Isabelle. In: Bertot, Y., Dowek, G., Hirschowitz, A., Paulin, C., Théry, L. (eds.) TPHOLs 1999. LNCS, vol. 1690. Springer, Heidelberg (1999)
[22] Klein, G., Nipkow, T.: A machine-checked model for a Java-like language, virtual machine and compiler. ACM Trans. Progr. Lang. Syst. 28(4), 619–695 (2006), http://doi.acm.org/10.1145/1146809.1146811

[23] Krauss, A.: Partial recursive functions in Higher-Order Logic. In: Furbach, U., Shankar, N. (eds.) IJCAR 2006. LNCS (LNAI), vol. 4130. Springer, Heidelberg (2006)

[24] Leinenbach, D., Petrova, E.: Pervasive compiler verification — from verified programs to verified systems. In: Workshop on Systems Software Verification (SSV 2008). Elsevier, Amsterdam (2008)

[25] Lochbihler, A.: Type safe nondeterminism — a formal semantics of Java threads. In: Foundations of Object-Oriented Languages (FOOL 2008) (2008)

[26] Müller, O., Nipkow, T., von Oheimb, D., Slotosch, O.: HOLCF = HOL + LCF. Journal of Functional Programming 9, 191–223 (1999)

[27] Nipkow, T.: Order-sorted polymorphism in Isabelle. In: Huet, G., Plotkin, G. (eds.) Logical Environments. Cambridge University Press, Cambridge (1993)

[28] Nipkow, T.: Structured proofs in Isar/HOL. In: Geuvers, H., Wiedijk, F. (eds.) TYPES 2002. LNCS, vol. 2646. Springer, Heidelberg (2003)

[29] Nipkow, T., Bauer, G., Schultz, P.: Flyspeck I: Tame graphs. In: Furbach, U., Shankar, N. (eds.) IJCAR 2006. LNCS (LNAI), vol. 4130, pp. 21–35. Springer, Heidelberg (2006)

[30] Nipkow, T., Paulson, L.C., Wenzel, M.: Isabelle/HOL. LNCS, vol. 2283. Springer, Heidelberg (2002)

[31] Obua, S.: Flyspeck II: The basic linear programs. Ph.D. thesis, Technische Universität München (2008)

[32] Paulson, L.C.: Natural deduction as higher-order resolution. Journal of Logic Programming 3 (1986)

[33] Paulson, L.C.: Isabelle: the next 700 theorem provers. In: Odifreddi, P. (ed.) Logic and Computer Science. Academic Press, London (1990)

[34] Paulson, L.C.: Set theory for verification: I. From foundations to functions. Journal of Automated Reasoning 11(3) (1993)

[35] Paulson, L.C.: A fixedpoint approach to implementing (co)inductive definitions. In: Bundy, A. (ed.) CADE 1994. LNCS, vol. 814. Springer, Heidelberg (1994)

[36] Paulson, L.C.: Set theory for verification: II. Induction and recursion. Journal of Automated Reasoning 15(2) (1995)

[37] Paulson, L.C.: Generic automatic proof tools. In: Veroff, R. (ed.) Automated Reasoning and its Applications: Essays in Honor of Larry Wos. MIT Press, Cambridge (1997)

[38] Paulson, L.C.: A generic tableau prover and its integration with Isabelle. Journal of Universal Computer Science 5(3) (1999)

[39] Paulson, L.C.: The relative consistency of the axiom of choice — mechanized using Isabelle/ZF. LMS Journal of Computation and Mathematics 6, 198–248 (2003)

[40] Paulson, L.C.: Organizing numerical theories using axiomatic type classes. Journal of Automated Reasoning 33(1) (2004)

[41] Paulson, L.C., Susanto, K.W.: Source-level proof reconstruction for interactive theorem proving. In: Schneider, K., Brandt, J. (eds.) TPHOLs 2007. LNCS, vol. 4732. Springer, Heidelberg (2007)

[42] Slind, K.: Function definition in higher order logic. In: von Wright, J., Harrison, J., Grundy, J. (eds.) TPHOLs 1996. LNCS, vol. 1125. Springer, Heidelberg (1996)

[43] Tuch, H., Klein, G., Norrish, M.: Types, bytes, and separation logic. In: Principles of Programming Languages (POPL 2007), pp. 97–108. ACM Press, New York (2007)

[44] Urban, C.: Nominal techniques in Isabelle/HOL. Journal of Automated Reasoning 40, 327–356 (2008)

[45] Urban, C., Cheney, J., Berghofer, S.: Mechanizing the metatheory of LF. In: 23rd IEEE Symp. Logic in Computer Science (LICS) (2008)

[46] Wasserrab, D., Nipkow, T., Snelting, G., Tip, F.: An operational semantics and type safety proof for multiple inheritance in C++. In: Object Oriented Programming, Systems, Languages, and Applications (OOPSLA 2006), pp. 345–362. ACM Press, New York (2006)

[47] Weber, T.: Bounded model generation for Isabelle/HOL. In: Ahrendt, W., Baumgartner, P., de Nivelle, H., Ranise, S., Tinelli, C. (eds.) Workshops Disproving and Pragmatics of Decision Procedures (PDPAR 2004), vol. 125, pp. 103–116. Elsevier, Amsterdam (2005)

[48] Wenzel, M.: Type classes and overloading in higher-order logic. In: Gunter, E.L., Felty, A.P. (eds.) TPHOLs 1997. LNCS, vol. 1275. Springer, Heidelberg (1997)

[49] Wenzel, M.: Isar — a generic interpretative approach to readable formal proof documents. In: Bertot, Y., Dowek, G., Hirschowitz, A., Paulin, C., Théry, L. (eds.) TPHOLs 1999. LNCS, vol. 1690. Springer, Heidelberg (1999)

[50] Wenzel, M.: Structured induction proofs in Isabelle/Isar. In: Borwein, J.M., Farmer, W.M. (eds.) MKM 2006. LNCS (LNAI), vol. 4108. Springer, Heidelberg (2006)

[51] Wenzel, M.: Isabelle/Isar — a generic framework for human-readable proof documents. In: R. Matuszewski, A. Zalewska (eds.) From Insight to Proof — Festschrift in Honour of Andrzej Trybulec, Studies in Logic, Grammar, and Rhetoric, vol. 10(23). University of Białystok (2007),
http://www.in.tum.de/~wenzelm/papers/isar-framework.pdf

[52] Wenzel, M., Paulson, L.C.: Isabelle/Isar. In: Wiedijk, F. (ed.) The Seventeen Provers of the World. LNCS (LNAI), vol. 3600. Springer, Heidelberg (2006)

[53] Wenzel, M., Wolff, B.: Building formal method tools in the Isabelle/Isar framework. In: Schneider, K., Brandt, J. (eds.) TPHOLs 2007. LNCS, vol. 4732. Springer, Heidelberg (2007)

[54] Wiedijk, F., Wenzel, M.: A comparison of the mathematical proof languages Mizar and Isar. Journal of Automated Reasoning 29(3-4) (2002)

A Compiled Implementation
of Normalization by Evaluation

Klaus Aehlig[1,*] , Florian Haftmann[2,**], and Tobias Nipkow[2]

[1] Department of Computer Science, Swansea University
[2] Institut für Informatik, Technische Universität München

Abstract. We present a novel compiled approach to Normalization by Evaluation (NBE) for ML-like languages. It supports efficient normalization of open λ-terms w.r.t. β-reduction and rewrite rules. We have implemented NBE and show both a detailed formal model of our implementation and its verification in Isabelle. Finally we discuss how NBE is turned into a proof rule in Isabelle.

1 Introduction

Symbolic normalization of terms w.r.t. user provided rewrite rules is one of the central tasks of any theorem prover. Several theorem provers (see §5) provide especially efficient normalizers which have been used to great effect [9,14] in carrying out massive computations during proofs. Existing implementations perform normalization of open terms either by compilation to an abstract machine or by Normalization by Evaluation, NBE for short. The idea of NBE is to carry out the computations by translating into some underlying functional language, evaluating there, and translating back. The key contributions of this paper are:

1. A novel compiled approach to NBE that exploits the pattern matching already available in a decent functional language, while allowing the normalization of open λ-terms w.r.t. β-reduction and a set of (possibly higher-order) rewrite rules.
2. A formal model and correctness proof[1] of our approach in Isabelle/HOL [15].

NBE is implemented and available at the user-level in Isabelle 2007, both to obtain the normal form t' of some given term t, and as a proof rule that yields the theorem $t = t'$.

Throughout the paper we refer to the underlying functional language as ML. This is only for brevity: any language in the ML family, including Haskell, is suitable. However, we assume that the language implementation provides its own evaluator at runtime, usually in the form of some compiler. The guiding

---

* Partially supported by grant EP/D03809X/1 of the British Engineering and Physical Sciences Research Council (EPSRC).
** Supported by DFG grant Ni 491/10-1.
[1] Available online at afp.sf.net

principle of our realization of NBE is to offload as much work as possible onto ML: not just substitution but also pattern matching. Thus the word 'compiled' in the title refers to both the translation from the theorem prover's λ-calculus into ML and from ML to some byte or machine code. The trusted basis of the theorem prover is not extended if the compiler used at runtime is the same as the one compiling the theorem prover.

2 Normalization by Evaluation in ML

Normalization by Evaluation uses the evaluation mechanism of an underlying metalanguage to normalize terms, typically of the λ-calculus. By means of an evaluation function $[\cdot]_\xi$, or, alternatively by compiling and running the compiled code, terms are embedded into this metalanguage. In other words, we now have a native function in the implementation language. Then, a function $\downarrow$, which acts as an "inverse of the evaluation functional" [5], serves to recover terms from the semantics. This process is also known as "type-directed partial evaluation" [7].

Normalization by Evaluation is best understood by assuming a semantics enjoying the following two properties.

– *Soundness:* if $r \to s$ then $[r]_\xi = [s]_\xi$, for any valuation ξ.
– *Reproduction:* if r is a term in normal form, then $\downarrow [r]_\uparrow = r$ with $\uparrow$ a special valuation.

These properties ensure that $\downarrow [r]_\uparrow$ actually yields a normal form of r if it exists. Indeed, let $r \to^* s$ with s normal; then $\downarrow [r]_\uparrow = \downarrow [s]_\uparrow = s$.

We implement untyped normalization by evaluation [1] in ML. To do so, we need to construct a model of the untyped λ-calculus, i.e., a data type containing its own function space. Moreover, in order to make the reproduction property possible, our model ought to include some syntactical elements in it, like constructors for free variables of our term language. Fortunately, ML allows data types containing their own function space. So we can simply define a universal type Univ like the following.

```
datatype Univ =
    Const of string * Univ list
  | Var of int * Univ list
  | Clo of int * (Univ list -> Univ) * Univ list
```

Note how the constructors of the data type allow to distinguish between basic types and proper functions of implementation language. In type-directed partial evaluation such a tagging is not needed, as the type of the argument already tells what to expect; on the other hand, this need of anticipating what argument will come restricts the implementation to a particular typing discipline, whereas our untyped approach is flexible enough to work with any form of rewrite calculus.

The data type Univ represents the embedding of the syntax and the embedding of the function space. There is no constructor for application. The reason is that semantical values of the λ-calculus correspond to normal terms, whereas an application of a function to some other value, in general, yields a redex. There-fore application is implemented by a function `apply: Univ -> Univ -> Univ` discussed below. The constructor `Const` serves to embed constructors of data types of the underlying theory; they are identified by the string argument. Nor-mal forms can have the shape $C t_1 \ldots t_k$ of a constructor C applied to several (normal) arguments. Therefore, we allow `Const` to come with a list of arguments, for convenience of the implementation in reverse order. In a similar manner, the constructor `Var` is used to represent expressions of the form $x t_1 \ldots t_k$ with x a variable.

The constructor `Clo` represents partially applied functions. More precisely, "`Clo` $(n, f, [a_k, \ldots, a_1])$" represents the $(n+k)$-ary function f applied to $a_1, \ldots, a_k$. This expression needs another n arguments before f can be evaluated. In the case of the pure λ-calculus, n would always be 1 and f would be a value obtained by using (Standard) ML's "`fn x => ...`" function abstraction. Of course, ML's un-derstanding of the function space is bigger than just the functions that can be ob-tained by evaluating a term in our language. For example, recursion can be used to construct representation for infinite terms. However, this will not be a problem for our implementation, for several reasons. First of all, we only claim that terms are normalised correctly—this suffices for our procedure to be admissible in a the-orem prover. During that normalisation process, only function that can be named by a (finite) term will occur as arguments to `Clo`. Moreover, only needing partial correctness, we will only ever be concerned with semantical values where our $\downarrow$-function terminates. But then, the fact that it did terminate, witnesses that the semantical value has a finite representation by one of our terms.

As mentioned, application is realised by an ML-function `apply`. With the discussed semantics in mind, it is easy to construct such a function: in the cases that $C t_1 \ldots t_k$ or $x t_1 \ldots t_k$ is applied to a value s, we just add it to the list. In the case of a partially applied function applied to some value s we either, in case more then one argument is still needed, collect this argument or, in case this was the last argument needed, we apply the function to its arguments.

```
fun apply (Clo (1, f, xs)) x = f (x :: xs)
  | apply (Clo (n, f, xs)) x = Clo (n - 1, f, x :: xs)
  | apply (Const (name, args)) x = Const (name, x :: args)
  | apply (Var (name, args)) x = Var (name, x :: args)
```

It should be noted that the first case in the above definition is the one that triggers the actual work: compiled versions of the functions of the theory are called. As discussed above, our semantical universe `Univ` allows only normal values. Therefore, this call carries out all the normalization work.

As an example, consider a function `append` defined in some Isabelle/HOL theory `T` based on the type `list` defined in theory `List`

```
fun append :: "'a list => 'a list => 'a list" where
  "append Nil         bs = bs" |
  "append (Cons a as) bs = Cons a (append as bs)"
```

and assume " `append (append as bs) cs = append as (append bs cs)` " was proved. Compiling these equations together with associativity of **append** yields the following ML code.

```
fun T_append [v_cs, Nbe.Const ("T.append", [v_bs, v_as])] =
        T_append [T_append [v_cs, v_bs], v_as]
  | T_append [v_bs, Nbe.Const ("List.Cons", [v_as, v_a])] =
        Nbe.Const ("List.Cons", [T_append [v_bs, v_as], v_a])
  | T_append [v_bs, Nbe.Const ("List.Nil", [])] =
        v_bs
  | T_append [v_a, v_b] =
        Nbe.Const ("T.append", [v_a, v_b])
```

The second and third clause of the function definition are in one-to-one correspondence with the definition of the function **append** in the theory. The arguments, both on the left and right side, are in reverse order; this is in accordance with our semantics that $f a_1 \dots a_n$ is implemented as "f [a_n,..., a_1]".

The last clause is a *default clause* fulfilling the need that the ML pattern matching be exhaustive. But our equations, in general, do not cover all cases. The constructor **Var** for variables is an example for a possible argument usually not covered by any rewrite rule. In this situation where we have all arguments for a function but no rewrite rule is applicable, no redex was generated by the last application—and neither will be by applying this expression to further arguments, as we have already exhausted the arity of the function. Therefore, we can use the **append** function as a constructor. Using (the names of) our compiled functions as additional constructors in our universal data type is a necessity of normalising open terms. In the presence of variables not every term reduces to one built up from only canonical constructors; instead, we might obtain normal forms with functions like **append**. Using them as additional constructors is the obvious way to represent these normal forms in our universal semantics.

Keeping this reproduction case in mind, we can understand the first clause. If the first argument is of the form **append**, in which case it cannot further be simplified, we can use associativity. Note that we are actually calling the append function, instead of using a constructor; in this way we ensure to produce a normal result.

Continuing the example, now assume that we want to normalise the expression "append [a,b] [c]". Then the following compiled version of this expression would be evaluated to obtain an element of **Univ**.

```
(Nbe.apply
  (Nbe.apply
    (Clo (2,T_append,[]))
    (Nbe.Const ("List.cons",
              [(Nbe.Const ("List.cons",
                          [(Nbe.Const ("List.nil", [])),
                           (Nbe.free "b")])),
               (Nbe.free "a")]))))
```

```
(Nbe.Const ("List.cons", [(Nbe.Const ("List.nil", [])),
                          (Nbe.free "c")]))))
```

As discussed, values of type `Univ` represent normal terms. Therefore we can easily implement the $\downarrow$-function, which will be called `term` in our implementation. The function `term` returns a normal term representing a given element of `Univ`. For values of the form "`Const` *name* $[v_n, \ldots, v_1]$" we take the constant C named by the string, recursively apply `term` to $v_1, \ldots, v_n$, obtaining $t_1, \ldots, t_n$, and build the application $C\, t_1 \ldots t_n$. Here, again, we keep in mind that arguments are in reverse order in the implementation. The definition in the case of a variable is similar. In the case $v =$ "`Clo` ..." of a closure we just carry out an eta expansion: the value denotes a function that needs at least another argument, so we can always write it as $\lambda x.\mathtt{term}(v\, x)$, with x a fresh syntactical variable. Naturally, this application of v to the fresh variable x is done via the function `apply` discussed above. In particular, this application might trigger a redex and therefore cause more computation to be carried out. For example, as normal form of "`append Nil`" we obtain—without adding any further equations!—the correct function "$\lambda u.\ u$".

Immediately from the definition we note that `term` can only output normal terms. Indeed, the `Const` construct is used only for constructors or functions where the arguments are of such a shape that no redex can occur. Expressions of the shape $x\, t_1 \ldots t_k$ and $\lambda x.t$ are always normal if $t, t_1, \ldots, t_k$ are; the latter we can assume by induction hypothesis. Note that we have shown the normality of the output essentially by considering ways to combine terms that preserve the normality. In fact, the normalisation property of normalisation by evaluation can be shown entirely by considering an appropriate typing discipline [8].

Compared to the expressivity of the underlying term language in Isabelle, our universal datatype is quite simple. This is due to the fact, that we consider an untyped term-rewriting mechanism. This simplicity, however, comes at a price: we have to translate back and forth between a typed and an untyped world. Forgetting the types to get to the untyped rewrite structure is, essentially, an easy task, even though some care has to be taken to ensure that the more advanced Isabelle features like type classes and overloading are compiled away correctly and the term to be normalised obeys the standard Hindley-Milner type discipline. More details of this transformation into standard typing discipline are described in §4.

From terms following this standard typing discipline the types are thrown away and the untyped normal form is computed, using the mechanism described earlier. Afterwards, the full type annotations are reconstructed. To this end, the types of all free variables have been stored before normalization; the most general types of the constants can be uniquely rediscovered from their names. The type of the whole expression is kept as well, given that the Isabelle object language enjoys subject reduction. Standard type inference will obtain the most general type annotations for all sub-terms such that all these constraints are met.

In most cases, these type reconstructions are unique, as follows from the structure of normal terms in the simply-typed lambda calculus. However, in the presence of polymorphic constants, the most general type could be more general than intended. For example, let `f` be a polymorphic constant of type "`('a => 'a) => bool`", say without any rewrite rule. Then the untyped normal form of "`f (λu::bool. u)`" would be "`f (λu. u)`" with most general type annotations "`f (λu::'a. u)`". To avoid such widening of types only those equations will be considered as being proved by normalization where the typing of the result is completely determined, i.e., those equations, where the most general type for the result does not introduce any new type variables. It should be noted that this, in particular, is always the case, if an expression evaluates to `True`.

3   Model and Verification

This section models the previous section in Isabelle/HOL and proves partial correctness of the ML level w.r.t. rewriting on the term level. In other words, we will show that, if NBE returns an output t' to an input t, then $t = t'$ could have also be obtained by term rewriting with equations that are consequences of the theory.

We do not attempt to handle questions of termination or uniqueness of normal forms. This would hardly be possible anyway, as arbitrary proven equations may be added as rewrite rules. Given this modest goal of only showing soundness, which however is enough to ensure conservativity of our extension of the theorem prover, we over-approximate the operational semantics of ML. That is, every reduction ML can make is also a possible reduction our model of ML can make. Conversely, our ML model is non-deterministic w.r.t. both the choice among the applicable clauses of a compiled function and the order in which to evaluate functions and arguments—any evaluation strategy is fine, even non left-linear equations are permitted in function definitions. This over-approximation shows that partial correctness of our implementation is quite independent of details of the implementation language. In particular, we could have chosen any functional language, including lazy ones like Haskell.

In the introduction it was mentioned that Normalization by Evaluation is best understood in terms of the mentioned properties "soundness of the semantics" (i.e., the semantics identifies enough terms) and "reproduction" (i.e., normal terms can be read off from the semantics). For showing partial correctness, however, the task is slightly different. First of all, we cannot really guarantee that our semantics identifies enough terms; there might be equalities that hold in the Isabelle theory under consideration that are not expressed as rewrite rules. Fortunately, this is not a problem. A failure of this property can only lead to two terms that are equal in the theory, but still have different normal forms. Then, the lack of this properties requires us to show a slightly stronger form of the reproduction property. We need to for *arbitrary* terms r that $\downarrow [r]_\uparrow$ is, if defined, a term that our theory equates with r. To show this property, we give a model of our implementation language and assign each internal state a "denoted term"; having this term denotation at hand we just have to show that each step our machine model makes either doesn't change the denoted term, or transforms it to a term of which our theory shows that it is equal.

3.1 Basic Notation

HOL conforms largely to everyday mathematical notation. This section introduces some non-standard notation and a few basic data types with their primitive operations.

The types of truth values and natural numbers are called *bool* and *nat*. The space of total functions is denoted by $\Rightarrow$. The notation $t :: \tau$ means that term t has type τ.

Sets over type α, type α *set*, follow the usual mathematical convention.

Lists over type α, type α *list*, come with the empty list $[]$, the infix constructor $\cdot$, the infix @ that appends two lists, and the standard functions *map* and *rev*.

3.2 Terms

We model bound variables by de Bruijn indices [6] and assume familiarity with this device, and in particular the usual lifting and substitution operations. Below we will not spell those out in detail but merely describe them informally—the details are straightforward. Because variables are de Bruijn indices, i.e. natural numbers, the types *vname* and *ml-vname* used below are merely abbreviations for *nat*. Type *cname* on the other hand is an arbitrary type of constant names, for example strings.

ML terms are modeled as a recursive **datatype**:

$$
\begin{aligned}
ml = \ & C_{ML} \ cname \\
| \ & V_{ML} \ ml\text{-}vname \\
| \ & A_{ML} \ ml \ (ml \ list) \\
| \ & Lam_{ML} \ ml \\
| \ & C_U \ cname \ (ml \ list) \\
| \ & V_U \ vname \ (ml \ list) \\
| \ & Clo \ ml \ (ml \ list) \ nat \\
| \ & apply \ ml \ ml
\end{aligned}
$$

The default type of variables u and v shall be *ml*.

The constructors come in three groups:

- The λ-calculus underlying ML is represented by C_{ML}, V_{ML}, A_{ML} and Lam_{ML}. Note that application A_{ML} applies an ML value to a list of ML values to cover both ordinary application (via singleton lists) and to model the fact that our compiled functions take lists as arguments. Constructor Lam_{ML} binds V_{ML}.
- Values of the datatype Univ (§2) are encoded by the constructors C_U, V_U and *Clo*.
- Constructor *apply* represents the ML function apply (§2).

Note that this does not model all of ML but just the fraction we need to express computations on elements of type Univ, i.e. encoded terms.

Capture-avoiding substitution $subst_{ML} \ \sigma \ u$, where $\sigma :: nat \Rightarrow ml$, replaces $V_{ML} \ i$ by $\sigma \ i$ in u. Notation $u[v/i]$ is a special case of $subst_{ML} \ \sigma \ u$ where σ

replaces V_{ML} i by v and decreases all ML variables $\geq i$ by 1. Lifting the free ML variables $\geq i$ is written $lift_{ML}$ i v. Predicate $closed_{ML}$ checks if an ML value has no free ML variables ($\geq$ a given de Bruijn index).

The term language of the logical level is an ordinary λ-calculus, again modeled as a recursive **datatype**:

$$tm = C\ cname \mid V\ vname \mid tm \cdot tm \mid \Lambda\ tm \mid term\ ml$$

The default type of variables r, s and t shall be tm.

This is the standard formalization of λ-terms (using de Bruijn), but augmented with $term$. It models the function $\texttt{term}$ from §2. The subset of terms not containing $term$ is called $pure$.

We abbreviate $(\cdots(t \cdot t_1) \cdot \cdots) \cdot t_n$ by $t \cdot\cdot [t_1,\ldots,t_n]$. We have the usual lifting and substitution functions for term variables. Capture-avoiding substitution $subst\ \sigma\ s$, where $\sigma :: nat \Rightarrow tm$, replaces $V\ i$ by $\sigma\ i$ in s and is only defined for pure terms. The special form $s[t/i]$ is defined in analogy with $u[v/i]$ above, only for term variables. Lifting the free term variables $\geq i$ is written $lift\ i$ and applies both to terms (where V is lifted) and ML values (where V_U is lifted).

In order to relate the encoding of terms in ML back to terms we define an auxiliary function $kernel :: ml \Rightarrow tm$ that maps closed ML terms to λ-terms. For succinctness $kernel$ is written as a postfix !; $map\ kernel\ vs$ is abbreviated to $vs!$. Note that postfix binds tighter than prefix, i.e. $f\ v!$ is $f\ (v!)$.

$$(C_{ML}\ nm)! = C\ nm$$
$$(A_{ML}\ v\ vs)! = v! \cdot\cdot (rev\ vs)!$$
$$(Lam_{ML}\ v)! = \Lambda\ ((lift\ 0\ v)[V_U\ 0\ []/0])!$$
$$(C_U\ nm\ vs)! = C\ nm \cdot\cdot (rev\ vs)!$$
$$(V_U\ x\ vs)! = V\ x \cdot\cdot (rev\ vs)!$$
$$(Clo\ f\ vs\ n)! = f! \cdot\cdot (rev\ vs)!$$
$$(apply\ v\ w)! = v! \cdot w!$$

The arguments lists vs need to be reversed because, as explained in §2, the representation of terms on the ML level reverses argument lists to allow $\texttt{apply}$ to add arguments to the front of the list.

The kernel of a tm, also written $t!$, replaces all subterms $term\ v$ of t by $v!$.

Note that ! is not structurally recursive in the Lam_{ML} case. Hence it is not obvious to Isabelle that ! is total, in contrast to all of our other functions. To allow its definition [13] we have shown that the (suitably defined) size of the argument decreases in each recursive call of !. In the Lam_{ML} case this is justified by proving that both lifting and substitution of $V_U\ i\ []$ for $V_{ML}\ i$ do not change the size of an ML term.

3.3   Reduction

We introduce two reduction relations: $\rightarrow$ on pure terms, the usual λ-calculus reductions, and $\Rightarrow$ on ML terms, which models evaluation in functional languages.

The reduction relation $\rightarrow$ on pure terms is defined by β-reduction: $\Lambda\ t \cdot s \rightarrow t[s/0]$, η-expansion: $t \rightarrow \Lambda\ (lift\ 0\ t \cdot V\ 0)$, rewriting:

$$\frac{(nm,\ ts,\ t) \in R}{C\ nm\ \cdot\cdot\ map\ (subst\ \sigma)\ ts\ \rightarrow\ subst\ \sigma\ t}$$

and context rules:

$$\frac{t \rightarrow t'}{\Lambda\ t \rightarrow \Lambda\ t'} \qquad \frac{s \rightarrow s'}{s \cdot t \rightarrow s' \cdot t} \qquad \frac{t \rightarrow t'}{s \cdot t \rightarrow s \cdot t'}$$

Note that $R :: (cname \times tm\ list \times tm)\ set$ is a global constant that models a (fixed) set of rewrite rules. The triple $(f,\ ts,\ t)$ models the rewrite rule $C\ f\ \cdot\cdot\ ts \rightarrow t$.

Just like $\rightarrow$ depends on R, $\Rightarrow$ depends on a compiled version of the rules, called $compR :: (cname \times ml\ list \times ml)\ set$. A triple $(f,\ vs,\ v)$ represents the ML equation with left-hand side $A_{ML}\ (C_{ML}\ f)\ vs$ and right-hand side v. The definition of $compR$ in terms of our compiler is given further below.

The ML reduction rules come in three groups. First we have β-reduction $A_{ML}\ (Lam_{ML}\ u)\ [v] \Rightarrow u[v/0]$ and invocation of a compiled function:

$$\frac{(nm,\ vs,\ v) \in compR \qquad \forall i.\ closed_{ML}\ 0\ (\sigma\ i)}{A_{ML}\ (C_{ML}\ nm)\ (map\ (subst_{ML}\ \sigma)\ vs) \Rightarrow subst_{ML}\ \sigma\ v}$$

This is simply one reduction step on the level of ML terms.

Then we have the reduction rules for function $apply$:

$$\frac{0 < n}{apply\ (Clo\ f\ vs\ (Suc\ n))\ v \Rightarrow Clo\ f\ (v \cdot vs)\ n}$$

$$apply\ (Clo\ f\ vs\ (Suc\ 0))\ v \Rightarrow A_{ML}\ f\ (v \cdot vs)$$

$$apply\ (C_U\ nm\ vs)\ v \Rightarrow C_U\ nm\ (v \cdot vs)$$

$$apply\ (V_U\ x\ vs)\ v \Rightarrow V_U\ x\ (v \cdot vs)$$

which directly realize the defining equations for `apply` in §2.

Finally we have all the context rules (not shown). They say that reduction can occur anywhere, except under a Lam_{ML}. Note that we do not fix lazy or eager evaluation but allow any strategy. Thus we cover different target languages. The price we pay is that we can only show partial correctness.

Because λ-calculus terms may contain $term$, they too reduce via $\Rightarrow$. These reduction rules realize the description of `term` in §2:

$$term\ (C_U\ nm\ vs) \Rightarrow C\ nm\ \cdot\cdot\ map\ term\ (rev\ vs)$$
$$term\ (V_U\ x\ vs) \Rightarrow V\ x\ \cdot\cdot\ map\ term\ (rev\ vs)$$
$$term\ (Clo\ vf\ vs\ n) \Rightarrow \Lambda\ (term\ (apply\ (lift\ 0\ (Clo\ vf\ vs\ n))\ (V_U\ 0\ [])))$$

The last clause formalizes η-expansion. By lifting, 0 becomes a fresh variable which the closure object is applied to and which is bound by the new Λ.

In addition we can reduce anywhere in a tm:

$$\frac{t \Rightarrow t'}{\Lambda\ t \Rightarrow \Lambda\ t'} \qquad \frac{s \Rightarrow s'}{s \cdot t \Rightarrow s' \cdot t} \qquad \frac{t \Rightarrow t'}{s \cdot t \Rightarrow s \cdot t'} \qquad \frac{v \Rightarrow v'}{term\ v \Rightarrow term\ v'}$$

3.4   Compilation

This section describes our compiler that takes a λ-calculus term and produces an ML term. Its type is $tm \Rightarrow (nat \Rightarrow ml) \Rightarrow ml$ and it is defined for pure terms only:

> $compile\ (V\ x)\ \sigma = \sigma\ x$
> $compile\ (C\ nm)\ \sigma = Clo\ (C_{ML}\ nm)\ []\ (arity\ nm)$
> $compile\ (s \cdot t)\ \sigma = apply\ (compile\ s\ \sigma)\ (compile\ t\ \sigma)$
> $compile\ (\Lambda\ t)\ \sigma = Clo\ (Lam_{ML}\ (compile\ t\ (V_{ML}\ 0\ \#\#\ \sigma)))\ []\ 1$

We explain the equations one by one.

1. In the variable case we look the result up in the additional argument σ. This is necessary to distinguish two situations. On the one hand the compiler is called to compile terms to be reduced. Free variables in those terms must be translated to V_U variables, their embedding in type Univ. Function *term* reverses this translation at the end of ML execution. On the other hand the compiler is also called to compile rewrite rules (R) to ML $(compR)$. In this case free variables must be translated to ML variables which are instantiated by pattern matching when that ML code is executed.
2. A constant becomes a closure with an empty argument list. The counter of missing arguments is set to *arity nm*, where *arity* is a global table mapping each constant to the number of arguments it expects. Note that our implementation takes care to create only closures with a non-zero counter—otherwise *apply* never fires. This does not show up in our verification because we only show partial correctness: even though the output would not be normal, it still would be a reduct of the input.
3. Term application becomes *apply*.
4. Term abstraction becomes a closure containing the translated ML function waiting for a single argument. The construction $V_{ML}\ 0\ \#\#\ \sigma$ is a new substitution that maps 0 to $V_{ML}\ 0$ and $i+1$ to $lift_{ML}\ 0\ (\sigma\ i)$. This is the de Bruijn way of moving under an abstraction.

Note that our actual compiler avoids building intermediate closures that are directly applied to an argument.

As explained above, the compiler serves two purposes: compiling terms to be executed (where the free variables are fixed) and compiling rules (where the free variables are considered open). These two instances are given separate names:

> $comp\text{-}open\ t = compile\ t\ V_{ML}$ $comp\text{-}fixed\ t = compile\ t\ (\lambda i.\ V_U\ i\ [])$

We can now define the set of compiled rewrite rules $compR$ as the union of the compilation of R and the default rules (§2) for each defined function symbol

$compR =$
$(\lambda(nm,\ ts,\ t).\ (nm,\ map\ comp\text{-}open\ (rev\ ts),\ comp\text{-}open\ t))\ `\ R\ \cup$
$(\lambda(nm,\ ts,\ t).\ \textsf{let}\ vs = map\ V_{ML}\ [0..{<}arity\ nm]\ \textsf{in}\ (nm,\ vs,\ C_U\ nm\ vs))\ `\ R$

where $f \, ' \, M$ is the image of a set under a function and $[m..{<}n]$ is the list $[m,\ldots,n-1]$. Since compilation moves from the term to the ML level, we need to reverse argument lists. On the left-hand sides of each compiled rule this is done explictly, on the right-hand side it happens implicitly by the interaction of *apply* with closures. For the default rewrite rules no reversal is necessary.

We can model the compiled rewrite rule as a set (rather than a list) because the original rewrite rules are already a set and impose no order. For partial correctness it is irrelevant in which order the clauses are tried. If the default rule is chosen, no reduction occurs, which is correct, too. Of course the actual implementation puts the default clause last. The implementation also ensures that in all clauses $f p_1 \ldots p_n = t$ for some function f, n is the same: additional parameters can always be added by extensionality.

3.5   Verification

The main theorem is partial correctness of compiled evaluation at the ML level w.r.t. term reduction:

Theorem 1. *If pure t, pure t' and term (comp-fixed t) $\Rightarrow*$ t' then t $\rightarrow*$ t'.*

Let us examine the key steps in the proof. The two inductive lemmas

Lemma 2. *If pure t and $\forall i.\ \sigma\ i = V_U\ i\ []$ then (compile t σ)! = t.*

Lemma 3. *If pure t and $\forall i.\ closed_{ML}\ n\ (\sigma\ i)$ then $closed_{ML}\ n\ (compile\ t\ \sigma)$.*

yield $(term\ (comp\text{-}fixed\ t))! = t$ and $closed_{ML}\ 0\ (term\ (comp\text{-}fixed\ t))$. Then

Theorem 4. *If t $\Rightarrow*$ t' and $closed_{ML}\ 0\ t$ then t! $\rightarrow*$ t'! $\wedge\ closed_{ML}\ 0\ t'$.*

yields the desired result $t \rightarrow* t'$ (because $pure\ t' \Longrightarrow t'! = t'$). Theorem 4 is proved by induction on $\Rightarrow*$ followed by induction on $\Rightarrow$. The inner induction, in the *term* case, requires the same theorem, but now on the ML level:

Theorem 5. *If v $\Rightarrow$ v' and $closed_{ML}\ 0\ v$ then v! $\rightarrow*$ v'! $\wedge\ closed_{ML}\ 0\ v'$.*

This is proved by induction on the reduction $\Rightarrow$ on ML terms. There are two nontrivial cases: β-reduction and application of a compiled rewrite rule. The former requires a delicate and involved lemma about the interaction of the kernel and substitution which is proved by induction on u (and whose proof requires an auxiliary notion of substitution):

Theorem 6. *If $closed_{ML}\ 0\ v$ and $closed_{ML}\ (Suc\ 0)\ u$ then $(u[v/0])! = ((lift\ 0\ u)[V_U\ 0\ []/0])![v!/0]$.*

The application of a compiled rewrite rule is justified by

Theorem 7. *If $(nm,\ vs,\ v) \in compR$ and $\forall i.\ closed_{ML}\ 0\ (\sigma\ i)$ then $C\ nm \cdot\cdot (map\ (subst_{ML}\ \sigma)\ (rev\ vs))! \rightarrow* (subst_{ML}\ \sigma\ v)!$.*

That is, taking the kernel of a compiled and instantiated rewrite rule yields a rewrite on the λ-term level. The conclusion is expressed with $\rightarrow *$ rather than $\rightarrow$ because the rule in *compR* may also be a default rule, in which case both sides become identical.

The proof of Theorem 7 requires one nontrivial inductive lemma:

Lemma 8. *If pure t and* $\forall i.\ closed_{ML}\ 0\ (\sigma\ i)$ *then* $(subst_{ML}\ \sigma\ (comp\text{-}open\ t))!$ $= subst\ (kernel \circ \sigma)\ t.$

In the proof of Theorem 7 this lemma is applied to *vs* and *v*, which are the output of *comp-open* by definition of *compR*. Hence we need that all rules in R are pure:

$$(nm,\ ts,\ t) \in R \implies (\forall\, t \in set\ ts.\ pure\ t) \land pure\ t$$

This is an axiom because R is otherwise arbitrary. It is trivially satisfied by our implementation because the inclusion of *term* as a constructor of λ-terms is an artefact of our model.

4   Realization in Isabelle

The implementation of our NBE approach in Isabelle/HOL is based on a generic code generator framework [12]. The following diagram and description explains how this is connected to the rest of Isabelle:

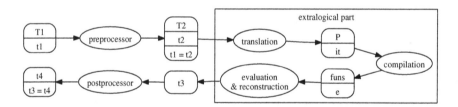

1. The input is an Isabelle term t_1 to be normalized w.r.t. a set of equational theorems T_1 (and β-reduction). Until evaluation both t_1 and T_1 are processed in parallel.
2. The framework allows one to configure arbitrary logical transformations on input t_1 (and T_1) and output t_3 (pre- and postprocessing). This is for the user's convenience and strictly on the level of theorems: both transformations yield equational theorems $t_1 = t_2$ and $t_3 = t_4$; together with the equation $t_2 = t_3$ stemming from the actual evaluation (this is where we have to trust the evaluator!), the desired $t_1 = t_4$ is obtained by transitivity and returned to the user.
3. The main task of the framework is to transform a set of equational theorems T_2 into a program P (and t_2 into it) in an abstract intermediate language capturing the essence of languages like SML or Haskell with an equational semantics. The intermediate term language is practically the same as the

Isabelle term language, and the equational semantics is preserved in the translation. The key changes are the replacement of an unordered *set* of equational theorems by a structured presentation with explicit dependencies, and, most importantly, the removal of overloading and the dictionary translation of type classes. For details see [12]. Inputs to NBE are in this intermediate language. Having compiled away type classes and overloading, NBE operates on terms following the Hindley-Milner type discipline, as assumed in §2.

4. *P* is compiled (via *comp-open*, see §3.4) to a series of SML function definitions `funs` and *it* (via *comp-fixed*) to an SML term `e`. Then `term` (`let funs in e end`) is given to the SML compiler, causing the evaluation of `e` and the translation of the result back into an Isabelle term; type reconstruction (see §2) on the result yields t_3.

We conducted a number of timing measurements to determine the relative performance of NBE w.r.t. two other normalization mechanisms available in Isabelle:

simp, the symbolic simplifier which operates on the level of Isabelle terms and theorems and produces a theorem purely by inference.

eval, the ground evaluator which compiles terms and theorems directly to SML, without support for open terms. It uses the same code generator framework but defines a native SML datatype for each Isabelle datatype, rather than operating on a universal datatype. For details see [12].

Our setup for this experiment ensures that all three evaluators use the same equational theorems and the same reduction strategy.

We measured the performance of three different programs: *eras* computes the first 100 prime numbers using the Sieve of Eratosthenes in a symbolic and naive implementation; *graph* computes the strongly connected components of a graph represented as a finite set of pairs; *sort* sorts a list of strings by insertion sort:[2]

	eras		*graph*		*sort*	
simp	4304	1384%	222717	11404%	1451169	22488%
nbe_s	339	109%	3312	170%	11174	173%
nbe	311	100%	1953	100%	6453	100%
eval	48	15%	292	15%	393	6%

Unsurprisingly, *nbe* turns out to be faster than *simp* and slower than *eval*. However the relative differences increase from left to right. In this order also the use of pattern matching in the examples increases. This shows the superiority of native pattern matching as exploited by *eval* over the pattern matching via strings in some universal datatype as required by *nbe*, which is in turn superior

---

[2] Absolute figures are in milliseconds using Isabelle 2007 with PolyML 5.1 on a Linux 2.6 AMD 1 GHz machine.

to pattern matching programmed in SML as in *simp*. This relevance of pattern matching motivated us to use integers (not strings) to identify constant names in patterns. Indeed, if we use an implementation using strings for constant names (nbe_s), there is a considerable loss of efficency.

There is a trade-off between performance and expressiveness. While *eval* is fast, it can evaluate only closed terms. Furthermore, if the result of *eval* is to be "read back" as an Isabelle term, it must only contain constructors and no function values. Finally, *eval* cannot cope with additional rewrite rules like associativity. With a comparably small performance penalty *nbe* can lift all these restrictions, while still outperforming the simplifier by 1–2 orders of magnitude.

5   Related Work

The work probably most closely related to ours is that of Berger, Eberl, and Schwichtenberg [3,4] who also integrated NBE into a proof assistant. However, their approach is based on a type-indexed semantics with constructors coinciding with those of the object language. Besides the administrative hassle, the commitment to a particular type system in the object language, and unneeded and unwanted η-expansions, the main disadvantage of this choice is that functions, like the **append** function in our example in §2, cannot serve the role as additional constructors. Note that in our example, this usage of an **append** constructor made it possible to effortlessly incorporate associativity into the definition of **T_append**, with pattern matching directly inherited from the implementation language.

The unavailability of the shape of a semantical object, unless it is built from a canonical constructor of some ground type, made it necessary in the approach by Berger et al. to revert to the term representation. This led to the artificial (at least from a user's point of view) and somewhat obscure distinction between so-called "computational rules" and "proper rewrite rules" where only the former are handled by NBE. The latter are carried out at a symbolic level (using pattern matching on the term representation). This mixture of computations on the term representation and in the implementation language requires a continuous changing between both representations. In fact, one full evaluation and reification is performed for each single usage of a rewrite rule.

Following Aehlig and Joachimski [1], our proof shows again that correctness of NBE is completely independent of any type system. In particular, no new version of NBE has to be invented each and every time it is applied to some term system with a different typing discipline. There simply is no need for logical relations in the proof.

Two other theorem proving systems provide specialized efficient normalisers for open λ-terms. Both of them are based on *abstract machines* and are therefore complementary to our compiled approach:

– Barras [2] extends the HOL [10] system with an abstract reduction machine for efficient rewriting. It is as general as our approach and even goes through the inference kernel. For efficiency reasons HOL's term language was extended with explicit substitutions.

- Grégoire and Leroy [11] present and verify a modification of the abstract machine underling OCaml. This modified abstract machine has become part of Coq's trusted proof kernel. The main difference is that they cannot deal with additional rewrite rules like associativity.

Compiled approaches to rewriting of first-order terms can also be found in other theorem provers, e.g. KIV [17].

6   Future Work

A small extension of the formalization is the straightforward proof normality of the output (see §2). More interesting are extensions of the class of permitted rewrite rules:

- Currently the implementation inherits ML's restriction to left-linear rules. It can be lifted to allow repeated variables on the left-hand side roughly as follows: make all variables distinct on the left-hand side but check for equality on the right-hand side. The details are more involved.
- More adventurous generalizations include ordered rewriting (where a rewrite rule only fires if certain ordering constraints are met) and conditional rewriting. The former should be easy to add, the latter would require a nontrivial generalization of the underlying code generator framework.

It would also be interesting to model λ-terms by different means than de Bruijn indices. Particularly prominent is the nominal approach [16] and its realisation by Urban [18] in Isabelle. As about one third of our proofs are primarily concerned with de Bruijn indices, it would be an interesting comparison to redo the verification in the nominal setup. Our preference for de Bruijn terms is due to the fact that the current implementation of nominal data types in Isabelle does not support nested data types, where recursion is through some other data type like *list*, which occurs in our model of ML terms.

References

1. Aehlig, K., Joachimski, F.: Operational aspects of untyped normalization by evaluation. Mathematical Structures in Computer Science 14(4), 587–611 (2004)
2. Barras, B.: Programming and computing in HOL. In: Aagaard, M.D., Harrison, J. (eds.) TPHOLs 2000. LNCS, vol. 1869, pp. 17–37. Springer, Heidelberg (2000)
3. Berger, U., Eberl, M., Schwichtenberg, H.: Term rewriting for normalization by evaluation. Information and Computation 183, 19–42 (2003)
4. Berger, U., Eberl, M., Schwichtenberg, H.: Normalization by evaluation. In: Möller, B., Tucker, J.V. (eds.) NADA 1997. LNCS, vol. 1546, pp. 117–137. Springer, Heidelberg (1998)
5. Berger, U., Schwichtenberg, H.: An inverse of the evaluation functional for typed λ–calculus. In: Vemuri, R. (ed.) Proceedings of the Sixth Annual IEEE Symposium on Logic in Computer Science (LICS 1991), pp. 203–211 (1991)

6. de Bruijn, N.G.: Lambda calculus notation with nameless dummies, a tool for automatic formula manipulation, with application to the Church–Rosser theorem. Indagationes Mathematicae 34, 381–392 (1972)
7. Danvy, O.: Type-directed partial evaluation. In: Proceedings of the Twenty-Third Annual ACM SIGPLAN-SIGACT Symposium on Priciples Of Programming Languages (POPL 1996) (1996)
8. Danvy, O., Rhiger, M., Rose, C.H.: Normalisation by evaluation with typed syntax. Journal of Functional Programming 11(6), 673–680 (2001)
9. Gonthier, G.: A computer-checked proof of the four-colour theorem, http://research.microsoft.com/ gonthier/4colproof.pdf
10. Gordon, M.J.C., Melham, T.F. (eds.): Introduction to HOL: a theorem-proving environment for higher order logic. Cambridge University Press, Cambridge (1993)
11. Grégoire, B., Leroy, X.: A compiled implementation of strong reduction. In: International Conference on Functional Programming 2002, pp. 235–246. ACM Press, New York (2002)
12. Haftmann, F., Nipkow, T.: A code generator framework for Isabelle/HOL. In: Schneider, K., Brandt, J. (eds.) Theorem Proving in Higher Order Logics: Emerging Trends Proceedings. Department of Computer Science, University of Kaiserslautern (2007)
13. Krauss, A.: Partial recursive functions in higher-order logic. In: Furbach, U., Shankar, N. (eds.) IJCAR 2006. LNCS (LNAI), vol. 4130, pp. 589–603. Springer, Heidelberg (2006)
14. Nipkow, T., Bauer, G., Schultz, P.: Flyspeck I: Tame graphs. In: Furbach, U., Shankar, N. (eds.) IJCAR 2006. LNCS (LNAI), vol. 4130, pp. 21–35. Springer, Heidelberg (2006)
15. Nipkow, T., Paulson, L.C., Wenzel, M.T.: Isabelle/HOL. LNCS, vol. 2283. Springer, Heidelberg (2002), http://www.in.tum.de/ nipkow/LNCS2283/
16. Pitts, A.M.: Nominal logic, a first order theory of names and binding. Information and Computation 186, 165–193 (2003)
17. Reif, W., Schellhorn, G., Stenzel, K., Balser, M.: tructured specifications and interactive proofs with KIV. In: Bibel, W., Schmitt, P. (eds.) Automated Deduction—A Basis for Applications. Systems and Implementation Techniques, vol. II, pp. 13–39. Kluwer Academic Publishers, Dordrecht (1998)
18. Urban, C., Tasson, C.: Nominal techniques in Isabelle/HOL. In: Nieuwenhuis, R. (ed.) CADE 2005. LNCS (LNAI), vol. 3632, pp. 38–53. Springer, Heidelberg (2005)

LCF-Style Propositional Simplification with BDDs and SAT Solvers

Hasan Amjad

Middlesex University School of Computing Science, London NW4 4BT, UK
Hasan.Amjad@cl.cam.ac.uk

Abstract. We improve, in both a logical and a practical sense, the simplification of the propositional structure of terms in interactive theorem provers. The method uses Binary Decision Diagrams (BDDs) and SAT solvers. We present experimental results to show that the time cost is acceptable.

1 Introduction

We consider the problem of simplifying the propositional structure of terms, in interactive theorem provers (ITPs) based on higher-order logic (HOL) or stronger type systems.

Most such ITPs use rewriting or equational reasoning (semi-automatic or manual) to do such simplification. Such tools include Coq [12], HOL4 [8], HOL Light [10], Isabelle/HOL [18], and PVS [17]. The PVS theorem prover is unique in this family, in additionally optionally allowing the use of Binary Decision Diagrams (BDDs) [1] for propositional simplification [2]. The propositional structure of the input term is encoded as a BDD, from which PVS can automatically extract a term in conjunctive normal form (CNF) that is logically equivalent to the input term. As BDDs usually achieve very compact encodings, the expectation is that the extracted term will be simpler than the input term.

BDDs are powerful tools for propositional reasoning, and have seen widespread adoption in the automated reasoning community. And yet BDDs are rarely if ever used by ITPs. In the current context, there are four broad objections to the use of BDDs for propositional simplification in ITPs:

1. *Not needed.* Current rewriting based implementations of simplifiers are sufficient for the terms that typically occur in interactive proof.
2. *Unsuitable.* The process of BDD-based simplification converts the term to CNF. This destroys the structure of the term and thus may often destroy any intuition that the human ITP user may have had about the term.
3. *Inefficient for LCF-style.* All the ITPs mentioned above, except PVS, are "LCF-style", or follow the "de Bruijn criterion". Roughly speaking, this means that they employ some high assurance facility for verifying their proofs. Typically, this is by translation of the proof to a very simple proof system – the implementation of which is easily understood and well tested

O. Ait Mohamed, C. Muñoz, and S. Tahar (Eds.): TPHOLs 2008, LNCS 5170, pp. 55–70, 2008.

– which accepts all correct proofs (and no incorrect ones) produced by the ITP. Verifying BDD operations in this fashion has been tried and found to be very costly [9,21].

4. *Classical.* BDDs are based on classical propositional logic, which limits their usefulness in non-classical contexts.

The last objection cannot be overcome using BDDs as they are currently implemented. Thus, we restrict ourselves to the classical setting. We address the remaining objections in this paper:

1. *Need.* We augment BDD-based simplification with new and known clause-form simplifications and prove that our simplification method can provide logical guarantees about the simplified term that are not provided by current rewrite-based simplifiers. We present experimental results that suggest that in practice our method always does better than rewrite-based simplifiers on a quantifiable measure of simplification quality.
2. *Suitability.* Our method improves on the BDD-based simplification of PVS by not completely flattening the input term. Instead, it selectively applies simplification to suitable sub-terms, thus largely retaining term structure.
3. *Efficiency.* We show that LCF-style BDD-based simplification is possible at not too great a cost, by verifying the BDD operations using a recent LCF-style integration [22] of SAT solvers [16,4] and ITPs. We present experimental results to support this claim.

The next section describes related work. We then give a brief account of the relevant aspects of normal forms, BDDs and SAT solvers, to keep the paper self-contained. Finally, we describe our work (§4) and present experimental results (§5). Henceforth, all discussion is restricted to classical purely propositional HOL terms, unless explicitly stated otherwise.

2   Related Work

There is a reasonable body of work on integrating BDDs in interactive provers. One of the earliest results combined higher-order logic with BDDs for symbolic trajectory evaluation [13]. A little later, temporal symbolic model checking was done in PVS [19]. These integrations trusted the underlying BDD engines. At about the same time, a serious attempt at using BDDs in an LCF-style manner [9] reported an approximate 100x slowdown. Later, a larger project added BDDs to the Coq theorem prover [21] and reported similar slowdowns, except that the faster programs were themselves extracted by reflection from the Coq representation, and could thus said to have higher assurance. The penalty for checking BDD proofs has thus more or less ensured that BDDs are not used internally by LCF-style theorem provers, in a non-trusted manner. There have of course been trusted integrations of BDDs with LCF-style provers [7].

This does not rule out the use of BDDs in interactive provers in general. BDDs are used in the ACL2 prover [14] to help with conditional rewriting and for deciding equality on bit vectors (see ACL2 System Documentation). The PVS theorem prover can use BDDs for propositional simplification, via its bddsimp function [2]. This was the inspiration for our work. Roughly speaking, when invoked on a goal with propositional structure, it uses BDDs to obtain the CNF of the goal, and each conjunct of the CNF becomes a separate subgoal.

LCF-style integrations of SAT solvers with interactive provers have a shorter history. The integration is trivial for the case where the solver returns a satisfying assignment: we simply substitute the assignments into the input term and check that the resulting ground term evaluates to true. This can be done efficiently. For the unsatisfiable case, the earliest work we know of is the LCF-style programming of Stålmarck's Algorithm in an ancestor of HOL4 [11]. This achieved good results but was never distributed due to licensing issues. Further work had to wait for the arrival of DPLL-based proof producing SAT solvers [25] and mature integrations were reported relatively recently [22].

3   Technical Background

We use $\Gamma \vdash t$ to denote that t is a theorem (under hypotheses Γ) in the mechanized object logic, i.e, the logic of the interactive prover. Quantification binds weaker than $\Leftrightarrow$ which binds weaker than all other propositional connectives. Propositional truth is denoted by $\top$ and falsity by $\bot$. All other notation is standard.

In pure propositional logic, there is no concept of variables. In HOL, variables of Boolean type do double duty as propositional letters. We will refer to propositional letters as variables, keeping in mind that quantification over these is not allowed in our setting.

3.1   Normal Forms

A *literal* is either a variable or its negation. Any term has a finite number of variables, so all involved literals can be encoded as numbers when working on a given term. We shall switch between the term and number representation of literals as convenient.

A *clause* is a disjunction of literals. Since both conjunction and disjunction are associative, commutative and idempotent (ACI), clauses can also be interpreted as sets of literals. If a literal occurs in a set, then we abuse notation and assume its underlying proposition also occurs in the set. We assume that any trivial clauses, i.e., containing both a literal and its negation, have been filtered out.

A term is in *conjunctive normal form* (CNF) if it is a conjunction of clauses. Any propositional term t can be transformed into a logically equivalent term in CNF. Again, by ACI, a CNF term can be interpreted as a set (of sets of literals), and we overload the notation accordingly. We will switch back and forth between the term and set interpretations, as convenience dictates.

Any term can be transformed to CNF, but the result can be exponentially larger than the original term. To avoid this, *definitional* CNF [20] introduces extra fresh[1] Boolean variable names as place-holders for subterms of the original

---

[1] Guaranteed not to already occur in the term.

problem. Conversion to definitional CNF is linear time in the worst case. We use $dCNF(t)$ to denote the definitional CNF of t.

The term $dCNF(t)$ is not logically equivalent to t, since there are valuations for the introduced *definitional variables* that can disrupt an otherwise satisfying assignment to the variables of t. However, it is equisatisfiable. This is expressed by the theorem

$$t \Leftrightarrow \exists V.dCNF(t) \tag{1}$$

where V is the set of all the definitional variables and the existential quantification is lifted to all $v \in V$ in the usual way.

3.2   BDDs

Reduced Ordered Binary Decision Diagrams (ROBDDs, shortened to BDDs) [1] are data structures for efficiently representing Boolean terms and Boolean operations on them. In theory, the problem is NP-complete. In practice, BDDs can often achieve very compact representatations. They are built by starting with the BDDs representing propositional variables and performing a bottom-up construction using the BDD operations corresponding to each propositional connective in the term.

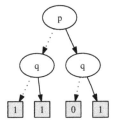   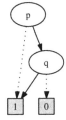

Fig. 1. Binary decision tree for $p \Rightarrow q$ **Fig. 2.** Corresponding ROBDD

For example, the decision tree of the term $p \Rightarrow q$ is given in Figure 1, and its BDD in 2. Dotted arcs indicate a valuation of $\bot$ and solid arcs a valuation of $\top$ to the parent node. A path from the root to the 1 node indicates an assignment that makes the formula true, and a path to the 0 node, a falsifying assignment.

To read off the CNF equivalent to a term t represented by a BDD, we treat every path from the root to the 0 node, with the signs of variables inverted, as a clause. So, for example, the CNF for Figure 2 is $\neg p \lor q$. This can be done by a single depth-first search of the BDD structure. We denote by $BDDCNF(t)$ the CNF term read off the BDD of a pure propositional term t. We state a standard result:

Proposition 1. $t \Leftrightarrow BDDCNF(t)$

A term is *tautology free* if no strict sub-term of it is true. A term is *contradiction free* if no strict sub-term of it is false. BDDs are canonical and built bottom-up. Hence sub-terms that are tautologies or contradictions are detected during construction, and are absorbed into the BDD. So the following result also holds:

Proposition 2. *BDDCNF(t) is tautology free and contradiction free.*

Representing BDDs efficiently in an LCF style prover causes too high a performance penalty (see §2 for details). Therefore, we assume that the results of BDD operations by themselves cannot produce LCF-style theorems in the object logic.

3.3   SAT Solvers

SAT solvers are algorithms for testing Boolean satisfiability. A SAT solver will accept a Boolean term in CNF and return a satisfying assignment to its variables. If the term is unsatisfiable, the solver will simply say so, though some SAT solvers will also return a resolution refutation proof from the clauses of the input CNF term [25]. Such proof-producing SAT solvers have been integrated with LCF-style ITPs [22]. We assume we have access to such an LCF-style integration.

Suppose we have a propositional term t, and we wish to check whether or not it is a tautology. This can be done by computing $dCNF(\neg t)$ (which can be done efficiently) and asking a SAT solver if that term is unsatisfiable. If so, we can derive $\vdash t$.

Thus, we can assume access to a black box procedure $SATprove(t)$ that returns $\vdash t$ iff t is a valid pure propositional term. This short description is sufficient for our purposes. A tutorial introduction to SAT solvers is available [15].

3.4   CNF Simplification

In general, SAT solvers require input in CNF. This has lead to much work on CNF simplification in the SAT community.

Equivalence-preserving CNF simplifications. Of special interest to us are equivalence-preserving simplifications, since these can be directly useful in term simplification where we must derive a logically equivalent but simpler term. Two methods, subsumption reduction (S-reduction for short) and decremental resolution reduction (DR-reduction), have been very effective in practice [3,24].

A clause C *subsumes* another clause D iff $C \subseteq D$, i.e., $C \Rightarrow D$. D is then called *subsumed* and C is called an *S-clause*. Given a clause set, any clauses subsumed by other clauses in the set are redundant and can be removed. This removal, an S-reduction, preserves equivalence, so we have that,

$$\forall CD.(C \Rightarrow D) \Rightarrow (C \wedge D \Leftrightarrow C) \tag{2}$$

A CNF term is considered *subsumption free* (S-free for short) if it has no S-clauses.

The *propositional resolution* rule is

$$\frac{C \vee p \qquad D \vee \neg p}{C \cup D}$$

where the resultant clause is called the *resolvent*, and written as $C \vee p \otimes D \vee \neg p$. p is the *pivot* literal, written as $pivot(C \vee p \otimes D \vee \neg p)$. The pivot occurs complementarily in the two input clauses; if needed, the sign will be clear from the context.

A resolution is *decremental* if the resolvent implies either of the two input clauses. If so, the implied input clause can be strengthened by removing its pivot literal (equivalent to replacement by the resolvent). This strengthening, a DR-reduction, also preserves equivalence, so we have,

$$\forall CD.(C \otimes D \Rightarrow D) \Rightarrow (C \wedge D \Leftrightarrow C \wedge C \otimes D) \tag{3}$$

A DR-reduction is possible iff one of the two input clauses of the corresponding resolution subsumes the other, modulo the pivot. The almost-subsuming clause is called a *DR-clause*. A CNF term is considered *decremental resolution free* (DR-free for short) if it has no DR-clauses.

We shall adapt the method of Een et al. [3] to our specific circumstances (§4.1). Their method turns a clause set S S-free, by checking whether any $C \in S$ is an S-clause. Let $L(p) = \{C|p \in C \wedge C \in S\}$, i.e, it gives all clauses in which the literal p occurs. Let $\#(C) = \bigoplus_{p \in C} 2^{p \bmod 64}$ where $\oplus$ is bitwise OR, i.e., $\#(C)$ computes a 64-bit hash of clause C. An overview of the algorithm is given in Figure 3, where & is bitwise AND, and ! is bitwise NOT. The test in line 5 is a fast semi-complete subset test: if true it guarantees that $C \not\subseteq D$, and avoids doing the full (and expensive) subset check. They then achieve DR-freedom as shown in Figure 4, where the test in line 3 uses the S-reduction routine.

1. **foreach** $C \in S$
2. $p \leftarrow$ the p such that $p \in C \wedge \forall q \in C.|L(p)| \leq |L(q)|$
3.   **foreach** $D \in L(p) - \{C\}$
4.   **if** $|C| > |D|$ **then continue**
5.   **if** $\#(C) \& !\#(D) \neq 0$ **then continue**
6.   **if** $C \subseteq D$ **then** $S \leftarrow S - \{D\}$

Fig. 3. S-reduction detection

1. **foreach** $C \in S$
2.   **foreach** $p \in C$
3.   **foreach** $D \in S - \{C\}$ such that $C[p \leftarrow \neg p] \subseteq D$
4. $D \leftarrow D - \{\neg p\}$

Fig. 4. DR-reduction detection

SAT-based CNF simplifications. A considerably more powerful class of CNF simplifications [6,23] removes redundant clauses by using information gleaned from invoking the SAT solver on the CNF term. They are all based on the fact that if a SAT solver produces a refutation from some CNF term t, then clauses of t not participating in the proof can be removed without affecting the unsatisfiability of t. The same cannot be said if t is satisfiable, since any clauses not used in finding a given satisfying assignment are not necessarily irrelevant to the truth value of t. We shall use the black box call $SATsimp(t)$ to denote the use of such off-the-shelf SAT-based simplifications, under the assumption that the CNF term t is unsatisfiable, and that the call returns some subset c of the clauses of t, such that c is unsatisfiable iff t is unsatisfiable.

4   Simplification

Roughly speaking, the core of our method works as follows:

1. Convert input term t to a BDD and read off the CNF equivalent c_0.
2. Further simplify c_0 using new and known CNF simplifications to obtain c_1.
3. Find redundant clauses in c_1 using SAT-based simplifications to obtain s.
4. If needed, use LCF-style SAT solver interface to prove $\vdash t \Leftrightarrow s$

Note that LCF-style proof is applied only in the final step, so the other phases can be optimised without regard for proof. The main challenges are: avoiding too large a c_0 in BDD construction; fast CNF simplification; using powerful SAT-based simplifications that work only on unsatisfiable terms, for arbitrary terms; achieving a useful term simplification. The last goal is intentionally vague, for now. We discuss this further in §4.3.

4.1   Faster CNF Simplification

The DR-reduction check in Figure 4 requires computing the hash of each $C[p \leftarrow \neg p]$ for each $p \in C$. We cannot compute them incrementally, because we cannot tell whether removing a literal from a clause turned the corresponding bit position in the hash to 0, without considering all the other literals of the clause.

We can improve on this since our goal is simplification of terms encountered in *interactive* proof: in ITPs, automatic or semi-automatic proof procedures very rarely use full-blown simplification internally, for efficiency reasons. This means that the terms we encounter will have considerably fewer variables than the typical SATLIB problem.

Suppose that instead of using a single 64-bit word for the hash, we use enough words so that $\#(-)$ maps each literal to a unique bit position in the hash, i.e., we turn $\#(-)$ into a perfect hashing function. Then the hash-based subset test (Figure 3, line 5) becomes complete. Further, if C is a subset of D modulo some $p \in C$ such that $\neg p \in D$, then $\#(C) \,\&\, !\#(D)$ will have exactly one bit switched on. This can also be detected in constant time (assuming fixed hash size). Figure 5 outlines a method that uses multiword hashes, and combines checking whether some $C \in S$ is an S-clause or a DR-clause. The recursion terminates because the number of literal occurrences in the underlying set S is always strictly smaller with each call. So either eventually there are no more reductions to be found, or we discover the empty clause $\bot$, which subsumes all clauses so S is reduced to $\{\bot\}$. This method is very fast, because it dispenses with the expensive subset checks altogether.

1. $EQsimp\_clause(C)$
2. $p \leftarrow$ any p such that $p \in C \wedge \forall q \in C.|L(p)| \leq |L(q)|$
3.   **foreach** $D \in L(p) - \{C\}$
4.   **if** $|C| > |D|$ **then continue**
5. $h \leftarrow \#(C) \,\&\, !\#(D)$
6.   **if** $h = 0$ **then** $S \leftarrow S - \{D\}$
7.   **else if** $h \& (h - 1) = 0$ **then**
8. $D \leftarrow D - \{pivot(C \otimes D)\}$
9. $EQsimp\_clause(D)$

Fig. 5. Combined S-reduction and DR-reduction detection

Our actual implementation is slightly more complex: we also need to check at various points that a clause under consideration has not already been removed from S due to the result of a previous call.

Our overall equivalence-preserving simplification procedure, $EQsimp$, takes as argument a CNF term S (actually a set of set of numbers) and then calls $EQsimp\_clause$ for each $C \in S$. The number of 64-bit words for the hash is determined once per $EQsimp$ call, and is given by dividing the number of variables by 32 and rounding up. This approach is not feasible for SATLIB problems with millions of variables, but in our experiments with interactive goals we rarely needed a hash size of more than four 64-bit words.

Proposition 3. $EQsimp(t)$ *is S-free and DR-free*

Proof $\#(-)$ is now a perfect hash, so bitwise operations on clause hashes coincide with logical operations on clauses. Hence, the test for S-reduction on line 6 of Figure 5 is sound and complete. For DR-reduction, if C subsumes D modulo $pivot(C \otimes D)$, the bit position for the occurrence of the pivot in C will be switched on in both $\#(C)$ and $!\#(D)$, and hence in their conjunction. The remaining bits in the conjunction will be off, as in the S-reduction test. Hence h will be "1-hot", i.e., will have exactly one bit switched on. The test in line 7 turns the most significant switched-on bit of h to off, thereby allowing 1-hot detection. So, the DR-reduction test is sound and complete.

$EQsimp(t)$ checks each clause for being an S-clause or DR-clause. Once checked, a clause cannot again become an S-clause or DR-clause unless it is strengthened in line 8, in which case we recheck it in line 9. Thus, when the algorithm terminates, no clause is an S-clause or a DR-clause. $\square$

Proposition 4. $t \Leftrightarrow EQsimp(t)$

Proof Immediate from (2) and (3). $\square$

4.2   Simplification Using SAT Solvers

Figure 6 gives an overview of our core algorithm, *simplify*, which takes a pure propositional term t and attempts to simplify it by BDD-based conversion to

$$1.\; simplify(t)$$
$$2.\qquad c_0 \leftarrow BDDCNF(t)$$
$$3.\qquad c_1 \leftarrow EQsimp(c_0)$$
$$4.\qquad s \leftarrow SATsimp(c_1 \wedge dCNF(\neg t)) \cap c_1$$
$$5.\qquad \textbf{return}\; SATprove(t \Leftrightarrow s)$$

Fig. 6. Core simplification method

CNF, application of CNF simplifications, and SAT-based simplifications, followed by a call to $SATprove$ if an LCF-style theorem is required. We have already seen all the simplifications except for SAT-based simplification, which will be the focus here.

We have $t \Leftrightarrow c_1$ by Propositions 1 and 4. However, the powerful $SATsimp$ works only for unsatisfiable terms, and in general we cannot expect to be so fortunate. To use it for simplifying arbitrary terms, we devise a specially crafted argument for $SATsimp$ and intersect the result with c_1. The correctness of this construction is central to the main theoretical result of the paper.

Theorem 5. $t \Leftrightarrow s$

Proof We have $t \Leftrightarrow c_1$. Now

$$(t \Leftrightarrow c_1) \Leftrightarrow (\neg t \vee c_1) \wedge (\neg c_1 \vee t) \tag{4}$$

and hence

$$c_1 \wedge \neg t \Leftrightarrow \bot \tag{5}$$

Then,

$$c_1 \wedge \neg t \Leftrightarrow \bot$$
$$\text{iff } c_1 \wedge (\exists V. dCNF(\neg t)) \Leftrightarrow \bot \quad \text{by (1)}$$
$$\text{iff } (\exists V. c_1 \wedge dCNF(\neg t)) \Leftrightarrow \bot \quad \text{no } v \in V \text{ occurs in } c_1$$
$$\text{iff } \forall V. \neg(c_1 \wedge dCNF(\neg t))$$

So $c_1 \wedge dCNF(\neg t)$ is unsatisfiable by (5), and of course it is in CNF, meeting the pre-conditions for the call to $SATsimp$, which returns some subset of its input clauses. Clearly $s \subseteq c_1$, hence $c_1 \Rightarrow s$. Then

$$\neg c_1 \wedge t \Rightarrow \bot \qquad\qquad \text{by (4) and } t \Leftrightarrow c_1$$
$$\text{iff } \neg s \wedge t \Rightarrow \bot \qquad\qquad \text{since } \neg s \Rightarrow \neg c_1 \tag{6}$$

Finally, s is that subset of c_1 that suffices for the proof of $c_1 \wedge dCNF(\neg t) \Leftrightarrow \bot$. So we have a proof of $s \wedge dCNF(\neg t) \Leftrightarrow \bot$. But $s \wedge dCNF(\neg t) \Leftrightarrow \bot$ iff $s \Rightarrow t$ by (1) together with some simple reasoning that uses the fact that no $v \in V$ occurs in s. Combining $s \Rightarrow t$ with (6) gives us $t \Leftrightarrow s$ as required. □

Thus, the call to $SATprove$ in line 5 succeeds, and we obtain $\vdash t \Leftrightarrow s$ as desired. Here s obeys the guarantees given by Propositions 2 and 3. Additionally, we know that s also does not contain any clauses of c_1 not needed by the SAT solver to prove $c_1 \Rightarrow t$. It may nevertheless contain redundant clauses: the solver does not guarantee to use the minimum number of clauses. So we do not phrase this as a guarantee.

4.3 Practicalities

Our hope is that s is "simpler" than t, though we have left this notion vague until now. ITP simplifiers perform many tasks, but in our setting we shall concentrate on the simplest one: given a pure propositional term, what does it mean to simplify it? We believe it means, first, to reduce the size without introducing new operators or variables, and second, to reduce the bracketing depth (modulo associativity) provided this does not conflict too much with the first goal.

These goals are quantifiable and approximate our intuition about propositional simplification to a reasonable degree. The restriction on the second goal implicitly acknowledges that flattening a term too much may lose structure that helps guide intuition. We shall measure term size and bracketing depth in the standard way. We denote the size of term t by $size(t)$.

By these criteria, the *simplify* procedure is too crude. It converts the term to possibly exponentially larger CNF, violating both goals. Instead, we parameterise *simplify* and use it as a subroutine in a control wrapper.

We introduce three parameters B_1, B_2 and F, of which the last two are real numbers, that enforce the following invariants on *simplify*:

1. No BDD has a node count exceeding B_1
2. $size(c_0) < size(t) \times B_2$
3. $size(s) < size(t) \times F$

If any invariant fails, *simplify* signals failure.

The first invariant imposes an upper bound on BDD size, and the second invariant ensures that the CNF generated from the BDD is not too big. The *BDDCNF* function is easily modified to enforce these invariants on-the-fly, rather than checking them after the full BDD or CNF term has been generated. However, our invariant checking for B_1 is not as low-level as it could be: it checks BDD size after each operation, so cannot avoid an exponential size explosion from a single BDD operation. This has not happened yet, but if it becomes a problem, we can simply impose a time limit on the BDD building procedure, rather than a size limit. The third invariant imposes an upper bound on $size(s)$.

All this is not enough, since invoking *simplify* on a term will now invariably report failure. Instead, our control wrapper invokes *simplify* once on every sub-term of t in a bottom-up manner. If the invocation succeeds, that sub-term is replaced by its simplified equivalent. Hence, the final simplified term is equivalent to t. Since *simplify* works semantically rather than by non-deterministic rewriting, there is no need to apply it to a given sub-term more than once.

This does mean that the relatively expensive *SATprove* call is made several times. We remedy the situation by modifying *simplify* to return the trivial theorem $t \Leftrightarrow s \vdash t \Leftrightarrow s$ on the last line, which takes negligible cost to generate. Recall that the control wrapper applies each sub-term simplification to t. So when the bottom up traversal has finished, we have a theorem of the form

$$t_0 \Leftrightarrow s_0, t_1 \Leftrightarrow s_1, \ldots \vdash t \Leftrightarrow s$$

where the t_i are those sub-terms of t which *simplify* succeeded in simplifying to s_i. We now make a single call to $SATprove(\bigwedge_i t_i \Leftrightarrow s_i)$, split the resulting theorem into its conjuncts, and use them to discharge every hypothesis in $t_i \Leftrightarrow s_i \vdash t \Leftrightarrow s$. Even though this $SATprove$ call deals with a larger term, in practice these terms are quite small by SAT solver standards, and do not really exercise the SAT interface. Thus, the only LCF-style proof is a single $SATprove$ call, followed by very low cost hypothesis discharges that number at most linear in $size(t)$ and considerably fewer in practice.

The guarantees of Proposition 2 and Proposition 3 now only hold locally, at sub-terms where *simplify* succeeded.

As observed earlier, a pleasant side effect of the invariants is that the term structure of t is not flattened beyond the reach of intuition. The degree of flattening can be controlled by the F parameter.

We claimed in the introduction (§1) to simplify the propositional structure of arbitrary terms, rather than pure propositional terms. This is easily done by replacing each atomic proposition with a fresh Boolean variable (but maintaining a bijection between the propositions and the variables), resulting in a pure propositional term that is logically equivalent (modulo the atomic propositions) to the original term. This is not new: PVS already does this, as does the HOL4 decision procedure for propositional tautologies.

As a quick example, consider the term

$$(((p \Rightarrow q) \Rightarrow p) \Rightarrow p) \wedge (p \wedge (p \Leftrightarrow q))$$

The HOL4, HOL Light, and Isabelle/HOL simplifiers (all in default setup) all fail to simplify this term, whereas our method returns $p \wedge q$ as expected. The top-level left conjunct is a tautology (Peirce's Law) that is famously hard for rewriting. The top-level right conjunct is interesting because all three simplifiers fail on it despite its simplicity.[2] Even the *BDDCNF* function by itself returns $p \wedge (\neg p \vee q)$ (assuming p is before q in the BDD ordering) and it requires a DR-reduction to obtain $p \wedge q$. PVS does return $p \wedge q$ but only because after extracting the CNF from the BDD, PVS performs a variant of unit propagation that, as a simplification strategy, is in general both weaker and slower than our reductions.

5   Experimental Results

In §1, we made three claims about the work: that it is not too slow, that it reduces term size better than existing rewrite-based simplifiers, and that it does

---

[2] They succeed if we make the simplifier aware of certain congruences for conjunction, which are not part of the default set of rewrites because they tend to slow down the rewriter and are not commonly useful.

so without annihilating term structure. Although we have shown that our simplified terms provide certain guarantees of semantic simplicity, these need not always translate into simpler syntax. Hence we present empirical evidence in support of the first two claims.

We compare our method against the HOL4 simplifier. The choice of HOL4 (rather than HOL Light, Isabelle/HOL or PVS) was made because:

1. We wish to evaluate the method in an LCF-style setting, ruling out PVS.
2. The HOL4 SAT interface's definitional CNF subroutine is faster than that of Isabelle/HOL. It avoids all proof by inlining definitions, as opposed to Isabelle/HOL where the definitional CNF produces an expensive LCF-style equivalance proof.
3. We understand the HOL4 simplifier better than the HOL Light simplifier, at the implementation level.

Our method was implemented to produce a valid LCF-style HOL4 theorem, in an interactive HOL4 proof environment. We used a trusted integration of HOL4 with the BuDDy BDD engine [7]. *EQsimp* was implemented by us in C++. For *SATsimp* we used the fixpoint technique of Zhang et al. [23]. For *SATprove*, we used an LCF-style integration of the MiniSat SAT solver [4] with HOL4 [22]. For the comparison, the HOL4 simplifier was invoked using SIMP_CONV bool_ss []. The test machine was an AMD Athlon X2 6400+, with 2GB of RAM. Memory consumption was not an issue for these experiments.

We used values of 10000 for B_1, 10.0 for B_2, and 1.1 for F throughout. The B_2 value reflects our intention that the CNF from the BDD should not be too large, but should be big enough to give the CNF simplifications something to work with. The F value reflects our intuition that a "simplified" term that is more than 10% bigger than its original size (without expanding any definitions, which never happens here) is best discarded. We have not experimented with any other values of the parameters.

Since existing propositional benchmark libraries like SATLIB and TPTP (BOO category) have problems already in CNF, we generated random propositional terms for testing, parameterised on a target term size. Randomness meant that for each term size parameter, actual term sizes varied slightly. Nine increasing target sizes were used. For each, the benchmarks were run 100 times. Table 1 presents the minimum, average and maximum term sizes and term bracketing depths for the input and simplified terms, and Table 2 gives the same statistics for execution times, for each size parameter.

As Table 1 shows, our method is always able to achieve a better reduction in term size. In the minimum simplified term sizes for our method, a 1 signifies a random term that was either a tautology or a contradiction, and this was caught by our method as per the guarantee of Proposition 2, and reduced to either ⊤ or ⊥. HOL4 invariably slightly increases the size of the term. In our method we can control this using the F parameter.

Table 1. Experimental results: term size and bracketing depth

| | Term Size | | | | | | | | Term Depth | | | | | | | | |
| | Input | | | HOL4 | | | BDD+SAT | | | Input | | | HOL4 | | | BDD+SAT | | |
	min	avg	max	min	avg	max	min	avg	max	min	avg	max	min	avg	max	min	avg	max
1	7	12.9	16	1	8.4	17	1	2.7	9	1	2.0	3	0	1.3	3	0	0.3	2
2	22	29.5	33	1	26.9	40	1	15.6	33	2	3.5	5	0	3.0	4	0	1.6	4
3	50	59.8	66	1	58.0	74	1	44.2	69	2	4.9	7	0	4.2	6	0	3.8	7
4	70	99.6	122	61	100.2	137	1	75.4	118	4	5.7	7	3	4.8	7	0	4.6	7
5	110	121.3	132	89	124.9	158	1	101.4	127	4	6.2	8	4	5.4	7	0	5.6	8
6	113	125.6	135	62	129.0	149	1	107.1	130	5	6.8	9	4	5.5	8	0	6.3	9
7	161	204.4	233	1	207.0	251	1	171.4	224	5	7.0	9	0	6.1	8	0	6.2	9
8	215	241.7	256	163	249.5	279	1	212.1	246	6	7.7	11	4	6.5	9	0	7.1	9
9	240	251.6	262	205	262.2	289	1	230.5	256	5	8.2	11	5	6.6	8	0	7.7	11

For bracketing depth, HOL4 achieves a marginally better score on larger terms. This may due to inwards movement of negations by HOL4, which reduces depth without badly affecting term size. More detailed analysis of this is needed.

Table 2. Experimental results: execution times (ms)

| | HOL4 | | | BDD+SAT | | |
	min	avg	max	min	avg	max
1	0	0.2	2	10	21.6	58
2	1	1.7	10	12	76.1	266
3	2	4.0	13	16	224.9	944
4	4	6.8	20	20	294.3	589
5	7	11.0	31	25	372.0	637
6	6	9.3	35	27	379.8	553
7	10	14.8	47	33	503.9	847
8	14	17.9	50	42	747.6	1620
9	14	18.7	53	40	876.2	3263

For execution time, (Table 2) HOL4 is about two orders of magnitude faster. However, we are not disappointed. First, in absolute terms our simplifier typically takes less than a second, which is acceptable in interactive proof. Second, our implementation is a mix of SML, C++, Perl and shell scripts, all communicating via disk files, brought on by the need to try out various third-party off-the-shelf implementations. Since the method invokes *simplify* for each subterm, the overhead of file creation, reading, writing, and spawning command shells adds up. This diagnosis is supported by profiling of individual modules: with the exception of *SATsimp*, they contribute very little to the time cost. Our current *SATsimp* works by invoking a SAT solver on the clause set file, extracting only the clauses used by the solver, and repeating the invocation on the new clause set file, until a fixpoint is reached. So it is also disk intensive. We fully expect the execution times to approach those of HOL4 with a little more engineering effort towards direct in-memory interfaces.

6   Conclusion

We have shown how BDDs and SAT solvers can be used for propositional term simplification. The method appears to improve on the simplification of pure propositional terms by rewrite-based simplifiers, at least on random terms with propositional structure sizes the same as those of terms typically encountered in interactive proof. Further we have proved that our simplified terms respect formal guarantees about semantic simplicity that cannot be furnished by rewrite-based simplifiers. Our treatment is tool independent, except that we require that the SAT solver is proof producing. We have also shown that the method can be used in an LCF-style framework with acceptable cost.

We are currently considering a more aggressive structure retention strategy that uses the definitions of definitional CNF to encode structural information, that can later be recovered from the CNF. However, given that the current method does a decent enough job, this is not very high on our list of priorities.

Even though we have restricted ourselves to propositional logic, in theory the results should be applicable to the use of SMT solvers rather than SAT solvers, and should allow us to do simplification in combinations of decidable theories. This awaits the development of a fast and mature LCF-style SMT interface, work on which is underway [5]. If so, our results could then be applied to the propositional structure of more expressive logics via Skolemization, as is done in PVS, since SMT solvers can reason about uninterpreted functions.

Possible direct applications of this work include isolating the cause of a failure to prove a putative tautology, and identification of dead code.[3]

We also plan to use *simplify* in a more fine grained manner, perhaps in conjunction with the rewriting engine of the theorem prover, as is done in ACL2: the HOL4 and Isabelle/HOL simplifiers do provide hooks for integrating other procedures with the rewrite engine. We plan to customize SAT-based simplifiers, for instance to try harder to exclude clauses containing no definitional variables. These plans will form the initial steps for future research.

References

1. Bryant, R.E.: Symbolic boolean manipulation with ordered binary decision diagrams. ACM Computing Surveys 24(3), 293–318 (1992)
2. Cyrluk, D., Rajan, S., Shankar, N., Srivas, M.K.: Effective theorem proving for hardware verification. In: Kumar, R., Kropf, T. (eds.) TPCD 1994. LNCS, vol. 901, pp. 203–222. Springer, Heidelberg (1995)
3. Eén, N., Biere, A.: Effective preprocessing in SAT through variable and clause elimination. In: Bacchus, F., Walsh, T. (eds.) SAT 2005. LNCS, vol. 3569, pp. 61–75. Springer, Heidelberg (2005)
4. Eén, N., Sörensson, N.: An extensible SAT-solver. In: Giunchiglia, E., Tacchella, A. (eds.) SAT 2003. LNCS, vol. 2919, pp. 502–518. Springer, Heidelberg (2004)

---

[3] Thanks to Larry Paulson and Laurent Théry respectively, for these ideas.

5. Fontaine, P., Marion, J.-Y., Merz, S., Nieto, L.P., Tiu, A.F.: Expressiveness + automation + soundness: Towards combining SMT solvers and interactive proof assistants. In: Hermanns, H., Palsberg, J. (eds.) TACAS 2006 and ETAPS 2006. LNCS, vol. 3920, pp. 167–181. Springer, Heidelberg (2006)
6. Gershman, R., Koifman, M., Strichman, O.: Deriving small unsatisfiable cores with dominators. In: Ball, T., Jones, R.B. (eds.) CAV 2006. LNCS, vol. 4144, pp. 109–122. Springer, Heidelberg (2006)
7. Gordon, M.J.C.: Programming combinations of deduction and BDD-based symbolic calculation. LMS Journal of Computation and Mathematics 5, 56–76 (2002)
8. Gordon, M.J.C., Melham, T.F. (eds.): Introduction to HOL: A theorem-proving environment for higher order logic. Cambridge University Press, Cambridge (1993)
9. Harrison, J.: Binary decision diagrams as a HOL derived rule. The Computer Journal 38(2), 162–170 (1995)
10. Harrison, J.: HOL Light. In: Srivas, M., Camilleri, A. (eds.) FMCAD 1996. LNCS, vol. 1166, pp. 265–269. Springer, Heidelberg (1996)
11. Harrison, J.: Stålmarck's algorithm as a HOL derived rule. In: von Wright, J., Harrison, J., Grundy, J. (eds.) TPHOLs 1996. LNCS, vol. 1125, pp. 221–234. Springer, Heidelberg (1996)
12. Huet, G., Kahn, G., Paulin-Mohring, C.: The Coq proof assistant: A tutorial: Version 7.2. Technical Report RT-0256, INRIA (February 2002)
13. Joyce, J.J., Seger, C.-J.H.: The HOL-Voss system: Model checking inside a general-purpose theorem prover. In: Joyce, J.J., Seger, C.-J.H. (eds.) HUG 1993. LNCS, vol. 780, pp. 185–198. Springer, Heidelberg (1994)
14. Kaufmann, M., Moore, J.: An industrial strength theorem prover for a logic based on Common Lisp. IEEE Transactions on Software Engineering 23(4), 203–213 (1997)
15. Mitchell, D.G.: A SAT solver primer. In: EATCS Bulletin. The Logic in Computer Science Column, EATCS, vol. 85, pp. 112–133 (February 2005)
16. Moskewicz, M.W., Madigan, C.F., Zhao, Y., Zhang, L., Malik, S.: Chaff: Engineering an efficient SAT solver. In: Proceedings of the 38th Design Automation Conference, pp. 530–535. ACM Press, New York (2001)
17. Owre, S., Rushby, J.M., Shankar, N.: PVS: A prototype verification system. In: Kapur, D. (ed.) CADE 1992. LNCS, vol. 607, pp. 748–752. Springer, Heidelberg (1992), http://pvs.csl.sri.com
18. Paulson, L.C.: Isabelle. LNCS, vol. 828. Springer, Heidelberg (1994)
19. Rajan, S., Shankar, N., Srivas, M.K.: An integration of model checking and automated proof checking. In: Wolper, P. (ed.) CAV 1995. LNCS, vol. 939, pp. 84–97. Springer, Heidelberg (1995)
20. Tseitin, G.S.: On the complexity of derivation in propositional calculus. In: Siekmann, J., Wrightson, G. (eds.) Automation Of Reasoning: Classical Papers On Computational Logic, Vol. II, 1967-1970, pp. 466–483. Springer, Heidelberg (1983); Slisenko, A.O. (eds.) Structures in Constructive Mathematics and Mathematical Logic Part II, pp. 115–125 (1968)
21. Verma, K.N., Goubault-Larrecq, J., Prasad, S., Arun-Kumar, S.: Reflecting BDDs in Coq. In: He, J., Sato, M. (eds.) ASIAN 2000. LNCS, vol. 1961, pp. 162–181. Springer, Heidelberg (2000)
22. Weber, T., Amjad, H.: Efficiently checking propositional refutations in HOL theorem provers. JAL (2007) (accepted for publication, July 2007) (to appear)

23. Zhang, L., Malik, S.: Extracting small unsatisfiable cores from unsatisfiable boolean formula. In: 6th SAT (2003) (presentation only)
24. Zhang, L.: On subsumption removal and on-the-fly CNF simplification. In: Bacchus, F., Walsh, T. (eds.) SAT 2005. LNCS, vol. 3569, pp. 482–489. Springer, Heidelberg (2005)
25. Zhang, L., Malik, S.: Validating SAT solvers using an independent resolution-based checker: Practical implementations and other applications. In: DATE, pp. 10880–10885. IEEE Computer Society Press, Los Alamitos (2003)

Nominal Inversion Principles

Stefan Berghofer and Christian Urban

Technische Universität München
Institut für Informatik, Boltzmannstraße 3, 85748 Garching, Germany

Abstract. When reasoning about inductively defined predicates, such as typing judgements or reduction relations, proofs are often done by inversion, that is by a case analysis on the last rule of a derivation. In HOL and other formal frameworks this case analysis involves solving equational constraints on the arguments of the inductively defined predicates. This is well-understood when the arguments consist of variables or injective term-constructors. However, when alpha-equivalence classes are involved, that is when term-constructors are not injective, these equational constraints give rise to annoying variable renamings. In this paper, we show that more convenient inversion principles can be derived where one does not have to deal with variable renamings. An interesting observation is that our result relies on the fact that inductive predicates must satisfy the variable convention compatibility condition, which was introduced to justify the admissibility of Barendregt's variable convention in rule inductions.

1 Introduction

Inductively defined predicates play an important role in formal methods; they are defined by a set of introduction rules and come equipped with rule induction and inversion principles. A typical example of an inductive predicate is beta-reduction defined by the four rules

$$\frac{}{App\ (Lam\ x.s_1)\ s_2 \longrightarrow_\beta s_1[x{:=}s_2]}b_1 \qquad \frac{s_1 \longrightarrow_\beta s_2}{App\ s_1\ t \longrightarrow_\beta App\ s_2\ t}b_2$$

$$\frac{s_1 \longrightarrow_\beta s_2}{App\ t\ s_1 \longrightarrow_\beta App\ t\ s_2}b_3 \qquad \frac{s_1 \longrightarrow_\beta s_2}{Lam\ x.s_1 \longrightarrow_\beta Lam\ x.s_2}b_4 \tag{1}$$

where $\_[\_{:=}\_]$ stands for capture-avoiding substitution. Another is the typing predicate for simply-typed lambda-terms defined by the rules

$$\frac{valid\ \Gamma \quad (x, T) \in \Gamma}{\Gamma \vdash Var\ x : T}t_1 \qquad \frac{\Gamma \vdash t_1 : T_1 \to T_2 \quad \Gamma \vdash t_2 : T_1}{\Gamma \vdash App\ t_1\ t_2 : T_2}t_2$$

$$\frac{(x, T_1){::}\Gamma \vdash t : T_2}{\Gamma \vdash Lam\ x.t : T_1 \to T_2}t_3 \tag{2}$$

where the typing contexts Γ are lists of (variable name,type)-pairs, $\in$ stands for list membership and $::$ for list-cons. The premise $valid\ \Gamma$ in the first typing rule is another inductive predicate which states that the typing context must not contain repeated occurrences of a variable name. This can be defined as follows:

$$\frac{}{valid\ []}v_1 \qquad \frac{valid\ \Gamma \quad x \mathbin{\#} \Gamma}{valid\ ((x, T){::}\Gamma)}v_2 \tag{3}$$

O. Ait Mohamed, C. Muñoz, and S. Tahar (Eds.): TPHOLs 2008, LNCS 5170, pp. 71–85, 2008.

where $[]$ stands for the empty typing context and $x \# \Gamma$ states that the variable name x does not occur in Γ.

The rule induction and inversion principles are the main thrust behind these definitions: they provide the infrastructure for convenient reasoning about inductive predicates. This is illustrated by the proof of the following lemma establishing that beta-reduction preserves typing.

Lemma 1 (Type Preservation). *If $\Gamma \vdash u : U$ and $u \longrightarrow_\beta u'$ then $\Gamma \vdash u' : U$.*

Type preservation can be proved by a rule induction on $\Gamma \vdash u : U$. This gives rise to three subgoals:

$$
\begin{aligned}
&(i) \quad Var\, x \longrightarrow_\beta u' \wedge \ldots \Rightarrow \Gamma \vdash u' : T \\
&(ii) \quad App\, t_1\, t_2 \longrightarrow_\beta u' \wedge \ldots \Rightarrow \Gamma \vdash u' : T_2 \\
&(iii) \quad Lam\, x.t \longrightarrow_\beta u' \wedge \ldots \Rightarrow \Gamma \vdash u' : T_1 \to T_2
\end{aligned}
$$

where we omitted some of the side-assumptions. The proof then proceeds by a case analysis, called *inversion*, of the assumptions about $\longrightarrow_\beta$.

In general, inversion is a reasoning principle that applies to any instance of an inductive predicate occurring in the assumptions; it relies on the observation that this instance must have been derived by at least one of the rules by which the inductive predicate is defined. In informal reasoning one therefore matches the assumption with the conclusion of every rule and tests whether the assumption and conclusion match. We will refer to this kind of informal reasoning as *inversion by matching* and describe it next.

In the case *(i)*, the assumption $Var\, x \longrightarrow_\beta u'$ matches with no conclusion in (1). Therefore this is an impossible case, which implies that the goal $\Gamma \vdash u' : T$ holds trivially.

In the case *(ii)*, the matching of $App\, t_1\, t_2 \longrightarrow_\beta u'$ with the conclusions in (1) succeeds in case of b_1, b_2 and b_3, and therefore three cases need to be considered. Let us first analyse the case corresponding to the rule

$$
\frac{s_1 \longrightarrow_\beta s_2}{App\, s_1\, t \longrightarrow_\beta App\, s_2\, t} b_2
$$

In this case we know for some s_2 that $u' = App\, s_2\, t_2$ (since t_1 matches with s_1, and t with t_2). By induction we can infer that $\Gamma \vdash s_2 : T_1 \to T_2$ and $\Gamma \vdash t_2 : T_1$ hold. Consequently, $\Gamma \vdash u' : T_2$ holds.

Continuing with our informal reasoning, the case of beta-reduction, i.e. $App\, (Lam\, x.s_1)\, s_2 \longrightarrow_\beta s_1[x:=s_2]$, goes as follows: For some term s_1, u' is equal to $s_1[x:=t_2]$ and t_1 equal to $Lam\, x.s_1$. The latter equation gives us that $\Gamma \vdash Lam\, x.s_1 : T_1 \to T_2$ and $\Gamma \vdash t_2 : T_1$ hold. To complete the proof we need the substitutivity lemma:

Lemma 2 (Type Substitutivity).
If $(x, U) :: \Gamma \vdash t : T$ and $\Gamma \vdash u : U$ then $\Gamma \vdash t[x:=u] : T$.

whose proof we omit. For this lemma to be useful, we have to invert the typing judgement $\Gamma \vdash Lam\, x.s_1 : T_1 \to T_2$. The informal inversion by matching gives us the desired result: this judgement matches with the conclusion of the rule t_3 and we obtain $(x, T_1) :: \Gamma \vdash s_1 : T_2$. So we can conclude in this case by using Lemma 2 (similarly in all remaining cases).

The point of these calculations is to show that the inversion by matching is very natural and convenient. It is also very typical in programming language research: similar proofs are described for System F$_{<:}$ in the POPLmark challenge (see Appendix of [2]). The contribution of this paper is to make this informal reasoning formal. The problem we have to solve for this arises from the fact that the examples above contain lambda-terms, where the term constructor *Lam* is not *injective*. By this we mean the property that in general one *cannot* infer from the equation

$$Lam\ x.t = Lam\ x'.t'$$

that

$$x = x' \quad \text{and} \quad t = t'$$

hold. This is in contrast to the injective term constructors *Var* and *App* where we have the implications

$$Var\ x = Var\ x' \quad \Rightarrow \quad x = x'$$
$$App\ t\ s = App\ t'\ s' \quad \Rightarrow \quad t = t' \wedge s = s'$$

Why the lack of injectivity leads to problems with formal inversion principles is explained in the next section. Section 3 characterises the form of rules in inductive definitions, Section 4 recalls some notions from the nominal logic work [7,9] and Section 5 describes the condition for variable-convention compatibility and gives the proof for our main result. Examples are described in Section 6 and Section 7 concludes and mentions related work.

2   Formal Inversion Principles

Unfortunately, the *formal* reasoning in systems such as HOL, Coq and LEGO is subtly different from the informal inversion by matching illustrated in the Introduction: instead of matching two instances of a relation, the formal inversion principles in these systems require equality constraints to be solved.

Consider the inversion principles given in Fig. 1, which are formally derived by Isabelle/HOL for beta-reduction and typing. Both inversion principles can be employed to prove a proposition P from the assumption $u_1 \longrightarrow_\beta u_2$ and $\Delta \vdash u : U$, respectively. Their general structure is as follows: each premise of the inversion rule corresponds to a rule of the inductive predicate. These premises are implications whose right-hand side is the proposition P, and whose left-hand side are conjunctions (note also in each case the outermost universal quantification ranging over the entire implication). The elements of these conjunctions can be divided into two parts: the first part consists of equality constraints expressing the equality between the arguments of the predicate to be inverted and the arguments of each conclusion in the inductive definition; the second part consists of the premises of the corresponding rule.

Returning to our running example of proving the type-preservation lemma, let us analyse how the formally derived inversion principles given in Fig. 1 behave. The case *(i)* in Lemma 1 required us to prove

$$Var\ x \longrightarrow_\beta u' \wedge \ldots \Rightarrow \Gamma \vdash u' : T$$

$$\frac{\begin{array}{l} \forall x\, s_2\, s_1.\ u_1 = App\ (Lam\ x.s_1)\ s_2 \wedge u_2 = s_1[x:=s_2] \Rightarrow P \\ \forall s_1\, s_2\, t.\ u_1 = App\ s_1\ t \wedge u_2 = App\ s_2\ t \wedge s_1 \longrightarrow_\beta s_2 \Rightarrow P \\ \forall s_1\, s_2\, t.\ u_1 = App\ t\ s_1 \wedge u_2 = App\ t\ s_2 \wedge s_1 \longrightarrow_\beta s_2 \Rightarrow P \\ \forall s_1\, s_2\, x.\ u_1 = Lam\ x.s_1 \wedge u_2 = Lam\ x.s_2 \wedge s_1 \longrightarrow_\beta s_2 \Rightarrow P \end{array}}{u_1 \longrightarrow_\beta u_2 \Rightarrow P} \tag{4}$$

$$\frac{\begin{array}{l} \forall \Gamma\, x\, T.\ \Delta = \Gamma \wedge u = Var\ x \wedge U = T \wedge valid\ \Gamma \wedge (x, T) \in \Gamma \Rightarrow P \\ \forall t_1\, T_1\, T_2\, t_2.\ \Delta = \Gamma \wedge u = App\ t_1\ t_2 \wedge U = T_2 \wedge \Gamma \vdash t_1 : T_1 \to T_2 \wedge \Gamma \vdash t_2 : T_1 \Rightarrow P \\ \forall x\, T_1\, \Gamma\, t\, T_2.\ \Delta = \Gamma \wedge u = Lam\ x.t \wedge U = T_1 \to T_2 \wedge (x, T_1)::\Gamma \vdash t : T_2 \Rightarrow P \end{array}}{\Delta \vdash u : U \Rightarrow P} \tag{5}$$

Fig. 1. Inversion principles derived by Isabelle/HOL for the inductive predicates beta-reduction and typing

If we use inversion principle for $\longrightarrow_\beta$ (i.e. (4)) and invert $Var\ x \longrightarrow_\beta u'$, we obtain the following four subgoals:

$$\begin{array}{l} \forall x'\, s_2\, s_1.\ Var\ x = App\ (Lam\ x'.s_1)\ s_2 \wedge u' = s_1[x':=s_2] \wedge \ldots \Rightarrow \Gamma \vdash u' : T \\ \forall s_1\, s_2\, t.\ Var\ x = App\ s_1\ t \wedge u' = App\ s_2\ t \wedge s_1 \longrightarrow_\beta s_2 \wedge \ldots \Rightarrow \Gamma \vdash u' : T \\ \forall s_1\, s_2\, t.\ Var\ x = App\ t\ s_1 \wedge u' = App\ t\ s_2 \wedge s_1 \longrightarrow_\beta s_2 \wedge \ldots \Rightarrow \Gamma \vdash u' : T \\ \forall s_1\, s_2\, x'.\ Var\ x = Lam\ x'.s_1 \wedge u' = Lam\ x'.s_2 \wedge s_1 \longrightarrow_\beta s_2 \wedge \ldots \Rightarrow \Gamma \vdash u' : T \end{array}$$

The left-hand sides of these subgoals all reduce to *False* because the term constructors are in conflict (*Var* can never be equal to *App*). Therefore we can quickly, like in the informal reasoning, discharge all subgoals.

In case *(ii)* where we invert $App\ t_1\ t_2 \longrightarrow_\beta u'$, we obtain the following four subgoals:

$$\begin{array}{l} \forall x\, s_2\, s_1.\ App\ t_1\ t_2 = App\ (Lam\ x.s_1)\ s_2 \wedge u' = s_1[x:=s_2] \wedge \ldots \Rightarrow \Gamma \vdash u' : T \\ \forall s_1\, s_2\, t.\ App\ t_1\ t_2 = App\ s_1\ t \wedge u' = App\ s_2\ t \wedge s_1 \longrightarrow_\beta s_2 \wedge \ldots \Rightarrow \Gamma \vdash u' : T \\ \forall s_1\, s_2\, t.\ App\ t_1\ t_2 = App\ t\ s_1 \wedge u' = App\ t\ s_2 \wedge s_1 \longrightarrow_\beta s_2 \wedge \ldots \Rightarrow \Gamma \vdash u' : T \\ \forall s_1\, s_2\, x.\ App\ t_1\ t_2 = Lam\ x.s_1 \wedge u' = Lam\ x.s_2 \wedge s_1 \longrightarrow_\beta s_2 \wedge \ldots \Rightarrow \Gamma \vdash u' : T \end{array}$$

The fourth subgoal can again be discharged because of the conflicting equality between *App* and *Lam*. The reasoning in the second and third is very similar with the informal inversion by matching, because the *App*-term constructor is injective and therefore we can infer

$$\begin{array}{l} App\ t_1\ t_2 = App\ s_1\ t \Rightarrow t_1 = s_1 \wedge t_2 = t, \text{ and} \\ App\ t_1\ t_2 = App\ t\ s_1 \Rightarrow t_1 = t \wedge t_2 = s_1 \end{array} \tag{6}$$

which are the same equations we would have got by the informal inversion by matching.

The first subgoal (corresponding to b_1) is more complicated: although we obtain by injectivity of *App* the equations $t_1 = Lam\ x.s_1$ and $t_2 = s_2$, we will encounter problems with inverting the typing judgement $\Gamma \vdash Lam\ x.s_1 : T_1 \to T_2$. That is, we will not be able to infer that $(x, T_1)::\Gamma \vdash s_1 : T_2$ holds. This is because *Lam* is not injective and we cannot reason as in (6).

We encounter the same problem with the reasoning in case *(iii)*. There we have to invert the reduction $Lam\ x.t \longrightarrow_\beta u'$ and obtain by using the first inversion principle from (4) the following four subgoals:

$\forall x'\, s_2\, s_1.\ Lam\ x.t = App\ (Lam\ x'.s_1)\ s_2 \wedge u' = s_1[x':=s_2] \Rightarrow \Gamma \vdash u' : T_1 \to T_2$

$\forall s_1\, s_2\, t.\ Lam\ x.t = App\ s_1\ t \wedge u' = App\ s_2\ t \wedge s_1 \longrightarrow_\beta s_2 \Rightarrow \Gamma \vdash u' : T_1 \to T_2$

$\forall s_1\, s_2\, t.\ Lam\ x.t = App\ t\ s_1 \wedge u' = App\ t\ s_2 \wedge s_1 \longrightarrow_\beta s_2 \Rightarrow \Gamma \vdash u' : T_1 \to T_2$

$\forall s_1\, s_2\, x'.\ Lam\ x.t = Lam\ x'.s_1 \wedge u' = Lam\ x'.s_2 \wedge s_1 \longrightarrow_\beta s_2 \Rightarrow \Gamma \vdash u' : T_1 \to T_2$

Again the first three cases reduce to *False*. However in the fourth case we end up with solving the equation

$$Lam\ x.t = Lam\ x'.s_1 \tag{7}$$

where the variables x' and s_1 are universally quantified (that is we cannot choose them). Since *Lam* is not injective, the only way to solve this equation is to unfold the definition of alpha-equivalence, which in the Nominal Datatype Package gives us the cases

(i) $x = x' \wedge t = s_1$ **or**

(ii) $x \neq x' \wedge t = (x\ x') \cdot s_1 \wedge x \# s_1$

where $(x\ x')$ is a permutative renaming of x and x', and $x \# s_1$ stands for x not occurring freely in s_1, see [7]. While the first case is easy to deal with (the induction hypothesis is immediately applicable), the second leads to the following proof state:

$$x \neq x' \wedge x \# s_1 \wedge s_1 \longrightarrow_\beta s_2 \wedge \ldots \Rightarrow \Gamma \vdash Lam\ x'.s_2 : T_1 \to T_2$$

with the induction hypothesis

$$\forall s'.\ (x\ x') \cdot s_1 \longrightarrow_\beta s' \Rightarrow (x, T_1)::\Gamma \vdash s' : T_2$$

Here the formal reasoning starts to hurt, as it is much harder than the informal inversion by matching. As one can see, the induction hypothesis is not directly applicable: we know $s_1 \longrightarrow_\beta s_2$ but we need that $(x\ x') \cdot s_1$ reduces to some term. Also the induction hypothesis gives us a typing-judgement involving the variable x, but we need one for x'. The most direct way to complete this case requires the following side lemmas:

Lemma 3.

(i) *If $s_1 \longrightarrow_\beta s_2$ then $(x\ x') \cdot s_1 \longrightarrow_\beta (x\ x') \cdot s_2$.*

(ii) *If $x \# s_1$ and $s_1 \longrightarrow_\beta s_2$ then $x \# s_2$.*

where, interestingly, the second is a property specific to beta-reduction.

Clearly, inverting $Lam\ x.t \longrightarrow_\beta u'$ in this way is not very convenient and the same difficulties arise if we try to invert $\Gamma \vdash Lam\ x.s_1 : T_1 \to T_2$ using (5) as needed in the *App*-case above. In contrast, inverting inductive predicates based on the locally nameless approach to binders (see [3]) is much simpler, because there all term constructors are injective—even *Lam*. We show in this paper that we can obtain stronger inversion principles (than given in Fig. 1), where they are stronger in the sense that we can avoid the renaming of the binder, as long as the binder is sufficiently fresh. In this way we can follow quite closely the informal reasoning of inversion by matching an assumption with all rules.

These strong inversion principles will depend on the inductive predicates to satisfy the *variable convention compatibility condition*, short *vc-condition*. The reason for this condition is that the informal reasoning (i.e. inversion by matching) can lead to faulty reasoning when alpha-equivalence classes are involved. Consider the following inductive definition of a two-place predicate (both arguments are alpha-equated lambda-terms)

$$\frac{}{Var\ x \hookrightarrow Var\ x} \qquad \frac{}{App\ t_1\ t_2 \hookrightarrow App\ t_1\ t_2} \qquad \frac{t \hookrightarrow t'}{Lam\ x.t \hookrightarrow t'} \tag{8}$$

Now choose two distinct variables, say x and y with $x \neq y$. A simple calculation shows that $Lam\ x.Var\ x \hookrightarrow Var\ x$ can be derived using the rules above. Therefore we can use it as an assumption. Since we are working with alpha-equated lambda terms, we have that $Lam\ x.Var\ x = Lam\ y.Var\ y$ and therefore also $Lam\ y.Var\ y \hookrightarrow Var\ x$ must hold. Next we apply the inversion principle naively to the latter instance of the relation, i.e. we invert by matching this instance with the conclusions of the rules shown in (8). Only the third rule matches, yielding the fact $Var\ y \hookrightarrow Var\ x$. Next we invert this instance of the relation: the first rule matches, enabling us to infer that $x = y$ holds. This, however, contradicts the assumption that x and y are distinct. The vc-condition will protect us from this kind of faulty reasoning.

3   Inductive Predicates

An inductive predicate, say R, is defined by a finite set of rules r_i

$$\frac{B_1}{R\ ts_1}r_1 \qquad \cdots \qquad \frac{B_n}{R\ ts_n}r_n \tag{9}$$

where in the premises the B_i are HOL-formulae possibly containing R and where in the conclusion the ts_i are the arguments of the predicate R. The ts_i are HOL-terms, which for the purposes of this paper we can assume to be either variables or constructed by term constructors. Again for the purposes of this paper HOL-formulae will be the ones given by the grammar

$$B ::= P\ ts \mid B_1 \wedge B_2 \mid B_1 \vee B_2 \mid B_1 \longrightarrow B_2 \mid \neg B \mid \forall x.\ B\ x \mid \exists x.\ B\ x$$

where P stands for atomic predicates and ts are the arguments of P. In (9) we have the usual assumption that the premises can contain the predicate R in positive position only (see [1]). However, the B_i can contain other predicates, these are usually called side-conditions. For example our typing rule t_1 has the side-condition concerning $\in$ and *valid* as premise.

In what follows it is convenient to have the notations $t[xs]$, where the xs contain all the variables of t, and $B[ys]$, where ys includes the free variables of B (in B some variables might be bound because of the universal and existential quantifiers). The meaning of a rule in (9) is then the implication

$$\forall xs_i.\ B_i[xs_i] \Rightarrow R\ ts[xs_i]$$

where each xs_i includes all free variables in r_i. That means every instantiation of the free variables in r_i will result in an instance of this rule. With the rules given in (9) comes the following inversion principle

$$\forall xs_1. \; ss = ts_1[xs_1] \wedge B_1[xs_1] \Rightarrow P \qquad \text{rule } r_1$$

$$\vdots$$

$$\frac{\forall xs_n. \; ss = ts_n[xs_n] \wedge B_n[xs_n] \Rightarrow P \qquad \text{rule } r_n}{R \; ss \Rightarrow P} \qquad (10)$$

where the ts_i correspond to the arguments in the conclusion of each rule and the B_i to the premises (not also that the xs_i do not include any of the free variables in ss and P). The inversion principles given for $\longrightarrow_\beta$ and the typing rues in Fig. 1 are instances of (10). We refer to this inversion principle as the *weak inversion principle*. As we have shown in Section 2: when applying the weak inversions to cases involving non-injective term constructors, we need to analyse cases involving annoying variable renamings. We will show later that a strong inversion principle can be derived from the weak one and using the strong one we can avoid the renamings.

4 Nominal Logic Work

Before we proceed, we introduce some necessary notions from the nominal logic work [7,9]. We assume that there are countably infinitely many names, which can be used as binders. We base our description on *permutation actions* and on the notion of *support*. The support of an object will, for the purposes of this paper, coincides with the set of free names of that object. For details and a proper definition of support see [8]. A name a is *fresh* w.r.t. an object, say t, provided that it is not free in t; we write this as $a \mathbin{\#} t$. Note that if t has finitely many free variables, then there exists a fresh variable w.r.t. t. We will also use the auxiliary notation $a \mathbin{\#} ts$, in which ts stands for a collection of objects $t_1, \ldots, t_n$, to mean $a \mathbin{\#} t_1, \ldots, a \mathbin{\#} t_n$. We further generalise this notation to a collection of names, namely $as \mathbin{\#} ts$, which means $a_1 \mathbin{\#} ts, \ldots, a_m \mathbin{\#} ts$.

Permutations are finite lists of swappings (i.e., pairs of variables). We write such permutations as $(a_1 \, b_1)(a_2 \, b_2) \cdots (a_n \, b_n)$; the empty list $[]$ stands for the identity permutation, list append (i.e. $\pi_1 \mathbin{@} \pi_2$) for the composition of two permutations and list reversal (i.e. π^{-1}) for the inverse of a permutation. We define the permutation action over the structure of types in HOL. The point of the permutation action is to push permutations inside the structure of every object, renaming names on the way. A permutation acting on names is therefore defined as follows:

$$[] \bullet a = a$$
$$(a, b){::}\pi \bullet c = \begin{cases} a & \text{if } \pi \bullet c = b \\ b & \text{if } \pi \bullet c = a \\ \pi \bullet c \text{ otherwise} \end{cases} \qquad (11)$$

The permutation action on lists, pairs and booleans is given by

$$\pi \bullet [] = []$$
$$\pi \bullet (x{::}xs) = \pi \bullet x{::}\pi \bullet xs$$
$$\pi \bullet (x, y) = (\pi \bullet x, \pi \bullet y)$$
$$\pi \bullet \textit{True} = \textit{True}$$
$$\pi \bullet \textit{False} = \textit{False}$$
$$(12)$$

Notice the last two lines imply the fact that for every HOL-formula B the equality $\pi \bullet B = B$ holds. This is because HOL is a classical logic and every formula is either true or false. For alpha-equated lambda-terms we have

$$
\begin{aligned}
\pi \bullet Var\, x &= Var\, (\pi \bullet x) \\
\pi \bullet App\, t_1\, t_2 &= App\, (\pi \bullet t_1)\, (\pi \bullet t_2) \\
\pi \bullet Lam\, x.t &= Lam\, (\pi \bullet x).(\pi \bullet t)
\end{aligned}
\tag{13}
$$

We can easily prove that the permutation actions in (11), (12) and (13) satisfy the following three properties:

$$
\begin{aligned}
&(i) \quad [] \bullet (\_) = (\_) \\
&(ii) \quad (\pi_1 \,@\, \pi_2) \bullet (\_) = \pi_1 \bullet \pi_2 \bullet (\_) \\
&(iii) \quad If\ \pi_1 \approx \pi_2\ then\ \pi_1 \bullet (\_) = \pi_2 \bullet (\_).
\end{aligned}
\tag{14}
$$

where in the last clause equality between two permutations, that is $\pi_1 \approx \pi_2$, is defined by the property that as $\pi_1 \bullet a = \pi_2 \bullet a$ holds for all names a. In the next section we need the following lemma about freshness and the permutation actions in (11), (12) and (13):

Proposition 1. *If* $a \,\#\, (\_)$ *and* $b \,\#\, (\_)$ *then* $(a\ b)\bullet(\_) = (\_)$.

The notion of *equivariance* is derived from the permutation actions:

Definition 1 (Equivariance [7]). *A HOL-term t, respectively a HOL-formula B, with free variables amongst xs is* equivariant *provided for all* π, *we have* $\pi \bullet t[xs] = t[\pi \bullet xs]$ *and* $\pi \bullet B[xs] = B[\pi \bullet xs]$.

From the definition of their permutation action, pairs, nil and list-cons are equivariant. For HOL-formulae we have:

$$
\begin{aligned}
\pi \bullet (A \wedge B) &= \pi \bullet A \wedge \pi \bullet B \\
\pi \bullet (A \vee B) &= \pi \bullet A \vee \pi \bullet B \\
\pi \bullet (A \longrightarrow B) &= \pi \bullet A \longrightarrow \pi \bullet B \\
\pi \bullet (\neg A) &= \neg\, \pi \bullet A \\
\pi \bullet (\forall x.\, P\, x) &= \forall x.\, \pi \bullet P\, (\pi^{-1} \bullet x) \\
\pi \bullet (\exists x.\, P\, x) &= \exists x.\, \pi \bullet P\, (\pi^{-1} \bullet x)
\end{aligned}
\tag{15}
$$

Therefore for all the structures we consider in this paper we can move permutations inside the structures until they reach variables, therefore all structures we consider in paper will be equivariant.

For proving our main result in the next section it is convenient to refine our notation $ts[xs]$ and $B[xs]$ for indicating the free variables of ts and B. The reason is that some of these variables stand for names and those names are potentially in *binding positions*. By binding position we mean the x in *Lam x.t*. In what follows the notation $ts[as;xs]$ and $B[as;xs]$ will be used to indicate that the variables in binding position of the ts are included in as and the other variables of the ts are either in as or in xs (similarly for HOL-formulae). We extend this notation also to rules: by writing $r[as;xs]$ we mean rules of the form

$$
\frac{B[as;xs]}{R\ ts[as;xs]}\ r_i[as;xs]
$$

However, unlike in the notation for HOL-terms and HOL-formulae, we mean in $r_i[as;xs]$ that the *as* stand *exactly* for the variables occurring somewhere in r_i in binding position and the *xs* stand for the rest of variables. To see how this notation works out in our examples, reconsider the definitions for the relations given in (1) and (2). Using our notation for these rules, we have

$$b_1[x;s_1,s_2] \qquad\qquad t_1[-;\Gamma,x,T]$$
$$b_2[-;s_1,s_2,t] \qquad\qquad t_2[-;\Gamma,t_1,t_2,T_1,T_2]$$
$$b_3[-;s_1,s_2,t] \qquad\qquad t_3[x;\Gamma,t,T_1,T_2]$$
$$b_4[x;s_1,s_2]$$

where '$-$' stands for no variable in binding position. An inductive definition for alpha-equivalence between lambda terms includes the two rules:

$$\frac{t_1 = t_2}{Lam\ x.t_1 = Lam\ x.t_2}\,a_1 \qquad\qquad \frac{x \neq y \quad t_1 = (x\ y)\bullet t_2 \quad x\ \#\ t_2}{Lam\ x.t_1 = Lam\ y.t_2}\,a_2$$

There our notation would be $a_1[x;t_1,t_2]$ and $a_2[x,y;t_1,t_2]$.

5   Strengthening of the Inversion Principle

In this section, we show how the "weak" inversion rules in (10) can be used to derive stronger inversion rules in which the equality constraints are formulated in such a way that they can be solved without having to rename variables.

We have seen in the example about $t \hookrightarrow t'$ from the Introduction that inversion principles involving alpha-equivalence classes require some care. In order to rule out the problematic case (and similar ones), we need to impose a condition on the rules of an inductive definition. It is interesting that the condition we impose is the same as the one introduced in [8] for justifying the admissibility of Barendregt's variable convention in rule inductions.

A rule is said to be *variable convention compatible*, or short *vc-compatible*, provided the following two properties are satisfied:

Definition 2 (Variable Convention Compatibility). *A rule $r[as;xs]$ with conclusion $R\ ts[as;xs]$ and premise $B[as;xs]$ is vc-compatible provided that:*

- *all HOL-terms and HOL-formulae occurring in r are equivariant, and*
- *the premise $B[as;xs]$ implies that as $\#$ ts$[as;xs]$ holds and that the as are distinct.*

Note that if rule r does not contain any variable in binding position, then the second condition is vacuously true. The first condition ensures that the relation R is equivariant. The equivariance property will allow us to push permutations inside HOL-terms and HOL-formulae until they reach free variables.

If every introduction rule in an inductive definition satisfies these conditions, then the inversion principle can be strengthened. The strengthened version looks as follows

$$\forall xs_1. \ (bs_1 \mathbin{\#} ss \wedge distinct(bs_1) \Rightarrow ss = ts_1[bs_1;xs_1] \wedge B_1[bs_1;xs_1]) \Rightarrow P \qquad \text{rule } r_1$$

$$\vdots$$

$$\underline{\forall xs_n. \ (bs_n \mathbin{\#} ss \wedge distinct(bs_n) \Rightarrow ss = ts_n[bs_n;xs_n] \wedge B_n[bs_n;xs_n]) \Rightarrow P} \qquad \text{rule } r_n$$
$$R \ ss \Rightarrow P$$

$$(16)$$

where for every rule $r_1,\ldots,r_n$ we have a case to analyse. In our notation the rules have the form $r_1[bs_1;xs_1],\ldots,r_n[bs_n;xs_n]$ where the bs_i are the variables in binding position. Note that in contrast to (10) the variables bs_i are no longer universally quantified, meaning that we are free to choose the names bs_i when we want to invoke the strong inversion principle. The only constraints we have is that the preconditions $bs_i \mathbin{\#} ss \wedge distinct(bs_i)$ need to be satisfied. This will be the case if the bs_i are sufficiently fresh.

We now prove the main result of this paper: if the rules of an inductive definition are vc-compatible, then the strong inversion principle in (16) holds.

Theorem 1. *For an inductive definition of the predicate R, involving vc-compatible rules only, a strong inversion principle exists deriving the implication $R \ ss \Rightarrow P$.*

Proof. We need to establish $R \ ss \Rightarrow P$ using the implications indicated in (16). To do so we will use the weak inversion rule from (10). For each rule $r_i[as_i;xs_i]$ of the form

$$\frac{B[as_i;xs_i]}{R \ ts_i[as_i;xs_i]}$$

we have to analyse one case of the form

$$\forall as_i \ xs_i. \ ss = ts_i[as_i;xs_i] \wedge B_i[as_i;xs_i] \Rightarrow P$$

To show P in these cases we have available the fact from (16), namely

$$\forall xs_i. \ (bs_i \mathbin{\#} ss \wedge distinct(bs_i) \Rightarrow ss = ts_i[bs_i;xs_i] \wedge B_i[bs_i;xs_i]) \Rightarrow P \qquad (17)$$

We first assume that

$$ss = ts_i[as_i;xs_i] \qquad (18)$$
$$B_i[as_i;xs_i] \qquad (19)$$

hold. Since $r_i[as_i;xs_i]$ is assumed to be vc-compatible, we further have that

$$(a) \ as_i \mathbin{\#} ts_i[as_i;xs_i] \quad \text{and} \quad (b) \ distinct(as_i) \qquad (20)$$

hold. The proof then proceeds by choosing for every name a in as_i a fresh name c such that for all the cs_i the following hold (cs_i is the collection of all those c):

$$(a) \ cs_i \mathbin{\#} ss \quad (b) \ cs_i \neq as_i \quad (c) \ cs_i \neq bs_i \quad (d) \ distinct(cs_i) \qquad (21)$$

Such a sequence cs_i always exists: the first three properties can be obtained since the terms ss, as_i and bs_i stand for finitely supported objects—so a free variable always exists; the last can be obtained by choosing the c one after another avoiding the ones that have already been chosen. We now build the permutation

$$\pi \stackrel{\text{def}}{=} (b_n \ c_n) \ldots (b_1 \ c_1) \ (a_n \ c_n) \ldots (a_1 \ c_1)$$

The point of π is that when applied to the as_i we get $\pi \bullet as_i = bs_i$. This follows from the properties in (20.b), (21.b-d) and the fact that we can assume $distinct(bs_i)$ holds (see below). We next instantiate in (17) the xs_i with $\pi \bullet xs_i$ giving us

$$(bs_i \# ss \wedge distinct(bs_i) \Rightarrow ss = ts_i[bs_i; \pi \bullet xs_i] \wedge B_i[bs_i; \pi \bullet xs_i]) \Rightarrow P$$

So in order to show P, it suffices to prove

$$ss = ts_i[bs_i; \pi \bullet xs_i] \wedge B_i[bs_i; \pi \bullet xs_i] \tag{22}$$

under the assumptions

$$(a) \ bs_i \# ss \qquad \text{and} \qquad (b) \ distinct(bs_i) \tag{23}$$

From (23.a) and (18) we obtain $bs_i \# ts_i[as_i; xs_i]$. Using this, (20.a) and Lemma 1, we have that $\pi \bullet ts_i[as_i; xs_i] = ts_i[as_i; xs_i]$. Since the rule is equivariant we have that $\pi \bullet ts_i[as_i; xs_i] = ts_i[bs_i; \pi \bullet xs_i]$ and thus also the first conjunct of (22). The reasoning for the other conjunct is as follows: using (19) and the fact that B_i is a boolean we have that $\pi \bullet B_i[as_i; xs_i]$ holds. Again by equivariance of the rule, we can move the permutation inside to obtain $B_i[bs_i; \pi \bullet xs_i]$—the second conjunct of (22). This concludes the proof. $\qquad\qquad\square$

Let us next describe how the stronger inversion principles simplify the formal reasoning in the type preservation lemma.

6 Examples

To use the strong inversion rules, we first have to make sure that the beta-reduction and typing relation are equivariant. For this we only have to observe that all constants (that is term constructors and functions) in the rules of $\longrightarrow_\beta$, typing and *valid* are equivariant. This follows either from the definition of the permutation action or is by a simple induction over the predicates (in our implementation Isabelle will infer this automatically). To show that the second condition in Definition 2 is satisfied we have to show that the binders are fresh w.r.t. the conclusions of the rule they appear in. That is a simple calculation for the rules

$$\frac{(x, T_1)::\Gamma \vdash t : T_2}{\Gamma \vdash Lam \ x.t : T_1 \rightarrow T_2} t_3 \qquad\qquad \frac{s_1 \longrightarrow_\beta s_2}{Lam \ x.s_1 \longrightarrow_\beta Lam \ x.s_2} b_4$$

In the first case we have to show that $x \# (\Gamma, Lam \ x.t, T_1 \rightarrow T_2)$ holds under the assumption that $(x, T_1)::\Gamma \vdash t : T_2$. Since we can show by a routine induction that typing judgements only include *valid* contexts, we have that $valid \ ((x, T_1)::\Gamma)$ holds.

$$\frac{\begin{array}{l} \forall s_2\ s_1.\ (y \mathbin{\#} (u_1, u_2) \Rightarrow u_1 = App\ (Lam\ y.s_1)\ s_2 \wedge u_2 = s_1[y:=s_2] \wedge y \mathbin{\#} s_2) \Rightarrow P \\ \forall s_1\ s_2\ t.\ u_1 = App\ s_1\ t \wedge u_2 = App\ s_2\ t \wedge s_1 \longrightarrow_\beta s_2 \Rightarrow P \\ \forall s_1\ s_2\ t.\ u_1 = App\ t\ s_1 \wedge u_2 = App\ t\ s_2 \wedge s_1 \longrightarrow_\beta s_2 \Rightarrow P \\ \forall s_1\ s_2.\ (x \mathbin{\#} (u_1, u_2) \Rightarrow u_1 = Lam\ x.s_1 \wedge u_2 = Lam\ x.s_2 \wedge s_1 \longrightarrow_\beta s_2) \Rightarrow P \end{array}}{u_1 \longrightarrow_\beta u_2 \Rightarrow P} \tag{24}$$

$$\frac{\begin{array}{l} \forall \Gamma\ x\ T.\ \Delta = \Gamma \wedge u = Var\ x \wedge U = T \wedge valid\ \Gamma \wedge (x, T) \in \Gamma \Rightarrow P \\ \forall t_1\ T_1\ T_2\ t_2.\ \Delta = \Gamma \wedge u = App\ t_1\ t_2 \wedge U = T_2 \wedge \Gamma \vdash t_1 : T_1 \to T_2 \wedge \Gamma \vdash t_2 : T_1 \Rightarrow P \\ \forall T_1\ \Gamma\ t\ T_2.\ (x \mathbin{\#} (\Delta, u, U) \Rightarrow \Delta = \Gamma \wedge u = Lam\ x.t \wedge U = T_1 \to T_2 \wedge (x, T_1){::}\Gamma \vdash t : T_2) \Rightarrow P \end{array}}{\Delta \vdash u : U \Rightarrow P} \tag{25}$$

Fig. 2. Strong inversion principles derived by the Nominal Datatype Package for the inductive predicates for beta reduction and typing

From this we can infer that $x \mathbin{\#} \Gamma$. We also know that $x \mathbin{\#} Lam\ x.t$ (since x is abstracted) and that $x \mathbin{\#} T_1 \to T_2$ (since types in the simply-typed lambda-calculus do not contain any variables). We can discharge the conditions in the other rule by similar arguments. However the condition will fail for the rule

$$\frac{}{App\ (Lam\ x.s_1)\ s_2 \longrightarrow_\beta s_1[x:=s_2]}b_1 \tag{26}$$

because we cannot determine whether $x \mathbin{\#} s_2$. However we can show that this beta-reduction rule is equivalent to the following more restricted rule

$$\frac{x \mathbin{\#} s_2}{App\ (Lam\ x.s_1)\ s_2 \longrightarrow_\beta s_1[x:=s_2]}b_1' \tag{27}$$

This is because we can choose a y such that $y \mathbin{\#} (s_1, s_2)$ and alpha-rename $App\ (Lam\ x.s_1)\ s_2$ to $App\ (Lam\ y.(y\ x)\bullet s_1)\ s_2$. Then apply the restricted rule to this term in order to obtain the reduct $((y\ x)\bullet s_1)[y:=s_2]$. By a structural induction over s_1, we can show that this term is equal to $s_1[x:=s_2]$ as desired. The point of this "manoeuvre" is that we can show that the restricted rule for beta-reduction does satisfy the vc-condition.

The result of these calculations is that there are strengthened inversion rules for beta-reduction and the typing-relation. They are given in Fig. 2. Using them for the type preservation lemma, the second and third case are the same as with the weak inversion rule (4). In the first and fourth case, however, the user does not need to show the claim for an arbitrary variable x', but for a sufficiently freshly chosen one (it has to be fresh w.r.t. (u_1, u_2)). In the strong inversion for the typing rule we have that the cases for variables and applications are the same as with the weak inversion rule (5). In the case of lambda abstractions, the user can choose a x so that $x \mathbin{\#} (\Delta, u, U)$. These choices will hugely simplify the formal reasoning. To give an impression of this fact we show next three lemmas in Isabelle/HOL proving special instances of inversion principles.

lemma *Ty-Lam-inversion*:
  assumes *ty*: $\Gamma \vdash Lam\ x.t : T$ **and** *fc*: $x \mathbin{\#} \Gamma$
  shows $\exists T_1\ T_2.\ T = T_1 \to T_2 \wedge (x, T_1){::}\Gamma \vdash t : T_2$
  using *ty fc* **by** (*cases rule: typing.strong-cases*) (*auto simp add: alpha*)

lemma *Beta-Lam-inversion*:
  assumes *red*: *Lam x.t* $\longrightarrow_\beta$ *s* **and** *fc*: $x\#s$
  shows $\exists\, t'.\ s = Lam\ x.t' \wedge t \longrightarrow_\beta t'$
  using *red fc* **by** (*cases rule*: *Beta.strong-cases*) (*auto simp add*: *alpha*)

lemma *Beta-App-inversion*:
  assumes *red*: *App* (*Lam x.t*) *s* $\longrightarrow_\beta$ *r* **and** *fc*: $x\#(s,r)$
  shows $(\exists\, t'.\ r = App\ (Lam\ x.t')\ s \wedge t \longrightarrow_\beta t') \vee$
         $(\exists\, s'.\ r = App\ (Lam\ x.t)\ s' \wedge s \longrightarrow_\beta s') \vee (r = t[x:=s])$
  using *red fc*
  by (*cases rule*: *Beta.strong-cases*) (*auto dest*: *Beta-Lam-inversion simp add*: *alpha*)

These lemmas are needed frequently in proofs about structural operational semantics. As seen in Section 2, it would have been quite painful to derive them using the weak inversion principles. We use the *alpha*-rule in the proofs above in order to rewrite the trivial alpha-equivalence *Lam x.t = Lam x.s* to $t = s$.

The Isar-proof of the complete type preservation lemma is given in Fig. 3. Lines 6 and 7 show the variable case. Lines 9-21 contain the steps for the case where a beta-reduction occurs (the other cases are automatic in Line 22). We first chose a fresh name x (Line 10); invert *App* $t_1\ t_2 \longrightarrow_\beta u'$ in Line 12 using the fresh x. In the only interesting case, we have that $\Gamma \vdash Lam\ x.s_1 : T_1 \rightarrow T_2$ holds (Line 15), which we can invert to $(x, T_1)::\Gamma \vdash s_1 : T_2$. To this we can apply the Lemma 2 (Line 20). In the lambda-case (Lines 24-31), we invert *Lam x.t* $\longrightarrow_\beta u'$. We know that x is fresh for u' by the strong induction (Line 5). We can apply the induction hypothesis in Line 28 and use the typing rule to conclude (Lines 30 and 31).

7   Conclusion and Related Work

As long as one is dealing with injective term constructors, the weak (or standard) inversion rules provided by Isabelle/HOL work similarly to the informal inversion by matching an assumption over the conclusions of inference rules. However, non-injective term constructors, such as *Lam* in the lambda-calculus, give rise to annoying variable renamings, and formal reasoning is quite different from and much more inconvenient than the informal inversion by matching. This was observed in [3], because in their locally nameless representation of binders, all term constructors are injective.

We have shown in this paper that if a binder is fresh with respect to the conclusion of the rule where the binder appears and the inductive predicate satisfies the vc-condition, then one can avoid the renamings. As a result the formal inversion principles are again as convenient the informal reasoning of inversion by matching—though the strong inversion principles only apply to vc-compatible inductive relations. In (8) we have shown that the informal inversion by matching can lead to faulty reasoning when the vc-condition is not satisfied. In our implementation this kind of faulty reasoning is prevented because the strong inversion principles are derived only when the user has verified the second part of the vc-condition (see Def. 2); the first part of that condition is verified automatically by observing that equivariant inductive predicates must be composed of equivariant components only.

```
1   lemma type-preservation:
2     assumes ty: Γ ⊢ u : U and red: u ⟶_β u′
3     shows Γ ⊢ u′ : U
4   using ty red
5   proof (nominal-induct avoiding: u′ rule: typing.strong-induct)
6     case (ty-Var Γ x T)
7     from ⟨Var x ⟶_β u′⟩ show Γ ⊢ u′ : T by (cases) (simp-all)
8   next
9     case (ty-App Γ t₁ T₁ T₂ t₂)
10    obtain x::name where fc: x # (Γ, App t₁ t₂, u′) by (rule exists-fresh-var)
11    from ⟨App t₁ t₂ ⟶_β u′⟩ show Γ ⊢ u′ : T₂ using fc
12    proof (cases rule: Beta.strong-cases[where x=x and xa=x])
13      case (Beta s₂ s₁)
14      then have eqs: t₁ = Lam x. s₁  t₂ = s₂  u′ = s₁[x:=s₂] using fc by (simp-all)
15      from ⟨Γ ⊢ t₁ : T₁ → T₂⟩ have Γ ⊢ Lam x. s₁ : T₁ → T₂ using eqs by simp
16      then have (x,T₁)::Γ ⊢ s₁ : T₂ using fc
17        by (cases rule: typing.strong-cases) (auto simp add: alpha)
18      moreover
19      from ⟨Γ ⊢ t₂ : T₁⟩ have Γ ⊢ s₂ : T₁ using eqs by simp
20      ultimately have Γ ⊢ s₁[x:=s₂] : T₂ by (rule type-substitutivity)
21      then show Γ ⊢ u′ : T₂ using eqs by simp
22    qed (auto intro: ty-App)
23  next
24    case (ty-Lam x T₁ Γ t T₂)
25    from ⟨Lam x. t ⟶_β u′⟩ ⟨x # u′⟩
26    obtain s₂ where t-red: t ⟶_β s₂ and eq: u′ = Lam x. s₂
27      by (cases rule: Beta.strong-cases) (auto simp add: alpha)
28    have ih: t ⟶_β s₂ ⟹ (x,T₁)::Γ ⊢ s₂ : T₂ by fact
29    with t-red have (x,T₁)::Γ ⊢ s₂ : T₂ by simp
30    then have Γ ⊢ Lam x. s₂ : T₁ → T₂ by (rule typing.ty-Lam)
31    with eq show Γ ⊢ u′ : T₁ → T₂ by simp
32  qed
```

Fig. 3. An Isar-proof of the type preservation lemma in Isabelle/HOL

What was surprising to us is that the strong inversion principles depend on the vc-condition that we introduced in previous work [8]. There, this condition was used to make sure that the variable convention in proofs by rule induction does not lead to faulty lemmas. An disadvantage of our approach is that in case of beta-reduction we have to use rule b_1' shown in (27) and so far we have no automatic method to derive from it the usual rule b_1 shown in (26).

The most closely related work to the one presented here is our own [8], where we study strong induction principles. Here we were concerned with inversion principles, which in our setting with non-injective term constructors are *not* a degenerated form of induction (as is usually the case). In contrast with that work [8], we also deal here with the case where rules include quantifiers. In the context of type theory, inversion principles have been studied by Cornes and Terrasse for the Coq proof assistant [4] and by McBride for the LEGO system [5]. McBride's implementation in LEGO uses an algorithm for solving equality constraints based on unification. The derivation of inversion principles for inductive sets in Isabelle's object logic HOL and ZF was first described by Paulson [6].

References

1. Aczel, P.: An Introduction to Inductive Definitions. In: Barwise, J. (ed.) Handbook of Mathematical Logic, pp. 739–782. Elsevier, Amsterdam (1977)
2. Aydemir, B.E., Bohannon, A., Fairbairn, M., Foster, J.N., Pierce, B.C., Sewell, P., Vytiniotis, D., Washburn, G., Weirich, S., Zdancewic, S.: Mechanized Metatheory for the Masses: The POPLMARK Challenge. In: Hurd, J., Melham, T. (eds.) TPHOLs 2005. LNCS, vol. 3603. Springer, Heidelberg (2005),
   http://www.cis.upenn.edu/plclub/wiki-static/poplmark.pdf
3. Aydemir, B.E., Charguéraud, A., Pierce, B.C., Pollack, R., Weirich, S.: Engineering formal metatheory. In: Necula, G.C., Wadler, P. (eds.) Proceedings of the 35th ACM SIGPLAN-SIGACT Symposium on Principles of Programming Languages, POPL 2008, San Francisco, California, USA, January 7-12, 2008, pp. 3–15. ACM Press, New York (2008)
4. Cornes, C., Terrasse, D.: Automating Inversion of Inductive Predicates in Coq. In: Berardi, S., Coppo, M. (eds.) TYPES 1995. LNCS, vol. 1158, pp. 85–104. Springer, Heidelberg (1996)
5. McBride, C.: Inverting Inductively Defined Relations in LEGO. In: Giménez, E. (ed.) TYPES 1996. LNCS, vol. 1512, pp. 236–253. Springer, Heidelberg (1998)
6. Paulson, L.C.: A fixedpoint approach to (co)inductive and (co)datatype definitions. In: Plotkin, G., Stirling, C., Tofte, M. (eds.) Proof, Language, and Interaction: Essays in Honor of Robin Milner, pp. 187–211. MIT Press, Cambridge (2000)
7. Pitts, A.M.: Nominal Logic, A First Order Theory of Names and Binding. Information and Computation 186, 165–193 (2003)
8. Urban, C., Berghofer, S., Norrish, M.: Barendregt's Variable Convention in Rule Inductions. In: Pfenning, F. (ed.) CADE 2007. LNCS (LNAI), vol. 4603, pp. 35–50. Springer, Heidelberg (2007)
9. Urban, C., Tasson, C.: Nominal Techniques in Isabelle/HOL. In: Nieuwenhuis, R. (ed.) CADE 2005. LNCS (LNAI), vol. 3632, pp. 38–53. Springer, Heidelberg (2005)

Canonical Big Operators

Yves Bertot[1], Georges Gonthier[2], Sidi Ould Biha[1], and Ioana Pasca[1]

[1] INRIA
[2] Microsoft Research
{Yves.Bertot,Sidi.Ould_Biha,Ioana.Pasca}@sophia.inria.fr
gonthier@microsoft.com

Abstract. In this paper, we present an approach to describe uniformly iterated "big" operations, like $\sum_{i=0}^{n} f(i)$ or $\max_{i \in I} f(i)$ and to provide lemmas that encapsulate all the commonly used reasoning steps on these constructs.

We show that these iterated operations can be handled generically using the syntactic notation and canonical structure facilities provided by the COQ system. We then show how these canonical big operations played a crucial enabling role in the study of various parts of linear algebra and multi-dimensional real analysis, as illustrated by the formal proofs of the properties of determinants, of the Cayley-Hamilton theorem and of Kantorovitch's theorem.

1 Introduction

One of the most versatile tools of the working mathematician is the "big operator" notation. At the stroke of a `\bigxx` LaTeX macro, she gets a bird's eye view of the algebra of her problem, revealing hidden symmetries, which she can immediately exploit using a rich set of partitioning, reindexing, and commutation operations.

So far, big operators have been missing from the toolbox of the formal mathematician, at least in their full generality, that is, allowing big of any operator indexed in any way, such as

$$\sum_{d|n} \phi(n/d)m^d \quad \text{or} \quad \bigoplus_{V_i \simeq W} V_i$$

We report here on the design of a generic big operator library for the COQ proof system [4,2]. This development was motivated and honed by the proof of several advanced results in algebra and analysis, which we also present.

This library is not just a collection of notations, although we do make good use of COQ's facilities in this respect. It contains a generic theory of big operators, including unique lemmas that perform complex operations such as reindexing and dependent commutation, for all operators, with minimal user input and under minimal assumptions.

Critically, the library relies on COQ's canonical structures (described below) for expressing structural and algebraic properties of indices and operators. This allows rewriting and resolution to infer such properties automatically, which

O. Ait Mohamed, C. Muñoz, and S. Tahar (Eds.): TPHOLs 2008, LNCS 5170, pp. 86–101, 2008.
© Springer-Verlag Berlin Heidelberg 2008

is essential for the library to be usable in practice. Although similar, neither dependent record subtyping nor axiomatic type classes would support this style of operator-centric inference.

The paper is organized as follows. In Section 2, we describe and use COQ's canonical structures to create the level foundation on which we will build our big operator theory, while in Section 3 we use COQ's syntactic notation facility to map a wide range of big operator forms to a single generic function; in Section 4 we develop a library of generic lemmas that can handle most of the common algebraic and logical operations on these forms. Finally, in Sections 5, 6, and 7, we put this library to work in the formalization of some classical results in algebra and analysis, including the Cauchy determinant formula, the Cayley-Hamilton theorem, and Kantorovitch's theorem.

2   Canonical Structures

Building a generic library that can accommodate a large variety of iterated operators requires more than notation — although the latter does play an important role, as we shall see in the next section. It calls for a logical framework that can express and classify the key properties of the two main components of big operators, namely, indexes and operations.

We implement this framework with COQ's Canonical Structure declaration, which we use in a new and nonstandard way. Although specific to COQ, the Canonical Structures are fairly close to record subtyping and type classes [15], so our approach could be ported to other systems, given some minor extensions.

For indices we actually reuse combinatorial structures that were developed for the Four Colour Theorem proof, and used in our finite group library [7], so the next section is a review of material from [7,6] that can also serve as an introduction to Canonical Structures.

2.1   Index Structures

We want to handle big operators indexed by arbitrary types. However, we need to compare and possibly enumerate indices to compute big operators, so the indices must have enriched types. In an object-oriented setting this could be achieved by subtyping; it is well-known that in higher-order logic nested dependent records (aka telescopes) can be used instead [11,16].

For example we can describe comparable ("equality") types as follows: [1]

```
Structure eqType : Type := EqType {
    sort :> Type;
  eqd : sort -> sort -> bool;
      _ : forall x y, (x == y) <-> (x = y)
} where "x == y" := (eqd x y).
```

---

[1] The actual code uses a mixin/class presentation and handles Coq technicalities like namespace management and reduction hints.

The :> symbol makes `sort` into a coercion, which means we can use a T : eqType as if it were a `Type` — type inference will insert the missing `sort` projection. The structure lets us define a unified notation for the generic comparison function `eqd`. Moreover every `eqType` contains an axiom stating that the comparison function *reflects* actual (Leibnitz) equality, so it is valid to rewrite x into y given x == y (i.e., (x == y)= true). Indeed, equality is decidable (and therefore proof-irrelevant [1]) for all `eqTypes`.

Unlike telescopes, COQ structures are not restricted to abstract types, and can be created for *existing* types. For example, if we can prove

Lemma eqnP : forall m n, eqn m n <-> m = n.

for an appropriate function `eqn : nat -> nat -> bool`, then we can make `nat`, a `Type`, behave as an `eqType` by declaring

Canonical Structure nat_eqType := EqType eqnP.

This creates a new `eqType` with sort ≡ `nat` and eqd ≡ `eqn` (both are inferred from the type of `eqnP`). Dually to coercions, this declaration allows `nat` to behave as `sort nat_eqType` during type inference. This lets COQ interpret 2 == n as (eqn 2 n), because it can solve the unification equation sort $?e \equiv_{\beta\delta\iota}$ nat and then evaluate eqd `nat_eqType` to `eqn`.

These details are crucial for the next section. However, the casual user can mostly gloss over them; he only cares that canonical structures let him use generic constructions and properties for specific instances, similarly to type classes.

The computation of a big operator must also enumerate the indices in its range. This is trivial if the range is an explicit sequence of type `seq I`, where `I` has an `eqType` structure, e.g., if the range is a `nat` interval. However, it is often more convenient to specify the range implicitly by a predicate, in which case the computation must be able to enumerate the entire index type, which must thus be *finite*. The following structure supports this capability:

```
Structure finType : Type := FinType {
  sort :> eqType;
  enum : seq sort;
  _ : forall x, count (fun y => y == x) enum = 1
}.
```

The axiom asserts that every value of the type occurs exactly once in `enum`.

This structure is very good for working with finite sets; we have a rich `finType`-based library of over 250 lemmas [7], which includes the construction of a type `ordinal n` (denoted I_(n)) of integers $0 \le i < n$, of function and set types over a `finType`, as well as canonical `eqType` and `finType` structures for all of these.

2.2 Operator Structures

Genericity over operations is more difficult to achieve, but probably more important than genericity over indices. We want to be able to use our library for all kinds of types and operations, from simple integer sums to GCDs of polynomials.

Using telescopes here would essentially amount to identifying big operators with generalized summations over a type with the following structure:

```
Structure additive_group : Type := AdditiveGroup {
  sort :> eqType;          zero : sort;
  opp : sort -> sort;      add : sort -> sort -> sort;
  _ : associative add;     _ : commutative add;
  _ : left_unit zero add;  _ : left_inverse zero opp add
}.
```

However this would be wrong, for several reasons:

1. It imposes strong axioms, which will not hold for many interesting operators, such as max over nat. Simply refining the telescope to take into account the many relevant axiom sets, from non-commutative monoids up to commutative rings, would lead to an uncomfortably deep hierarchy. The latter would cause type inference to generate terms bloated with long projection chains.
2. It makes the representation of the summation depend on the *proofs* that the operator satisfies algebraic properties, and by extension on any data on which these proofs might depend. This artificial dependency could only be broken by breaking the summation abstraction thereby losing all generic notations and properties.
3. It is only parametric in the operator *type*, not the operator itself, so we could only have at most one generic big operator per type. This is inadequate even for abstract rings, where we want both sums and products, and woefully inadequate for integers, which have sum, product, min, max, GCD and LCM!

The proper solution to this parametricity problem lies in the observation that unlike type classes or telescopes, canonical structures can be used to enrich not only types, but arbitrary values. Indeed, this was already the case for finType and additive_group, which both enriched the eqType structure — a *record*.

This fact allows us to define iteration for arbitrary operators, because we can use a structure to meet the algebraic requirements of our lemmas. For instance we define

```
Structure law T unit : Type := Law {
  operator : T -> T -> T;   mul1m : left_unit unit operator;
  _ : associative operator; mulm1 : right_unit unit operator
}.
```

and then

```
Canonical Structure andb_monoid := Law andbA andTb andbT.
Canonical Structure addn_monoid := Law addnA add0n addn0.
Canonical Structure gcdn_monoid := Law gcdnA gcd0n gcdn0.
...
```

This simple series of canonical structure declarations lets CoQ know that boolean conjunction, integer addition and *GCD*, etc, are monoidal laws, so that it can automatically discharge this condition when a lemma or rewrite rule is used.

We define similar structures for abelian monoids and semirings; note that nesting depth (issue 1 above) is not a problem here as these structures appear in the proof terms only, not in the big operator expression.

3 Notations

To capture the commonalities between all possible big operators, we provide a host of notations that are independent from the operator being used, the operator and the value for the empty range being given as parameters. Thus the notation has the following shape:

\big [op / nil]_ (*index and range description*) F

3.1 Range Descriptions

The part called *index and range description* is responsible for giving the name of the bound variable and stating the set over which this variable is supposed to range. There are mainly three ways to give the range: take a list of values that have to be covered, take a specific integer interval, or take the entire type of the bound variable (which must then be a finType) or a subset thereof. We use the notation $(i <- r)$ to range over a list, the notations $(m <= i < n)$ or $(i < n)$ to range over an interval, the notations (i) or $(i : t)$ to range over the entire index type, and the notation $(i \in A)$ to range over a subset. In all cases, the variable i is bound in F.

On top of these variants, we choose to add the possibility to filter the range with a predicate, meaning that the big operator takes only the elements of the range that satisfy the predicate. This is simply written by adding | P at the end of the index and range description. Again, the variable i is bound in the formula P. Thus, the following notation represents the addition of all squares of even numbers between 0 and $n - 1$.

\big[addn/0]_(i < n | even i) i^2

Since natural numbers in an interval can easily be enumerated, all notations reduce to the same function, where the range is a list of values that do not need to belong to a finite type and a filtering predicate is always provided. This notation is implemented by the following code:

```
Definition reducebig R I op nil r (P : pred I) (F : I -> R) : R :=
  foldr (fun i x => if P i then op (F i) x else x) nil r.

Notation "\big [ op / nil ]_ ( i <- r | P ) F" :=
  (reducebig op nil r (fun i => P%B) (fun i => F)) : big_scope.
```

It is a simple structural recursive function which follows the structure of the list r and tests whether P is satisfied on the first element to decide whether the value of F on this element is combined with the value computed for the rest; at the end of the list the nil value is used.

3.2   Operator Inference

We then define other notation that is specialized for the case where the operator satisfies a particular structure. For a variable ranging on a type, the various patterns are as follows:

- \sum_(i) F is used when the result type is nat, nil is 0 and the operator is nat addition, or when the operator is the add field of an additive_group structure (see 2.2) whose zero is nil. Thus, when the type of formula F is sort of a canonical additive_group structure, this operation is automatically understood as the iteration of its additive law.
- \prod_(i) F is used when the result type is nat, nil is 1 and the operator is nat multiplication, or when the operator is the multiplication of a ring or group with unit nil.
- \max_(i) F is used when the operator is the nat binary max and nil is 0.

Note that the denoted term is always of the form reducebig ..., so generic lemmas apply uniformly regardless of which notation is used to for the range and operator.

4   Main Lemmas

Canonical structures play a crucial role when organizing the large library of lemmas that we provide to reason about big operations (there are around 80 lemmas). A first collection of lemmas helps reasoning about big operations without any assumption on the operator being iterated. Other collections of lemmas are for operators that respect a plain monoid structure (with only associativity and a neutral element), an abelian monoid structure (with commutativity) or a semi-ring structure (where two operators interact through distributivity).

For instance, lemmas applicable to a monoid operator handle a big operator where op has the form operator l and require l to have the type law, while lemmas applicable to an abelian monoid operator handle a big operator where op has the form operator (law_of_abelian l) and require l to have the type abelian_law. For a given operator op that is both associative and commutative and has a neutral element, two canonical structures are constructed, one with type law and the other with type abelian_law. The user always writes \big[op/nil]_...; when a lemma requiring associativity is applied the corresponding canonical structure is automatically inferred. Thus, we have a single notation that is independent of properties satisfied by operators; lemmas refer to properties through records, and we use canonical structures to reconcile the two approaches at the time we use the lemmas. This will be apparent as we study in more detail some of the lemmas.

4.1   Lemmas for Plain Operators

Operations like replacing the general term or predicate of a big operation by an equivalent one or unmapping its index range do not require any property of the operator.

To cope with rewriting in parts of a big operation, we provide a variety of congruence lemmas. Here is one example, which can be used to express that we can rewrite in the predicate and the formula parts of a big operation.

```
Lemma eq_big : forall (r : seq I) (P1 P2 : pred I) F1 F2,
  P1 =1 P2 -> {in P1, F1 =1 F2} ->
  \big[op/nil]_(i <- r | P1 i) F1 i =
  \big[op/nil]_(i <- r | P2 i) F2 i.
```

This lemma expresses that two big operations can be proved equal even though their predicates and formulas may appear to be different. The first premise `P1 =1 P2` expresses that it suffices that the predicates are extensionally equal (the two functions are equal on every argument), the second premise `{in P1, F1 =1 F2}` that it suffices that the two formulas are extensionnally equal on the subset of the type determined by the predicate `P1`.

Other collections of lemmas concern rewritings that occur simultaneously in the range and in some other part of the big operation. For instance, a combined rewriting in the range and the formula makes it possible to change all elements in the range list, compensating by a composition in the formula and the filtering predicate:

```
Lemma big_maps : forall (J : eqType) (h : J -> I) r F P,
  \big[op/nil]_(i <- maps h r | P i) F i =
  \big[op/nil]_(j <- r | P (h j)) F (h j).
```

We also have lemmas that make it possible to change the length of the range: we can assert that a sum up to n_1 is equal to a sum up to n_2, with $n_1 \leq n_2$, if the predicate filters out all numbers that are larger than or equal to n_1 and smaller than n_2:

```
Lemma big_nat_widen : forall m n1 n2 P F, n1 <= n2 ->
  \big[op/nil]_(m <= i < n1 | P i) F i =
  \big[op/nil]_(m <= i < n2 | P i && (i < n1)) F i.
```

4.2   Plain Monoid Re-indexing

When the iterated operation is associative and the nil value is the neutral element, nicer decomposition lemmas can be obtained. To express that the operator is a monoid law we use a notation *%M. We also use a specific notation for the nil value, but this is only to enhance readability. For instance, we can state a lemma that helps decomposing a list range in two sub-lists, where ++ stands for the concatenation of lists:

```
Lemma big_cat : forall I (r1 r2 : seq I) P F,
  \big[*%M/1]_(i <- r1 ++ r2 | P i) F i =
  \big[*%M/1]_(i <- r1 | P i) F i *
    \big[*%M/1]_(i <- r2 | P i) F i.
```

which would be written in standard mathematics:

$$\prod_{i \in r_1 \cup r_2, P_i} F_i = \prod_{i \in r_1, P_i} F_i * \prod_{i \in r_2, P_i} F_i$$

We actually provide half a dozen lemmas that are specific, to monoidal laws.

4.3 Abelian Monoid Re-indexing

To handle commutative monoidal operators, we redefine our notation *%M to express that it has to be the operator of an abelian monoidal law. This is done with the following notation declaration:

```
Notation Local "*%M" := (operator (law_of_abelian op)).
```

In this case, permuting elements in the range or grouping them according to a partition becomes possible. Here are two of the main lemmas, concerned with partitioning an index set and with swapping nested sum operators.

To describe partitions, we use an auxiliary function and view each subset in the partition as the inverse image of one element:

```
Lemma partition_big :
  forall (I J : finType) (P : pred I) p (Q : pred J) F,
  (forall i, P i -> Q (p i)) ->
  \big[*%M/1]_(i | P i) F i =
  \big[*%M/1]_(j | Q j) \big[*%M/1]_(i | P i && (p i == j)) F i.
```

$$(\forall i, P_i \to Q_{p(i)}) \to \prod_{i \in I, P_i} F_i = \prod_{j \in J, Q_j} \prod_{\substack{i \in I \\ P_i \wedge p(i)=j}} F_i$$

To permute nested sum operators, we start by showing that that nested big operations can be reduced to a single big operation where pairs of indices are enumerated. Through a re-indexing operation on the pairs, we then obtain a variety of commutation lemmas, of which we show only the simplest one:

```
Lemma exchange_big : forall (I J : finType) P Q F,
  \big[*%M/1]_(i : I | P i) \big[*%M/1]_(j : J | Q j) F i j =
  \big[*%M/1]_(j | Q j) \big[*%M/1]_(i | P i) F i j.
```

$$\prod_{i \in I, P_i} \prod_{j \in J, Q_j} F_{i,j} = \prod_{j \in J, Q_j} \prod_{i \in I, P_i} F_{i,j}$$

4.4 Distributivity

Distributivity plays a role when several operators interact, usually in a semi-ring structure. Here we adapt our notation so that *%M refers to the multiplication operation of a semi-ring and +%M refers to the addition of the same semi-ring. Here is a first simple lemma:

```
Lemma big_distrl : forall I (r : seq I) alpha P F,
  (\big[+%M/0]_(i <- r | P i) F i) * alpha =
  \big[+%M/0]_(i <- r | P i) (F i * alpha).
```

In general, big products of big sums can be transformed into big sums of big products: this is another form of swapping lemma that gives rise to pairs of indices. Here is one of our lemmas to handle this, where {ffun I -> J} describes the set of all functional graphs in I*J (a finite type that actually describes all functions from the finite type I to the finite type J):

```
Lemma bigA_distr_bigA :
  forall (I J : finType) F,
  \big[*%M/1]_(i : I) \big[+%M/0]_(j : J) F i j =
  \big[+%M/0]_(f : {ffun I -> J}) \big[*%M/1]_(i) F i (f i).
```

It is remarkable that none of these lemmas requires a proof that C1, the value of empty "big products", actually be the neutral element for multiplication.

5   Some Results on Determinants

The first motivating example for our big operator library was the study of determinants; it uses many key features of the library, including the compact notation, generic indexing, and reindexing and swapping lemmas.

5.1   The Leibnitz Formula

While in practice determinants are best computed from a triangular decomposition, or by using Laplace's formula to expand with respect to a fixed row, it is impractical to derive any of the theoretical properties of determinants from such expressions because of their lack of symmetry. In contrast, the highly symmetrical but impractical Leibnitz formula calls for summing over permutations; our generic library handles this quite gracefully:

```
Definition determinant n (A : M_(n)) :=
  \sum_(s : S_(n)) (-1)^+s * \prod_(i) A i (s i).
```

The actual CoQ proofs that this definition yields a multilinear alternate form are only 7 and 14 lines long, respectively; the proof of the Laplace formula is 80 lines (most of which compute the parity of a cyclic permutation), but then we only need 16 lines to prove the Cramer rule:

$$A.\text{adj } A = \text{adj } A.A = \det A.\text{Id} \qquad (1)$$

5.2   The Cauchy Formula

The Cauchy formula simply states that the determinant commutes with matrix product. It is fairly tricky to establish rigorously for abstract rings; here is a self-contained proof, for $n \times n$ matrices:

$$\det AB = \sum_{\sigma \in S_n} (-1)^{\sigma} \prod_i \left(\sum_j A_{ij} B_{j\sigma(i)} \right)$$

$$= \sum_{\phi: [1,n] \to [1,n]} \sum_{\sigma \in S_n} (-1)^{\sigma} \prod_i A_{i\phi(i)} B_{\phi(i)\sigma(i)}$$

$$= \sum_{\phi \notin S_n} \sum_{\sigma \in S_n} (-1)^{\sigma} \prod_i A_{i\phi(i)} B_{\phi(i)\sigma(i)} + \sum_{\phi \in S_n} \sum_{\sigma \in S_n} (-1)^{\sigma} \prod_i A_{i\phi(i)} B_{\phi(i)\sigma(i)}$$

$$= \sum_{\phi \notin S_n} \left(\prod_i A_{i\phi(i)} \right) \sum_{\sigma \in S_n} (-1)^{\sigma} \prod_i B_{\phi(i)\sigma(i)}$$

$$+ \sum_{\phi \in S_n} (-1)^{\phi} \left(\prod_i A_{i\phi(i)} \right) \sum_{\sigma \in S_n} (-1)^{\phi^{-1}\sigma} \prod_k B_{k\sigma(\phi^{-1}(k))}$$

$$= \sum_{\phi \notin S_n} \left(\prod_i A_{i\phi(i)} \right) \det \left(B_{\phi(i)j} \right)_{ij} + (\det A) \sum_{\tau \in S_n} (-1)^{\tau} \prod_k B_{k\tau(k)}$$

$$= 0 + (\det A)(\det B)$$

The first step swaps the iterated product of the Leibnitz formula with the sum in the general term of the matrix product, generating a sum over all functions from indices to indices. This is split into a sum over non-injective functions and a sum over permutations. The former is rearranged into a weighted sum of determinants of matrices with repeated rows, while the latter is reindexed, using the group properties of permutations, to become the desired product of determinants.

Remarkably, the formal CoQ proof is only 25 lines long, and actually shorter than the above proof sketch, because all of the required sum manipulations are directly supported by the library, and our previous work on finite groups [7] supplies the all required permutation facts.

6   The Cayley-Hamilton Theorem

After proving the Cramer rule, the next step was formalizing the Cayley-Hamilton theorem [3]. For a commutative ring R and a square matrix A on R, this theorem states that A is a root of its characteristic polynomial $p_A(x) = \det(xI_n - A)$.

To prove the Cayley-Hamilton theorem, we apply the Cramer rule (1) to the $(xI_n - A) \in M_n(R[x])$ and we obtain:

$$\text{adj}\,(xI_n - A) * (xI_n - A) = \det(xI_n - A) * I_n = p_A(x) * I_n \qquad (2)$$

This is an equality in $M_n(R[x])$. However the ring $M_n(R[x])$ of matrices with polynomial coefficients and the ring of polynomials with matrix coefficients $(M_n(R))[x]$ are isomorphic. For example, the following equality exhibits the correspondence:

$$\begin{pmatrix} x^2+1 & x-2 \\ -x+3 & 2x-4 \end{pmatrix} = x^2 \begin{pmatrix} 1 & 0 \\ 0 & 0 \end{pmatrix} + x \begin{pmatrix} 0 & 1 \\ -1 & 2 \end{pmatrix} + \begin{pmatrix} 1 & -2 \\ 3 & -4 \end{pmatrix}$$

We call $\phi : M_n(R[x]) \rightarrow (M_n(R))[x]$ the isomorphism from one ring to the other. In $(M_n(R))[x]$, the equality (2) is written :

$$\phi(\text{adj } (xI_n - A)) * (x - A) = p_A^\phi(x) \tag{3}$$

where $p_A^\phi(x)$ is in fact the polynomial with scalar matrix coefficients obtained by applying ϕ to $p_A(x) * I_n$. This shows that $(x - A)$ is a factor of $p_A^\phi(x)$ in $(M_n(R))[x]$, so $p_A^\phi(A) = O_n$.

To formalize this proof, we developed a library to describe polynomials.

6.1   Polynomials

A polynomial is formally defined by the list of its coefficients a_i which are elements of a ring R :

$$a_n x^n + a_{n-1} x^{n-1} + \cdots + a_1 x + a_0$$

It is natural to represent a polynomial with the list $(a_0, \ldots, a_n)$; however, it is also handy to use polynomials as functions of type $nat \rightarrow R$, which return 0 almost everywhere.

We can easily change from one representation to the other by using a lemma that states that two polynomials are equal if their functional representations are extensionally equal. The list representation is used to define the operations on polynomials by induction on the coefficient list. With the representation as function of type $nat \rightarrow R$, we can reuse lemmas on big operators to prove algebraic properties of polynomials, in a style that is close to standard mathematics. For example the following property of the coefficients of the product of two polynomials is expressed using a big sum.

```
Lemma coef_mul_poly : forall p1 p2 i,
  coef (p1 * p2) i = \sum_(j < i.+1) coef p1 j * coef p2 (i - j).
```

With this new point of view, we prove the associativity of polynomial multiplication by simply reusing re-indexation and distributivity lemmas for big operators.

In the following, the notation \poly_(i < n) E, where E is an expression on i, corresponds to the polynomial $\sum_{i<n} E_i x^i$. The notations \X, \C c and p.[c] correspond respectively to the monomial x, the constant polynomial c and the evaluation of a polynomial p in a value c.

In the polynomials library we give a proof of the factor theorem :

```
Theorem factor_theorem : forall p c,
  (exists q, p = q * (\X - \C c)) <-> (p.[c] = 0).
```

We proved the equivalence, but we only need the implication from left to right for the Cayley-Hamilton theorem. This proof is only 12 lines long, thanks to the lemmas on big operators.

6.2 Proving the Cayley-Hamilton Theorem

The morphism between the ring of matrices of polynomials and the ring of polynomials of matrices is the central part of the proof. It is best described using big operators:

```
Definition phi (A : M(R[X])) : M(R)[X] :=
  \poly_(k < \max_(i) \max_(j) size (A i j))
    \matrix_(i, j) coef (A i j) k.
```

In this formula, the notation `\matrix_(i, j) E` denotes the matrix whose coefficient at position `(i, j)` is described by the expression `E`. The length of the resulting polynomial is the maximum size of coefficient lists in the input matrix, described with the `\max` operator. Big operator lemmas are also instrumental in the proofs of morphism properties for ϕ.

The characteristic polynomial is defined as follow :

```
Definition char_poly (A : M(R)) : R[X] := \det(\Z \X - matrixC A).
```

In this formula `\Z` stands for the scalar multiplication by the identity matrix. We also define `Zpoly` as the canonical injection from the ring of polynomials with scalar coefficients into the ring of polynomials with matrix coefficient. With these definitions the Cayley-Hamilton theorem has the following statement.

```
Theorem Cayley_Hamilton : forall A, (Zpoly (char_poly A)).[A] = 0.
```

The main proof is done in three lines.

7 Multivariate Real Analysis and Kantorovitch's Theorem

We also conducted an experiment in giving a complete formalization for Kantorovitch's theorem [14]. This theorem in numerical analysis gives sufficient conditions for the convergence of Newton's method for finding the root of a function $f : \mathbb{R}^p \to \mathbb{R}^p$. The main challenge was to find a representation for vectors in $\mathbb{R}^p$ and formalize multivariate analysis concepts.

After a careful analysis, the choice was made to base this formalization on the Reals from standard CoQ and on the SSREFLECT extension. This choice turned out to be adequate for vectors, matrices, and the use of big operations to abstract over dimensions.

We provide a canonical structure of field for $\mathbb{R}$ and encode vectors as functional graphs from a finite type of dimension p to $\mathbb{R}$. In practice, this gives both a view of vectors as lists and vectors as functions over the index type, thus facilitating the description of component-wise operations.

We then simply formalize a norm on vectors as a big operation. The norm is $\|x\| = \max_i |x_i|$, which in CoQ can be expressed as:

```
Definition norm (v : Rvec p) := \big[Rmax/0]_(i < p) Rabs (v i).
```

With this definition, a lemma stating the positivity of the norm

```
Lemma norm_pos : forall v, 0 <= norm v.
```

is easily proved by applying a generic lemma named `big_prop`. This induction scheme states that a property which is closed with respect to the operator, satisfied by the `nil` value, and by the formula for every index is also satisfied by the result of big operation. In this case, the property is positiveness. Other required properties for norms have about the same level of complexity.

Nevertheless, the use of the maximum as an indexed operation posed some difficulties. As stated before, the lemmas on big operations are organized in a sort of hierarchy following the algebraic structure given by the operator. In the case of the maximum, we have associativity and commutativity, but we do not have a neutral element on the type of real numbers. Since we work only with positive numbers (and the maximum on this subset has 0 for neutral element), we would like to be able to use the lemmas that deal with an abelian monoid structure.

There are two possible solutions for this problem. The first is to have a new type for positive reals. We can define the canonical structure of abelian monoid on this new type, manipulate the indexed operation as desired and inject the result in the original type. The second solution is to define a new operator that gives the type the desired structure. This operator has to be equal to the original one on the target subset (here, the positive reals). Such a change of operator is covered in the library by a lemma called `eq_big_op`.

We adopted the second approach, as we had this definition at hand:

$$\mathrm{max}'\ x\ y = \begin{cases} \mathrm{max}\ x\ y & \text{if } x > 0 \text{ or } y > 0; \\ \mathrm{min}\ x\ y & \text{if } x, y \le 0 \end{cases}$$

However, we could easily have fallen back to the first solution if such a construction had not been available.

Another interesting example regards the decomposition of a vector to prove a multi-dimensional variant of the mean value theorem.

$$f(x_1, \ldots, x_p) - f(y_1, \ldots, y_p) = f(x_1, \ldots, x_p) - f(y_1, x_2, \ldots, x_p) +$$
$$f(y_1, x_2, \ldots, x_p) - f(y_1, y_2, x_3, \ldots, x_p) + \ldots + f(y_1, \ldots, y_{p-1}, x_p) - f(y_1, \ldots, y_p)$$
$$= \sum_{i=1}^{p} (x_i - y_i) \frac{\partial f(y_1, \ldots, y_{i-1}, c_i, x_{i+1}, \ldots, x_p)}{\partial x_i}$$

This simple and elegant proof goes through naturally in our formalization.

During the development we also needed a formalization of matrices in order to represent, for example, the Jacobian of a partially derivable function. We used the `matrix` library developed during the formalization of the Cayley-Hamilton theorem, which we enriched with additional concepts, like the norm of a matrix, compatible with our vector norm: $\|A\| = \max_i \sum_j |a_{ij}|$.

Most of the lemmas we have described so far are concerned with equality, but results about norms also exhibit the need for lemmas concerned with binary relations. For instance, we use a lemma named `big_rel` which states that if a relation R is reflexive and transitive, R satisfies some stability condition with respect to the operator, and formulas F and G are related by R for every index, then the big operation on F is related by R with the big operation on G. Such a lemma is instrumental in the proof of the following results:

$$\|AB\| \leq \|A\|\|B\| \quad \text{and} \quad \|A\| < 1 \rightarrow \det(I_p - A) \neq 0$$

The first result relies on `big_rel` and distributivity lemmas, while the second result relies on the convergence of a series of matrices. One of the intermediate lemmas for the second result is expressed as follows:

```
Lemma mat_norm_sum : forall (A : nat -> MR(p)) n,
  norm (\sum_(i <= n) A i) <= \sum_(i <= n) norm (A i).
```

This lemma is a direct consequence of one of the generic lemmas from the library, named `big_morph_rel`: it suffices to show that `norm` has a morphism-like property with respect to the relation `<=`, addition of matrices (on the left), and addition of real numbers (on the right).

8   Conclusion

This work is based on the SSREFLECT extension of CoQ [6]. This extension relies extensively on canonical structures and reflexion. The work described in this paper is available on Internet at: `www-sop.inria.fr/marelle/bigops`.

8.1   Related Work

The HOL-Light system [8] also provides generic iterated operations and applications to multi-dimensional spaces. Separate work of T. Hales and J. Harrison [9] provide formalizations of euclidean space.

HOL-Light lacks dependent types but does not restrict itself to constructive logic. As a result, finite types cannot be described as records like our `finType` and iteration is actually defined on subsets of infinite types. Properties are mainly provided for abelian monoidal laws with a neutral element. This approach is less generic than ours, but it is already strong enough for many results. In particular, the system library contains results for real matrices and determinants similar to ours, but its applications do not go all the way to the Cayley-Hamilton theorem. Our work lives in a different setting: the main part is done in constructive logic with dependent types and we use enumerations for finite types which allows us to define big operations for plain operators.

Work by Gamboa, Cowles, and van Baalen also describes matrix computations in ACL2 [5]; they don't make a systematic use of big operations and determinants are described through a process of gaussian elimination, but almost no properties are proved.

In the CoQ system N. Magaud implemented vectors and matrices as dependent lists [12] but this is mainly an exercise in dependent types. J. Stein [17] and S. Obua [13] also describe linear algebra using big operations with monoid laws, for instance for matrix multiplication, but do not study determinants.

8.2 Overview and Perspectives

A commonly held opinion is that the formalization of mathematics is a long and difficult process for two reasons: first, more detail is required than in standard mathematical proofs and second, the formalized corpus is too small as a foundation, so that many lemmas have to be re-proved before addressing significant results. This opinion overlooks an important area where progress can be made, the area of infra-structure. Infra-structure can help in formalizing mathematics if statements and proofs can be expressed concisely and if the details can be collected automatically. This paper brings a contribution to the infra-structure aspect of formalizing mathematics.

We also bring a collection of lemmas organized in a way that increases their reusability drastically and we illustrate the gain with big operators for proofs on the properties of determinants and matrices. We feel we can approach new landmarks that were hitherto considered out of reach like the Cayley-Hamilton theorem. In the Mathematical Components project [7] we also reuse big operations to study generated groups.

Two questions come to mind: if this library on big operators has such a positive and structuring impact, what is the infrastructure behind the it that makes it so powerful? What is the next concept that deserves a systematic treatment and will have the same structuring effect?

To answer the first question, we propose to consider canonical structures as the key advance. First proposed by Saïbi in his study of category theory [10], these structures are instrumental here as they take over the automatic search for relevant information attached to each operator. Also, we propose to use canonical structures to attach properties to *operators*, while usual approaches attach properties to *types*. We can now write big operations simply, the required properties are inferred from the canonical structure declarations when applying lemmas.

We can't actually answer the second question yet, but we believe that big operations have opened the road to a re-newed study of linear algebra, with notions like bases, linear combinations, and so on, or of algorithms in other parts of algebra, like the algorithm of sub-resultants, the proof of which already relied on an abstract notion of determinants.

References

1. Barbanera, F., Berardi, S.: Proof-irrelevance out of Excluded-middle and Choice in the Calculus of Constructions. Journal of Functional Programming 6(3), 519–525 (1996)
2. Bertot, Y., Castéran, P.: Interactive Theorem Proving and Program Development, Coq'Art: the Calculus of Inductive Constructions. Springer, Heidelberg (2004)
3. Biha, S.O.: Formalisation des mathématiques: une preuve du théorème de Cayley-Hamilton. In: Journées Francophones des Langages Applicatifs, pp. 1–14 (2008)
4. Coq development team. The Coq Proof Assistant Reference Manual, version 8.1 (2006)

5. Cowles, J., Gamboa, R., Baalen, J.V.: Using ACL2 Arrays to Formalize Matrix Algebra. In: ACL2 Workshop (2003)
6. Gonthier, G., Mahboubi, A.: A small scale reflection extension for the Coq system. INRIA Technical report, http://hal.inria.fr/inria-00258384
7. Gonthier, G., Mahboubi, A., Rideau, L., Tassi, E., Théry, L.: A modular formalisation of finite group theory. In: Schneider, K., Brandt, J. (eds.) TPHOLs 2007. LNCS, vol. 4732, pp. 86–101. Springer, Heidelberg (2007)
8. Harrison, J.: HOL Light: A Tutorial Introduction. In: FMCAD, pp. 265–269 (1996)
9. Harrison, J.: A HOL Theory of Euclidian Space. In: Hurd, J., Melham, T. (eds.) TPHOLs 2005. LNCS, vol. 3603, pp. 114–129. Springer, Heidelberg (2005)
10. Huet, G., Saïbi, A.: Constructive category theory. In: Proof, language, and interaction: essays in honour of Robin Milner, pp. 239–275. MIT Press, Cambridge (2000)
11. Kammuller, F.: Modular Structures as Dependent Types in Isabelle. In: Altenkirch, T., Naraschewski, W., Reus, B. (eds.) TYPES 1998. LNCS, vol. 1657, pp. 121–132. Springer, Heidelberg (1999)
12. Magaud, N.: Ring properties for square matrices, http://coq.inria.fr/contribs-eng.html
13. Obua, S.: Proving Bounds for Real Linear Programs in Isabelle/HOL. In: Theorem Proving in Higher-Order Logics, pp. 227–244 (2005)
14. Paşca, I.: A Formal Verification for Kantorovitch's Theorem. In: Journées Francophones des Langages Applicatifs, pp. 15–29 (2008)
15. Paulson, L.C.: Organizing Numerical Theories Using Axiomatic Type Classes. J. Autom. Reason. 33(1), 29–49 (2004)
16. Pollack, R.: Dependently Typed Records for Representing Mathematical Structure. In: Aagaard, M.D., Harrison, J. (eds.) TPHOLs 2000. LNCS, vol. 1869, pp. 462–479. Springer, Heidelberg (2000)
17. Stein, J.: Documentation for the formalization of Linerar Agebra, http://www.cs.ru.nl/~jasper/

A Type of Partial Recursive Functions

Ana Bove[1] and Venanzio Capretta[2]

[1] Department of Computer Science and Engineering,
Chalmers University of Technology, 412 96 Göteborg, Sweden
Tel.: +46-31-7721020, Fax: +46-31-7723663
bove@chalmers.se
[2] Computer Science Institute (ICIS), Radboud University Nijmegen
Tel.: +31-24-3652631, Fax: +31-24-3652525
venanzio@cs.ru.nl

Abstract. Our goal is to define a type of partial recursive functions in constructive type theory. In a series of previous articles, we studied two different formulations of partial functions and general recursion. We could obtain a type only by extending the theory with either an impredicative universe or with coinductive definitions. Here we present a new type constructor that eludes such entities of dubious constructive credentials. We start by showing how to break down a recursive function definition into three components: the first component generates the arguments of the recursive calls, the second evaluates them, and the last computes the output from the results of the recursive calls. We use this dissection as the basis for the introduction rule of the new type constructor. Every partial recursive function is associated with an inductive domain predicate; evaluation of the function requires a proof that the input values satisfy the predicate. We give a constructive justification for the new construct by interpreting it into the base type theory. This shows that the extended theory is consistent and constructive.

1 Introduction

Our research investigates the formalisation of partial functions and general recursion in constructive type theory. In a series of previous articles, we expound two different ways of achieving that goal.

In our first approach [3,4,2,5], we define an inductive (domain) predicate that characterises the inputs on which a function terminates. The constructors of this predicate are automatically determined from the recursive equations defining the function. The function itself is formalised by structural recursion on an extra argument, a proof that the input values satisfy the domain predicate.

In our second approach [9,6], refined by Megacz [19], we associate to each data type a coinductive type of partial elements. Computations are modelled by (possibly) infinite structures. Partial and general recursive functions are implemented by corecursion on these types.

It is desirable that all partial functions with the same source and target are elements of the same type, rather than each function having an ad hoc type. In three of the articles mentioned above, we succeed in defining such a type, but

O. Ait Mohamed, C. Muñoz, and S. Tahar (Eds.): TPHOLs 2008, LNCS 5170, pp. 102–117, 2008.

there is always a cost to pay. In [5], we need to work in an impredicative type theory. Alternatively, we could work with a hierarchy of universes and accept that we are not able to formalise all recursive function definitions. In [9,6], we need support for coinductive definitions.

Other approaches to the definition of a type of partial recursive functions can be found in the literature; we summarise those we believe are more relevant.

Constable and Mendler [13] introduce a type of partial functions as a new type constructor in the Nuprl system [12]. Given a partial function, one can compute its domain predicate, which contains basically the same information as our domain predicates. A difference is that when defining a partial function in the Nuprl system, the actual definition of the partial function does not depend on its domain predicate. This would not be possible in the intuitionistic type theories in which we work. In other words, in Nuprl, a partial function from A to B maps an element a of A into an element of B provided there is some proof p that a belongs to the domain of the function. In our case, the formalisation of the partial function would map both the a and the p into B.

In [14], Constable and Smith develop a partial type theory for the Nuprl type system in which, for each type of the underlying total theory, there exists another type which might contain diverging terms. Together with this type of partial elements, a termination predicate and an induction principle to reason about partial functions are introduced.

Audebaud [1] uses the above idea [14] to define a conservative extension of the Calculus of Constructions [15] with fixed point terms and a type of partial objects. Strong normalisation still holds for terms with no fixed points. From the computational point of view, an equivalent of the Kleene theorem for partial recursive functions is obtained, but logical aspects need more examination.

Based on our work [3], Setzer [22,23] defines a type (of *codes*) of partial functions. From the code of a partial function, one can extract the domain of the function and the function itself, and one can evaluate the function on a certain argument. Nested functions and higher-order functions can also be coded as elements of the type of partial functions.

The approach illustrated in the present article is based on an analysis of a recursive definition into three components. The first component determines the arguments of the recursive calls, the second computes the recursive calls, and the third combines the results of the recursive calls to produce the output. This decomposition was described in [10] and is analogous to the separation of an hylomorphism into the composition of an anamorphism and a catamorphism in [20]. The method works in standard intuitionistic type theory and gives simple and direct formalisations, but we must pay a price: nested recursion and higher-order partiality are not possible anymore. The class of recursive functions that can be coded is still Turing-complete, but some function definitions may need to be rewritten to fit the pattern.

We formalised our approach in the proof assistant Coq [24]. The file containing the formalisation can be obtained from the following web page: www.cs.ru.nl/~venanzio/Coq/rec_fun_type/

The paper is organised as follows. Section 2 gives a closer look at functions and shows how a function can be split into three components, following the description in [10,11]. Section 3 gives the formal rules of the type of partial recursive functions. Section 4 formalises the signatures that describe the structure of the recursive calls in a way that ensures that the rules of the new type are predicative. Section 5 shows the consistency of the type theory extended with the new type by modelling it inside the base theory. Section 6 presents a modification of the rules to allow a general recursive pattern in which the input can also be used directly in the computation of the output, and not just in the generation of the recursive arguments. Finally, Section 7 summarises the achieved results.

2   A Closer Look at Recursive Functions

A common informal way to define a partial recursive function is by a sequence of recursive equations, as it could be implemented in a pure functional language like Haskell [17]:

$$f : A \rightarrow B$$
$$f \; p_0 = e_0[(f \; a_{00}), \ldots, (f \; a_{0k_0})]$$
$$\vdots$$
$$f \; p_n = e_n[(f \; a_{n0}), \ldots, (f \; a_{nk_n})].$$

Here $p_0, \ldots, p_n$ are *patterns*, that is, general terms containing variables and constructors. The notation $e[\cdots]$ denotes an expression e containing occurrences of the terms inside the square brackets. We restrict the form of the definition by requiring that the only occurrences of f are the displayed ones; specifically, f does not occur in any of the a_{ij}'s. This means that we are excluding nested recursive definitions.

When we apply a function to a concrete argument, the argument is matched against the patterns to find the equation that must be evaluated to give the result. We will not enter here into details about this process, nor about issues like exhaustiveness or overlapping patterns. What is important in what follows is that, when we give an actual input a to the function f, the system compares a with each pattern until it finds one, say p_i, that matches a. Then, it computes the new recursive arguments $a_{i0}, \ldots, a_{ik_i}$ on which to apply the function, and recursively computes f on these arguments. Finally, the system feeds the results of the recursive calls to the expression e_i in order to obtain the final output.

Next, we first illustrate with an example the issues involved in the process just described and then we present a generalisation of the example that allows a dissection of functions into three well-defined components. In the following section, we base our new type of *partial recursive functions* on this dissection.

A First Example

The example is a recursive functional program to translate Gödel's coding system for trees into Cantor's.

Kurt Gödel devised a system to encode expressions of a formal language into natural numbers by exploiting the uniqueness of prime factorisation [16]. His idea was that a complex expression can be encoded by giving the codes of its immediate n subexpressions as exponents to the first n prime numbers. As an instance of this process let us consider the coding of unlabelled well-founded trees of arbitrary branching degree. Let a tree t be represented by a node with a finite number of subtrees, that is, something of the form $\mathsf{node}(t_0, \ldots, t_k)$. Gödel's representation recursively encodes each tree t_i into a natural number g_i and then uses these g_i's as exponents of prime numbers to obtain the code of t as follows: if p_i is the $(i+1)$th prime number, the code of t is $p_0^{g_0} \cdots \cdot p_k^{g_k}$. A leaf is a node with no branches, $\mathsf{node}()$, and has code 1.

If we are interested in trees with small branching degree, high prime numbers are never used. In this case it is more convenient to adopt a different encoding that uses Cantor's pairing function. According to Cantor, a pair $\langle n_1, n_2 \rangle$ of natural numbers can be encoded by the number

$$\mathsf{pair}\ \langle n_1, n_2 \rangle = (n_1 + n_2) \cdot (n_1 + n_2 + 1)/2 + n_2.$$

This is a bijection between pairs and single numbers. Longer vectors can be encoded by repeated use of the pairing function: $\mathsf{tuple}_0 \langle \rangle = 0$, $\mathsf{tuple}_1 \langle n_0 \rangle = n_0$ and, recursively, $\mathsf{tuple}_{k+1} \langle n_0, \ldots, n_k \rangle = \mathsf{pair}\ \langle n_0, \mathsf{tuple}_k \langle n_1, \ldots, n_k \rangle \rangle$. We can now represent trees by an encoding similar to Gödel's: if c_i is the Cantor code of the tree t_i, then $\mathsf{pair}\ \langle k+1, \mathsf{tuple}_{k+1} \langle c_0, \ldots, c_k \rangle \rangle$ is the code of $\mathsf{node}(t_0, \ldots, t_k)$.

We adopt the convention of writing $\langle x_0, \ldots, x_{k-1} \rangle$ for a tuple of length k, understanding it to denote the empty tuple $\langle \rangle$ if $k = 0$. Similarly, a product $p_0^{i_0} \cdots p_{k-1}^{i_{k-1}}$ is understood to be equal to 1 if $k = 0$.

We define a translation function that maps Gödel's encoding of trees into Cantor's encoding:

$$\mathsf{trans\_code} \colon \mathbb{N} \to \mathbb{N}$$
$$\mathsf{trans\_code}\ x = \mathsf{pair}\ \langle k, \mathsf{tuple}_k \langle c_0, \ldots, c_{k-1} \rangle \rangle$$
$$\text{where}\quad c_i = \mathsf{trans\_code}\ g_i \quad \text{for } i = 0, \ldots, k-1$$
$$\text{with } k, g_0, \ldots, g_{k-1} \text{ such that } x = p_0^{g_0} \cdots p_{k-1}^{g_{k-1}}.$$

Note that the function is undefined on 0, since 0 cannot be written as a product of powers of primes. This kind of partiality is relatively easy to deal with, since it is decidable, unlike the partiality arising from non-termination, which is the real topic of our work. We assume that one of the following solutions is adopted: state that $\mathsf{trans\_code}\ 0 = 0$, turn this kind of partiality into a non-terminating loop by adding the equation $\mathsf{trans\_code}\ 0 = \mathsf{trans\_code}\ 0$, or extend the target type to $\mathbb{N} + \mathsf{Unit}$ (the operator $+$ denotes the disjoint union on types and Unit is the type with only one element) to generate an exception on non-matchable inputs. So we brush aside this technical issue for the rest of the example.

We can analyse this translation algorithm by splitting it into three components, following the categorical description in [10,11].

The first step consists in computing, from the input x, the index $k - 1$ of the largest prime divisor of x and the exponents $g_0, \ldots, g_{k-1}$ of the prime numbers

$p_0, \ldots, p_{k-1}$ in x (with $k = 0$ if $x = 1$). We give a name and a type to this function:

$$\alpha_{\text{trans\_code}} \colon \mathbb{N} \to \sum_{k:\mathbb{N}} \mathbb{N}^k$$
$$\alpha_{\text{trans\_code}} \ x = \langle k, \langle g_0, \ldots, g_{k-1} \rangle \rangle$$
$$\text{with } k, g_0, \ldots, g_{k-1} \text{ such that } x = p_0^{g_0} \cdots p_{k-1}^{g_{k-1}}.$$

The sum type $\sum_{k:\mathbb{N}} \mathbb{N}^k$ consists of elements of the form $\langle k, g \rangle$, where k is the length of the tuple and g is a tuple of k natural numbers.

The second step consists in applying the function trans_code recursively to the elements of the vector obtained in the first step. We indicate the lifting of the function to tuples by putting an arrow over it:

$$\overrightarrow{\text{trans\_code}} \colon \sum_{k:\mathbb{N}} \mathbb{N}^k \to \sum_{k:\mathbb{N}} \mathbb{N}^k$$
$$\overrightarrow{\text{trans\_code}} \ \langle k, \langle g_0, \ldots, g_{k-1} \rangle \rangle = \langle k, \langle \text{trans\_code} \ g_0, \ldots, \text{trans\_code} \ g_{k-1} \rangle \rangle.$$

The third and final step of the translation algorithm is the computation of the output from the results of the recursive calls:

$$\beta_{\text{trans\_code}} \colon \sum_{k:\mathbb{N}} \mathbb{N}^k \to \mathbb{N}$$
$$\beta_{\text{trans\_code}} \ \langle k, \langle c_0, \ldots, c_{k-1} \rangle \rangle = \text{pair} \ \langle k, \text{tuple}_k \langle c_0, \ldots, c_{k-1} \rangle \rangle.$$

The function trans_code is now specified by giving the fixed point equation

$$\text{trans\_code} = \beta_{\text{trans\_code}} \circ \overrightarrow{\text{trans\_code}} \circ \alpha_{\text{trans\_code}}.$$

The General Framework

The explanation given above suggests an analysis into three steps, already studied from the categorical point of view in [10,11].

Recall the general form of the definition of a function f given by several (recursive) equations and presented at the beginning of this section. We adopt a uniform formulation that abstracts away from the actual matching algorithm: the structure of the recursive calls is given by a type operator (a functor in categorical terms) which, in the case of the function f, is as follows:

$$F \colon \text{Set} \to \text{Set}$$
$$FX = X^{k_0} + \cdots + X^{k_n}.$$

The form of the functor may also be more complex, possibly containing type parameters and dependent families, for example, for the trans_code function the functor is $F_{\text{trans\_code}} \ X = \sum_{k:\mathbb{N}} X^k$.

The first step in the computation of a function is represented by a map $\alpha : A \to FA$ (in categorical terms, an F-coalgebra). Specifically, for f we have:

$$\alpha \colon A \to A^{k_0} + \cdots + A^{k_n}$$
$$\alpha \ p_0 = \text{in}_0 \langle a_{00}, \ldots, a_{0k_0} \rangle$$
$$\vdots$$
$$\alpha \ p_n = \text{in}_n \langle a_{n0}, \ldots, a_{nk_n} \rangle,$$

where in_i is the ith injection: if $n = 0$ then in_0 is just the identity, and for $n > 0$,

$$\text{in}_i \ y = \text{inl} \, \overbrace{(\text{inr} \, (\cdots (\text{inr}(y)) \cdots))}^{i \text{ times}} \text{ if } 0 \leqslant i < n \text{ and } \text{in}_n \ y = \overbrace{\text{inr} \, (\cdots (\text{inr}(y)) \cdots)}^{n \text{ times}}.$$

By the function α above we mean the following: when α is applied to a particular argument, this is matched against the different patterns and when the first pattern matching the argument is found, the tuple with the recursive arguments is computed and returned.

The second step in the computation of f, the evaluation of the recursive calls, is the lifting of f by the functor F, $Ff \colon FA \to FB$.

The last step is a mapping $\beta \colon FB \to B$ (in categorical jargon, an F-algebra). It is computed by applying the appropriate e_i from the recursive equations.

In short, the analysis can be expressed by the following diagram:

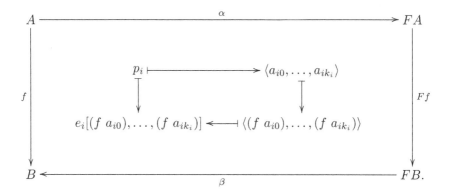

The function f can now be given by the single equation $f = \beta \circ Ff \circ \alpha$, in place of the set of recursive equations presented at the beginning of this section.

A Final Example

Let us illustrate this analysis again with another example: a generalisation of the Fibonacci sequence that depends on three numerical parameters a, b, and c, and on a function parameter $g \colon \mathbb{N} \to \mathbb{N}$ (the actual Fibonacci sequence is obtained for $a = b = 1$, $c = 0$, and $g = \text{id}_\mathbb{N}$):

$$\begin{aligned}
&\mathfrak{fib} : \mathbb{N} \to \mathbb{N} \\
&\mathfrak{fib} \ 0 = a \\
&\mathfrak{fib} \ 1 = b + c \cdot \mathfrak{fib} \ (g \ 0) \\
&\mathfrak{fib} \ (m + 2) = \mathfrak{fib} \ (g \ m) + \mathfrak{fib} \ (g \ (m + 1)).
\end{aligned}$$

The interest in this generalisation lies in the fact that, for some choices of the parameters, $\mathfrak{fib}$ will be a total function (for example, for the choices that give the actual Fibonacci sequence), while, for other choices, $\mathfrak{fib}$ will be partial (for example, if one chooses $a = b = 1$ and $c = 0$, but $g = (+1)$, the successor function over natural numbers). For the $\mathfrak{fib}$ function, F, α, and β are as follows:

$$F_{\mathsf{fib}} \, X = \mathsf{Unit} + X + X^2$$

$$\begin{aligned}
\alpha_{\mathsf{fib}} \ 0 &= \mathsf{in}_0(\mathsf{tt}) & \beta_{\mathsf{fib}} \ \mathsf{in}_0(\mathsf{tt}) &= a \\
\alpha_{\mathsf{fib}} \ 1 &= \mathsf{in}_1(g \ 0) & \beta_{\mathsf{fib}} \ \mathsf{in}_1(x) &= b + c \cdot x \\
\alpha_{\mathsf{fib}} \ (m+2) &= \mathsf{in}_2\langle g \ m, \ g \ (m+1)\rangle & \beta_{\mathsf{fib}} \ \mathsf{in}_2\langle y, z\rangle &= y + z
\end{aligned}$$

where tt is the only element of the type Unit (which we identify with X^0).

3   The Type of Partial Recursive Functions

Inspired by the previous analysis, we introduce a new type constructor for partial recursive functions in which the coalgebra-algebra pair is used in the introduction rule. For the elimination rule, we define a domain predicate similar to the one in the Bove/Capretta method [3]. For full generality, the method we used in the previous section must be adapted by defining a *lifted universal quantifier*: for every predicate $P: X \to \mathsf{Prop}$ and functor F, we define $\bigwedge_{F,P} : FX \to \mathsf{Prop}$ as the conjunction of the statement of P on every element of type X occurring in an element of FX; its formal definition will be given in the next section.

We give now the formal rules for the type of partial recursive functions:

- Formation:

$$\frac{A: \mathsf{Set} \quad B: \mathsf{Set}}{A \rightharpoonup B: \mathsf{Set}};$$

- Introduction:

$$\frac{\alpha: A \to FA \quad \beta: FB \to B}{\mathsf{rec}(\alpha, \beta): A \rightharpoonup B} \ F \text{ a functor;}$$

- Domain predicate:

$$\frac{f: A \rightharpoonup B}{\mathsf{Dom}_f: A \to \mathsf{Prop}}, \qquad \frac{\alpha: A \to FA \quad \beta: FB \to B \quad a: A \quad h: \bigwedge_{F, \mathsf{Dom}_{\mathsf{rec}(\alpha, \beta)}}(\alpha \ a)}{\mathsf{dom}_{\alpha, \beta}(a, h): \mathsf{Dom}_{\mathsf{rec}(\alpha, \beta)} \ a};$$

- Application:

$$\frac{f: A \rightharpoonup B \quad a: A \quad h: \mathsf{Dom}_f \ a}{\mathsf{app}_f \ a \ h: B};$$

- Reduction:

$$\mathsf{app}_{\mathsf{rec}(\alpha, \beta)} \ a \ \mathsf{dom}_{\alpha, \beta}(a, h) \rightsquigarrow \beta \left(\overline{\mathsf{rec}(\alpha, \beta)} \ (\alpha \ a) \ h\right)$$

where $\overline{f}$ is the lifting of the function $f : A \rightharpoonup B$ by the functor F:

$$\overline{f} : \forall t: FA, \bigwedge_{F, \mathsf{Dom}_f} t \to FB;$$

its formal definition, which is more complex than the lifting of trans_code in the previous section because of the presence of the domain predicate, will also be given in the next section. We could also see $\overline{f}$ as a recursive function $\overline{f}: F\alpha \rightharpoonup F\beta$ by setting $\overline{f} = \mathsf{rec}(F\alpha, F\beta)$. This definition would be equivalent to the one given.

With these rules, the functions trans_code and fib can simply be defined as:

$$\text{trans\_code}: \mathbb{N} \rightharpoonup \mathbb{N} \qquad\qquad\qquad \text{fib}: \mathbb{N} \rightharpoonup \mathbb{N}$$
$$\text{trans\_code} = \text{rec}(\alpha_{\text{trans\_code}}, \beta_{\text{trans\_code}}) \qquad \text{fib} = \text{rec}(\alpha_{\text{fib}}, \beta_{\text{fib}}).$$

4 A Predicative Reflection

The rules given in the previous section do not take F as an explicit parameter, but in an intensional type theory F should always be given explicitly. Since every function has its own different functor F, we should really write $\text{rec}(F, \alpha, \beta)$ in place of just $\text{rec}(\alpha, \beta)$.

So, how should we formalise our functors? A functor is a higher-order object of type Set $\rightarrow$ Set. (If we want to be really formal, we should give a more complex dependent type to F, also containing a proof of functoriality.) It is therefore clear that with a solution like this the type $A \rightarrow B$ would be inherently impredicative. One option to circumvent this impasse consists in stratifying partial functions over the hierarchy of predicative universes $U_0, U_1, U_2, \ldots$ by defining $A \rightharpoonup_i B: U_{i+1}$ as the type of partial functions $\text{rec}(F, \alpha, \beta)$ where $F: U_i \rightarrow U_i$.

An alternative solution, adopted here, is to be stricter about the functors we allow. Observe that each equation in the definition of a recursive function always contains a finite number of recursive calls, therefore we can limit ourselves to finitary functors. The class of these functors is small enough to be encoded by a small type, similarly to the encoding of the larger class of strictly positive functors given in [8]. In other words, a functor can be given by a *signature* in the same way that algebraic structures are defined in Universal Algebras [7].

If the functor is of the form $FX = X^{k_0} + \cdots + X^{k_n}$, then the signature may consist of just the list of exponents $[k_0, \ldots, k_n]$. However, some functions have a more general signature: they may have an *a priori* unknown number of potential cases; this is the case for trans_code.

Therefore, we define a signature to be a mapping $\sigma: \mathbb{N} \rightarrow \mathbb{N}$ that specifies, for every possible number of recursive arguments, how many equations contain that number of recursive arguments. Intuitively, for each *arity* n, the signature σ gives us the number $(\sigma\, n)$ of cases (equations) of the function that generate that number of recursive calls. In other words, the function will have $(\sigma\, 0)$ cases with no recursive calls, $(\sigma\, 1)$ cases with a single recursive call, $(\sigma\, 2)$ cases with two recursive calls, and, in general, $(\sigma\, i)$ cases with i recursive calls. In the definition of the functor, the coefficient $(\sigma\, i)$ means that we take $(\sigma\, i)$ copies of X^i.

We define an operator $\_ \star \_$ such that $m \star Y$ gives the sum of m copies of Y:

$$\_ \star \_: \mathbb{N} \rightarrow \text{Set} \rightarrow \text{Set}$$
$$0 \star Y = \emptyset$$
$$1 \star Y = Y$$
$$(m+1) \star Y = Y + (m \star Y).$$

A signature σ represents the following functor F_σ:

$$F_\sigma\, X = (\sigma\, 0) \star \text{Unit} + (\sigma\, 1) \star X + (\sigma\, 2) \star X^2 + \cdots$$

or, more formally,

$$F_\sigma X = \sum_{n:\mathbb{N}} (\sigma\ n) \star X^n.$$

The corresponding signatures in the examples trans_code and fib are:

$$\sigma_{\text{trans\_code}}\ n = 1 \quad \text{and} \quad \sigma_{\text{fib}}\ n = \begin{cases} 1 \text{ if } n=0,1,2 \\ 0 \text{ otherwise.} \end{cases}$$

When we adopt this restrained class of functors, the definitions of the α and β components of a function look a bit different from what we have seen before. For example, for the fib function, we have now that:

$$F_{\text{fib}}\ X = \sum_{n:\mathbb{N}} (\sigma_{\text{fib}}\ n) \star X^n$$

$$\alpha_{\text{fib}}\ 0 = \langle 0, \text{tt} \rangle \qquad\qquad \beta_{\text{fib}}\ \langle 0, \text{tt} \rangle = a$$
$$\alpha_{\text{fib}}\ 1 = \langle 1, g\ 0 \rangle \qquad\qquad \beta_{\text{fib}}\ \langle 1, x \rangle = b + c \cdot x$$
$$\alpha_{\text{fib}}\ (m+2) = \langle 2, \langle g\ m, g\ (m+1) \rangle \rangle \qquad \beta_{\text{fib}}\ \langle 2, \langle x, y \rangle \rangle = y + z.$$

This gives us a version of fib equivalent to the one given in Section 2. Recall that in_0 is simply the identity, so we do not write it in the equations above.

We now revise the rules given in the previous section to use signatures as parameters in place of functors. The introduction rule for partial recursive functions becomes:

$$\frac{\sigma:\mathbb{N} \to \mathbb{N} \quad \alpha:A \to F_\sigma\ A \quad \beta:F_\sigma\ B \to B}{\text{rec}(\sigma, \alpha, \beta): A \rightharpoonup B}.$$

Similarly, in the rest of the rules, we should simply add the parameter σ and replace F with F_σ.

We can now be more precise about the definition of the operator $\bigwedge_{F,P}$ when the functor F is given by a signature. We actually substitute the functor argument F with σ and define, for $P: X \to$ Prop:

$$\bigwedge_{\sigma,P}: F_\sigma\ X \to \text{Prop}$$
$$\bigwedge_{\sigma,P}\ \langle 0, \text{in}_i\ \text{tt} \rangle = \text{True}$$
$$\bigwedge_{\sigma,P}\ \langle n+1, \text{in}_i\ \langle x_0, \ldots, x_n \rangle \rangle = (P\ x_0) \wedge \cdots \wedge (P\ x_n)$$

with $0 \leqslant i < \sigma\ n$ and True the trivially true proposition (isomorphic to Unit).

The introduction rule for the domain predicate becomes:

$$\frac{\sigma:\mathbb{N} \to \mathbb{N} \quad \alpha:A \to F_\sigma\ A \quad \beta:F_\sigma\ B \to B \quad a:A \quad h:\bigwedge_{\sigma,\text{Dom}_{\text{rec}(\sigma,\alpha,\beta)}} (\alpha\ a)}{\text{dom}_{\sigma,\alpha,\beta}(a, h): \text{Dom}_{\text{rec}(\sigma,\alpha,\beta)}\ a}.$$

Finally, we specify how to lift a function $f: A \rightharpoonup B$ by a functor specified by a signature:

$$\overline{f}^\sigma: \forall t: F_\sigma\ A, \bigwedge_{\sigma,\text{Dom}_f} t \to F_\sigma\ B$$
$$\overline{f}^\sigma\ \langle 0, \text{in}_i\ \text{tt} \rangle\ \text{tt} = \langle 0, \text{in}_i\ \text{tt} \rangle$$
$$\overline{f}^\sigma\ \langle n+1, \text{in}_i\ \langle x_0, \ldots, x_n \rangle \rangle\ \langle h_0, \ldots, h_n \rangle =$$
$$\langle n+1, \text{in}_i\ \langle \text{app}_f\ x_0\ h_0, \ldots, \text{app}_f\ x_n\ h_n \rangle \rangle$$

with $0 \leqslant i < \sigma\ n$ and tt the only constructor of True.

Observe that the definitions of app_f and $\overline{f}^\sigma$ are mutually dependent. The recursion is well-founded, since the recursive calls are on structurally smaller arguments. Systems like Coq provide support for mutual recursion.

We mentioned before that $\mathsf{rec}(\sigma, \alpha, \beta)$ can itself be alternatively defined as a partial recursive function by $\mathsf{rec}(\sigma, F_\sigma \alpha, F_\sigma \beta)$. This possibility relies on the equivalence of $\mathsf{Dom}_{\mathsf{rec}(\sigma, F_\sigma \alpha, F_\sigma \beta)}$ t with $\bigwedge_{\sigma, \mathsf{Dom}_{\mathsf{rec}}(\sigma, \alpha, \beta)}$ t, which can be shown by induction on Dom. We leave this verification to the reader and stick to our original definition for the rest of the article.

5 Consistency of the Extended Type Theory

We show here that if we extend a consistent type theory with the new type of partial functions, we obtain a consistent new theory. We achieve this goal by translating the constructors for the new type of partial recursive functions into the base theory. The base theory must be expressive enough to provide the needed operators, specifically it must have Σ-types and inductive dependent families. Our reference system is Martin-Löf's type theory [18,21], but most versions of dependent type theory will work as well.

Let $\mathsf{\Pi}$ be a consistent and normalising type system and let $\mathsf{P\Pi}$ be its extension with the type of partial functions presented in Sections 3 and 4.

We first define an interpretation function $[\![\_]\!] : \mathsf{P\Pi} \to \mathsf{\Pi}$. The translation is defined by recursion on the structure of the terms and types of $\mathsf{P\Pi}$. The constructors that are already present in $\mathsf{\Pi}$ are translated into themselves; so we only need to specify how to interpret the new constructors related to the type of partial functions.

We first define a type which we will use to interpret the type of partial functions:

$$A \rightharpoonup^* B := \sum_{\sigma : \mathbb{N} \to \mathbb{N}} (A \to F_\sigma A) \times (F_\sigma B \to B).$$

Let now $f : A \rightharpoonup^* B$. We use the abbreviations $\sigma_f = \pi_1 f$, $\alpha_f = \pi_1 (\pi_2 f)$, and $\beta_f = \pi_2 (\pi_2 f)$, with π_1 and π_2 the first and second projection from a pair, respectively, and we represent elements of the above Σ-type as triples $\langle \sigma, \alpha, \beta \rangle$ rather than nested pairs $\langle \sigma, \langle \alpha, \beta \rangle \rangle$.

We then define an inductive predicate which we will use to interpret the domain of f:

$$\mathsf{Dom}_f^* : A \to \mathsf{Prop}$$
$$\mathsf{dom}_f^* : \forall a : A, \bigwedge_{\sigma_f, \mathsf{Dom}_f^*} (\alpha_f \, a) \to \mathsf{Dom}_f^* \, a.$$

Finally, we define an application operator for f by recursion on Dom_f^*:

$$\mathsf{app}_f^* : \forall a : A, \mathsf{Dom}_f^* \, a \to B$$
$$\mathsf{app}_f^* \, a \, \mathsf{dom}_f^*(a, h) = \beta_f \, (\overline{f}^{*\,\sigma_f} \, (\alpha_f \, a) \, h).$$

The definition of $\overline{f}^{*\,\sigma_f}$ is exactly that of $\overline{f}^{\sigma_f}$ given at the end of Section 4 but with calls to app^* in place of calls to app, and with Dom^* in place of Dom in the type of its last argument.

The translation proceeds now formally by structural induction on the types and terms of PTT. All the elements that are already present in TT are left unchanged by the translation. Therefore, the translation of type and term variables is the identity, $[\![X]\!] = X$ and $[\![x]\!] = x$; standard type constructors, like products and sums, and their constructors, that is, abstractions and pairs, are translated straightforwardly, hence $[\![\Pi x\!:\! A.B]\!] = \Pi x\!:\! [\![A]\!].[\![B]\!]$, $[\![\Sigma_{x:A} B]\!] = \Sigma_{x:[\![A]\!]} [\![B]\!]$, $[\![\lambda x.a]\!] = \lambda x.[\![a]\!]$ and $[\![\langle x,y \rangle]\!] = \langle [\![x]\!], [\![y]\!] \rangle$; and so on for all operations already present in TT, for example, application is translated as $[\![(d\ e)]\!] = ([\![d]\!]\ [\![e]\!])$ and projections as $[\![\pi_i\ p]\!] = \pi_i\ [\![p]\!]$ for $i = 1, 2$.

The new constructions for partial recursive functions are translated using the starred definitions presented above:

$$[\![A \rightharpoonup B]\!] = [\![A]\!] \rightharpoonup^* [\![B]\!]$$
$$[\![\mathsf{Dom}_f]\!] = \mathsf{Dom}^*_{[\![f]\!]}$$
$$[\![\mathsf{rec}\,(\sigma, \alpha, \beta)]\!] = \langle [\![\sigma]\!], [\![\alpha]\!], [\![\beta]\!] \rangle$$
$$[\![\mathsf{dom}_{\sigma,\alpha,\beta}(a, h)]\!] = \mathsf{dom}^*_{\langle [\![\sigma]\!], [\![\alpha]\!], [\![\beta]\!] \rangle}([\![a]\!], [\![h]\!])$$
$$[\![\mathsf{app}_f\ a\ h]\!] = \mathsf{app}^*_{[\![f]\!]}\ [\![a]\!]\ [\![h]\!]$$

It is worth noticing that any term of TT is returned unchanged from the interpretation $[\![\_]\!]$ (this can be proved by simple structural induction on the terms of TT). In addition, the functor associated to a signature, the lifted universal quantifier, and the lifting operator on functions commute with the interpretation function.

Lemma 1. *For terms of the appropriate type, we have that:*

$$[\![F_\sigma]\!] = F_{[\![\sigma]\!]}, \quad [\![\bigwedge_{\sigma,P}]\!] = \bigwedge_{[\![\sigma]\!],[\![P]\!]}, \quad and \quad [\![\overline{f}^\sigma]\!] = \overline{[\![f]\!]}^{*\,[\![\sigma]\!]}.$$

Proof. For the first two statements, it is sufficient to check that the definitions of these operators use only constructions from the base type theory and none of the new ones defined for partial recursive functions.

The proof of the third statement follows from from the fact that $\overline{f}^\sigma$ and $\overline{[\![f]\!]}^{*\,[\![\sigma]\!]}$ are defined basically in the same way. The difference between the definitions is that first one uses app while the second uses app*, but then we have that $[\![\mathsf{app}_f\ a\ h]\!] = \mathsf{app}^*_{[\![f]\!]}\ [\![a]\!]\ [\![h]\!]$. □

In addition, the interpretation is sound, as we show in the following lemma.

Lemma 2 (Soundness of $[\![\_]\!]$). *Let Γ be a context, t a term and T a type. If $\Gamma \vdash_{\mathsf{PTT}} t\!:\! T$ then $[\![\Gamma]\!] \vdash_{\mathsf{TT}} [\![t]\!]\!:\! [\![T]\!]$.*

Proof. By induction on the derivation of $\Gamma \vdash_{\mathsf{PTT}} t\!:\! T$. We consider here only the new rules of the extended system.

– Formation:

$$\frac{A\!:\!\mathsf{Set} \quad B\!:\!\mathsf{Set}}{A \rightharpoonup B\!:\!\mathsf{Set}}.$$

By inductive hypotheses (IH), both $[\![A]\!]$ and $[\![B]\!]$ are types. Hence, it is easy to see that $[\![A \rightharpoonup B]\!] = \Sigma_{\sigma:\mathbb{N}\to\mathbb{N}} ([\![A]\!] \to F_\sigma [\![A]\!]) \times (F_\sigma [\![B]\!] \to [\![B]\!])$ is a type.

- Introduction:

$$\frac{\sigma: \mathbb{N} \to \mathbb{N} \quad \alpha: A \to F_\sigma\, A \quad \beta: F_\sigma\, B \to B}{\mathsf{rec}(\sigma, \alpha, \beta): A \rightharpoonup B}.$$

By IH, $[\![\sigma]\!]: \mathbb{N} \to \mathbb{N}$, $[\![\alpha]\!]: [\![A]\!] \to F_{[\![\sigma]\!]} [\![A]\!] = [\![A]\!] \to F_{[\![\sigma]\!]} [\![A]\!]$ and $[\![\beta]\!]: F_{[\![\sigma]\!]} [\![B]\!] \to [\![B]\!]$. It is easy to check now that $[\![\mathsf{rec}(\sigma, \alpha, \beta)]\!] = \langle [\![\sigma]\!], [\![\alpha]\!], [\![\beta]\!] \rangle$ has type $[\![A \rightharpoonup B]\!]$.

- Domain predicate:

$$\frac{f: A \rightharpoonup B}{\mathsf{Dom}_f: A \to \mathsf{Prop}} \qquad \frac{\sigma: \cdots \quad \alpha: \cdots \quad \beta: \cdots \quad a: A \quad h: \bigwedge_{\sigma, \mathsf{Dom}_{\mathsf{rec}(\sigma,\alpha,\beta)}} (\alpha\, a)}{\mathsf{dom}_{\sigma,\alpha,\beta}\, (a, h): \mathsf{Dom}_{\mathsf{rec}(\sigma,\alpha,\beta)}\, a}.$$

By IH we have that $[\![f]\!]: [\![A \rightharpoonup B]\!]$, $[\![\sigma]\!], [\![\alpha]\!]$ and $[\![\beta]\!]$ have the types shown above, $[\![a]\!]: [\![A]\!]$, and $[\![h]\!]: [\![\bigwedge_{\sigma, \mathsf{Dom}_{\mathsf{rec}(\sigma \cdot \alpha, \beta)}} (\alpha\, a)]\!] = \bigwedge_{[\![\sigma]\!], \mathsf{Dom}^*_{[\![\mathsf{rec}(\sigma,\alpha,\beta)]\!]}} ([\![\alpha]\!] [\![a]\!])$. By definition $[\![\mathsf{Dom}_f]\!] = \mathsf{Dom}^*_{[\![f]\!]}$ has type $[\![A]\!] \to \mathsf{Prop}$. We know that $[\![\mathsf{dom}_{\sigma,\alpha,\beta}\, (a, h)]\!] = \mathsf{dom}^*_{[\![\mathsf{rec}(\sigma,\alpha,\beta)]\!]} ([\![a]\!], [\![h]\!])$, which has type $\mathsf{Dom}^*_{[\![\mathsf{rec}(\sigma,\alpha,\beta)]\!]} [\![a]\!]$ as required.

- Application:

$$\frac{f: A \rightharpoonup B \quad a: A \quad h: \mathsf{Dom}_f\, a}{\mathsf{app}_f\, a\, h: B}.$$

Let $[\![f]\!]: [\![A \rightharpoonup B]\!]$, $[\![a]\!]: [\![A]\!]$, and $[\![h]\!]: \mathsf{Dom}^*_{[\![f]\!]} [\![a]\!]$ by IH. Then, by definition, $[\![\mathsf{app}_f\, a\, h]\!] = \mathsf{app}^*_{[\![f]\!]} [\![a]\!] [\![h]\!]$, which has type $[\![B]\!]$. $\square$

We can also check that the interpretation of the reduction rule is sound. Be aware of the overloading of the symbol $\rightsquigarrow$, used to denote reduction both in PTT and TT. It should be clear from the context in which theory we are performing the reduction. We denote the reflexive-transitive closure of $\rightsquigarrow$ in TT by $\rightsquigarrow^*$.

Lemma 3 (Preservation of reductions). *Let Γ, t and T such that $\Gamma \vdash_{\mathsf{PTT}} t: T$. If $t \rightsquigarrow t'$ in PTT for a certain term t', then $[\![t]\!] \rightsquigarrow^* [\![t']\!]$ in TT.*

Proof. We just need to check that the new reduction rule for application of a partial recursive function is interpreted correctly:

$$\mathsf{app}_{\mathsf{rec}(\sigma,\alpha,\beta)}\, a\, \mathsf{dom}_{\sigma,\alpha,\beta}(a, h) \rightsquigarrow \beta\, \overline{(\mathsf{rec}(\sigma, \alpha, \beta)}^\sigma\, (\alpha\, a)\, h).$$

The interpretation of the left-hand side of the rule can be reduced using the definition of app^*:

$$[\![\mathsf{app}_{\mathsf{rec}(\sigma,\alpha,\beta)}\, a\, \mathsf{dom}_{\sigma,\alpha,\beta}(a, h))]\!] = \mathsf{app}^*_{[\![\mathsf{rec}(\sigma,\alpha,\beta)]\!]} [\![a]\!]\, \mathsf{dom}^*_{[\![\mathsf{rec}(\sigma,\alpha,\beta)]\!]} ([\![a]\!], [\![h]\!])$$
$$\rightsquigarrow [\![\beta]\!]\, (\overline{[\![\mathsf{rec}(\sigma, \alpha, \beta)]\!]}^{*\,[\![\sigma]\!]} ([\![\alpha]\!] [\![a]\!])\, [\![h]\!]).$$

The interpretation of the right-hand side gives us the following:

$$[\![\beta \overline{(\mathsf{rec}(\sigma, \alpha, \beta)}^\sigma (\alpha\ a)\ h)]\!] = [\![\beta]\!] ([\![\overline{\mathsf{rec}(\sigma, \alpha, \beta)}^\sigma]\!] ([\![\alpha]\!]\ [\![a]\!])\ [\![h]\!]).$$

Now, the correctness of the translation of the reduction rule for application is a consequence of Lemma 1. □

Conservativity and consistency follow now straightforwardly.

Theorem 1 (Conservativity of PTT). *Let Γ and T be a context and a type, respectively, in TT, and let t be a term in PTT. If $\Gamma \vdash_{\mathsf{PTT}} t : T$ then $\Gamma \vdash_{\mathsf{TT}} [\![t]\!] : T$.*

Proof. By Lemma 2 we have that $[\![\Gamma]\!] \vdash_{\mathsf{TT}} [\![t]\!] : [\![T]\!]$. Every term of TT is returned unchanged by the interpretation $[\![\_]\!]$; hence, $\Gamma = [\![\Gamma]\!]$ and $T = [\![T]\!]$. □

Theorem 2 (Consistency of PTT). PTT *is consistent.*

Proof. Let us assume that $\vdash_{\mathsf{PTT}} t : \bot$ for some term t. Then, by Theorem 1, we have that $\vdash_{\mathsf{TT}} [\![t]\!] : \bot$, which is absurd since TT is consistent. □

The results of this section show that we can implement types of partial recursive functions satisfying our formal rules in standard type theory. Our formalisation is then to be used as a useful module to develop recursion theory in a system like Coq, as we exemplified in the Coq formalisation that accompanies this article. A direct implementation of the extended system could also have its advantages, avoiding the overhead of the interpretation and improving efficiency by direct computation of the reduction rule for partial functions.

6   Full Recursion

So far we have considered recursive definitions in which the argument was used only to generate recursive calls. This is a scheme of definition by iteration. A more general scheme, giving full recursion, allows the argument to be used directly and not just inside the recursive calls. The most simple example is the recursive equation in the definition of the factorial function, $\mathsf{fact}\ (n + 1) = (n + 1) \cdot (\mathsf{fact}\ n)$, where the argument $(n + 1)$ is used as a factor of the multiplication, beside generating the recursive call to n. The factorial is a structurally recursive function and standard type theory can be used to define it with no problem. So we generalise the factorial function as we have generalised the Fibonacci function in order to consider a function that displays the same phenomenon, but requires at the same time general recursion. Let $a : \mathbb{N}$ and $g : \mathbb{N} \to \mathbb{N}$ be fixed parameters, then we define:

$$\mathsf{fact} : \mathbb{N} \to \mathbb{N}$$
$$\mathsf{fact}\ 0 = a$$
$$\mathsf{fact}\ (n + 1) = (n + 1) \cdot \mathsf{fact}\ (g\ n).$$

If $a = 1$ and g is the identity, we get the usual factorial function, but for some choices of g, for example the successor function, fact will be a partial function.

This function does not fall into the scheme described in the previous sections because we have passed the input $(n + 1)$ directly as an argument of the multiplication. That is, the β component uses not only the result of the recursive call, but also the input itself. The type of β must then be changed: it involves not only the functor F_σ, but also the type of the input. The introduction rule for partial recursive functions becomes now:

$$\frac{\sigma: \mathbb{N} \to \mathbb{N} \quad \alpha: A \to F_\sigma\, A \quad \beta: A \times F_\sigma\, B \to B}{\mathsf{rec}(\sigma, \alpha, \beta): A \rightharpoonup B}.$$

Observe that the α component remains the same: α still determines just the structure of the recursive calls and says nothing about the extra parameter. The change in the type of β should of course be made in all the other rules as well.

The reduction rule must also be modified to feed the input as an extra argument to β:

$$\mathsf{app}_{\mathsf{rec}(\sigma, \alpha, \beta)}\ a\ \mathsf{dom}_{\sigma,\alpha,\beta}(a, h) \rightsquigarrow \beta\,\langle a, \overline{\mathsf{rec}(\sigma, \alpha, \beta)}^\sigma\,(\alpha\ a)\ h\rangle.$$

The interpretation and the proof of soundness can be easily modified to take this generalisation into account.

In the case of the $\mathfrak{fact}$ function, the components σ, α and β are as follows:

$$\sigma_{\mathfrak{fact}}\ n = \begin{cases} 1 \text{ for } n = 0, 1 \\ 0 \text{ otherwise} \end{cases} \quad \begin{aligned} &\alpha_{\mathfrak{fact}}\ 0 = \langle 0, \mathsf{tt}\rangle \\ &\alpha_{\mathfrak{fact}}\ (n+1) = \langle 1, (g\,n)\rangle \end{aligned} \quad \begin{aligned} &\beta_{\mathfrak{fact}}\ \langle x, \langle 0, \mathsf{tt}\rangle\rangle = a \\ &\beta_{\mathfrak{fact}}\ \langle x, \langle 1, y\rangle\rangle = x \cdot y. \end{aligned}$$

7   Conclusions

The work presented here is a contribution to the formalisation of general recursion in intensional type theories. For every pair of types A and B, we define a type of partial functions $A \rightharpoonup B$. Its introduction rule allows us to define a function by prescribing the structure of recursive calls and the operations that produce the output from the recursive results. Every function is coupled with a domain predicate inductively defined to be true whenever it holds for the recursive arguments. Application is allowed only for elements in the domain and the reduction rule performs recursion over the proof of the domain predicate.

The fact that all recursive functions between two given types A and B can be collected in a single type facilitates the definition of higher-order programs. Indeed, the source type A can itself be a type of partial functions, $A = X \rightharpoonup Y$. However, the mechanism does not immediately allow the inheritance of partiality from a higher-order partial argument. For example, the formalisation of the *map* function on lists (which is itself structurally recursive) raises the problem of how the partiality of the mapped argument function is reflected on the list argument. The method explained here does not directly give an answer to this problem.

Another problematic issue is that nested recursive functions are not allowed: the α component must directly identify the recursive arguments, without recursively calling the function that we are defining.

Partiality may come not only from infinite recursion but also from a sequence of recursive equations whose patterns are not exhaustive. We did not provide a direct method to deal with the fact that the α component must be total. However, as we already mentioned before, there are easy fixes for this since we can add equations to make the patterns exhaustive by doing one of three things: assign a default value to the new patterns, make the function loop on them, or add an *undefined* element to the target type B, that is, using $B + \mathsf{Unit}$ in place of B.

We proved the conservativity of the new constructors with respect to a base type theory. Thus the extended system is sound whenever the base theory is.

We maintain that this new type could be of use in the formalisation of general recursive algorithms and in the proof of their correctness. It can be readily implemented in type-theory based proof assistants like Coq.

References

1. Audebaud, P.: Partial objects in the calculus of constructions. In: Kahn, G. (ed.) Proceedings of the Sixth Annual IEEE Symp. on Logic in Computer Science, LICS 1991, July 1991, pp. 86–95. IEEE Computer Society Press, Los Alamitos (1991)
2. Bove, A.: General recursion in type theory. In: Geuvers, H., Wiedijk, F. (eds.) TYPES 2002. LNCS, vol. 2646, pp. 39–58. Springer, Heidelberg (2003)
3. Bove, A., Capretta, V.: Nested general recursion and partiality in type theory. In: Boulton, R.J., Jackson, P.B. (eds.) TPHOLs 2001. LNCS, vol. 2152, pp. 121–135. Springer, Heidelberg (2001)
4. Bove, A., Capretta, V.: Modelling general recursion in type theory. Mathematical Structures in Computer Science 15(4), 671–708 (2005)
5. Bove, A., Capretta, V.: Recursive functions with higher order domains. In: Urzyczyn, P. (ed.) TLCA 2005. LNCS, vol. 3461, pp. 116–130. Springer, Heidelberg (2005)
6. Bove, A., Capretta, V.: Computation by prophecy. In: Della Rocca, S.R. (ed.) TLCA 2007. LNCS, vol. 4583, pp. 70–83. Springer, Heidelberg (2007)
7. Capretta, V.: Universal algebra in type theory. In: Bertot, Y., Dowek, G., Hirschowitz, A., Paulin, C., Théry, L. (eds.) TPHOLs 1999. LNCS, vol. 1690, pp. 131–148. Springer, Heidelberg (1999)
8. Capretta, V.: Recursive families of inductive types. In: Aagaard, M.D., Harrison, J. (eds.) TPHOLs 2000. LNCS, vol. 1869, pp. 73–89. Springer, Heidelberg (2000)
9. Capretta, V.: General recursion via coinductive types. Logical Methods in Computer Science 1(2), 1–18 (2005)
10. Capretta, V., Uustalu, T., Vene, V.: Recursive coalgebras from comonads. In: Proceedings of the Workshop on Coalgebraic Methods in Computer Science (CMCS 2004). Electronic Notes in Theoretical Computer Science, vol. 106, pp. 43–61 (2004)
11. Capretta, V., Uustalu, T., Vene, V.: Recursive coalgebras from comonads. Information and Computation 204(4), 437–468 (2006)
12. Constable, R.L.: Constructive mathematics as a programming logic I: Some principles of theory. Annals of Mathematics, vol. 24. Elsevier Science Publishers, North-Holland, Amsterdam (1985)
13. Constable, R.L., Mendler, N.P.: Recursive definitions in type theory. In: Parikh, R. (ed.) Logic of Programs 1985. LNCS, vol. 193, pp. 61–78. Springer, Heidelberg (1985)

14. Constable, R.L., Smith, S.F.: Partial objects in constructive type theory. In: Logic in Computer Science, Ithaca, New York, pp. 183–193. IEEE, Los Alamitos (1987)
15. Coquand, T., Huet, G.: The Calculus of Constructions. Information and Computation 76, 95–120 (1988)
16. Gödel, K.: Über formal unentscheidbare sätze der Principia Mathematica und verwandter systeme. Monatshefte für Mathematik und Physik 38, 173–198 (1931)
17. Jones, S.P.: Haskell 98 Language and Libraries: The Revised Report, April 2003. Cambridge University Press, Cambridge (2003)
18. Martin-Löf, P.: Intuitionistic Type Theory. Bibliopolis, 1984. Notes by Giovanni Sambin of a series of lectures given in Padua (June 1980)
19. Megacz, A.: A coinductive monad for Prop-bounded recursion. In: Stump, A., Xi, H. (eds.) PLPV 2007: Proceedings of the 2007 workshop on Programming languages meets program verification, pp. 11–20. ACM Press, New York (2007)
20. Meijer, E., Fokkinga, M.M., Paterson, R.: Functional programming with bananas, lenses, envelopes and barbed wire. In: Hughes, J. (ed.) FPCA 1991. LNCS, vol. 523, pp. 124–144. Springer, Heidelberg (1991)
21. Nordström, B., Petersson, K., Smith, J.M.: Programming in Martin-Löf's Type Theory. An Introduction. International Series of Monographs on Computer Scence, vol. 7. Oxford University Press, Oxford (1990)
22. Setzer, A.: Partial recursive functions in Martin-Löf Type Theory. In: Beckmann, A., Berger, U., Löwe, B., Tucker, J.V. (eds.) CiE 2006. LNCS, vol. 3988, p. 505. Springer, Heidelberg (2006)
23. Setzer, A.: A data type of partial recursive functions in Martin-Löf Type Theory, http://www.cs.swan.ac.uk/ csetzer/articles/setzerDataTypePar RecPostProceedings.ps
24. The Coq Development Team. LogiCal Project. The Coq Proof Assistant. Reference Manual. Version 8. INRIA (2004), http://pauillac.inria.fr/coq/coq-eng.html

Formal Reasoning About Causality Analysis

Jens Brandt and Klaus Schneider

University of Kaiserslautern
Embedded Systems Group, Department of Computer Science
P.O. Box 3049, 67653 Kaiserslautern, Germany
http://es.cs.uni-kl.de

Abstract. Systems that can immediately react to their inputs may suffer from cyclic dependencies between their actions and the corresponding trigger conditions. For this reason, causality analysis has to be employed to check the constructiveness of the programs which implies the existence of unique and consistent behaviours. In this paper, we describe the embedding of various views of causality analysis into the HOL4 theorem prover to check their equivalence. In particular, we show the equivalence between the classical analysis procedure, which is based on a fixpoint computation, and a formulation as a (bounded) model checking problem.

1 Introduction

For the modelling of embedded systems, or more generally, the modelling of systems with concurrent actions whose execution may consume time, many *models of computation* [8,10,12] have been considered. Among these models of computation are asynchronous models like dataflow process networks and Hoare's CSP, discrete-event models as used by most hardware description languages including VHDL, Verilog and SystemC, and synchronous models as used by synchronous hardware circuits, synchronous programming languages [3] and classical automata theory.

Most of these models of computation define a *causality* relation between actions and their trigger conditions. Roughly speaking, causality determines a logical sequentiality between trigger conditions and subsequent reactions according to the stimuli of the environment. Causality can be achieved in various ways: for example, the notion of δ-time has been introduced in hardware description languages to circumvent causality problems, and tagged tokens have been introduced in dataflow computers to solve this problem.

In synchronous models of computation, however, causality is not given by construction. There is neither a finer notion of time as given by δ-time nor are there tagged values to establish a causality relation. Instead, the execution of a synchronous system is partitioned into macro steps that consist of finitely many micro steps. The values of the variables remain constant during all micro steps within a macro step and change synchronously when proceeding to the next macro step. Time is given as a logical time in the number of macro steps. Hence, causality analysis is much more difficult than in other models of computation.

O. Ait Mohamed, C. Muñoz, and S. Tahar (Eds.): TPHOLs 2008, LNCS 5170, pp. 118–133, 2008.
© Springer-Verlag Berlin Heidelberg 2008

For this reason, special procedures for causality analysis have been developed for synchronous languages [4,22]. It is well-known that a synchronous program is causally correct (constructive) if and only if for all inputs, there is a dynamic schedule of the micro step actions in all macro steps (hence, the program can be executed on a sequential machine). Moreover, it is known that causality analysis is equivalent to the stability analysis of combinational feedback loops in hardware circuits [11,9,17,13,22,4,15,16,14] under Brzozowski and Seger's unbounded delay model [5,22].

Since the causality of a program depends on its syntax (or equivalently on the particular structure of a hardware circuit) and not only on its semantics, the causality analysis furthermore depends on the translation to semantically equivalent hardware circuits or sequential programs. We have studied such variants in depth in our previous research [19,20,21]. As it turned out that the problem is very subtle and depends on details of the used definitions, we believe that the research in this area would greatly benefit from a solid formal treatment using an interactive theorem prover like HOL4. Hence, this paper is the first step towards such a formal treatment, and it already presents an equivalence proof between two kinds of causality analyses, namely one that could be used at run-time with concrete inputs and a static one, which is used at compile-time with symbolic inputs (so that a symbolic analysis is obtained). The symbolic causality analysis is the second added value of this paper, since symbolic methods only exist for the special case of Boolean event variables so far.

In this paper, we do not consider a particular language like our synchronous programming language Quartz [18]. Instead, we consider a language-independent formalisation of synchronous systems that is based on so-called guarded actions to obtain a theory that is as general as possible. A guarded action is thereby a pair $(\gamma, \mathcal{C})$ where the guard γ is a Boolean condition, and where the action $\mathcal{C}$ is an atomic action of the system under consideration. Throughout this paper, we only consider immediate assignments of the form $y = \tau$, where y is an output variable, and τ is an expression that is type-consistent with y. Further kinds of actions like assumptions, assertions, and delayed assignments [18] do not further complicate the causality analysis [19]. Moreover, we do not take into account the reachability of states, which is necessary in a full causality analysis by a compiler. Again, this would not further complicate the algorithms. Finally, we only consider variables of type Boolean, while a real programming language offers typically further types. However, types do not complicate our symbolic analysis either.

Guarded actions as the ones considered here have been widely used in computer science so far, and entire programming languages like Unity [6,1,2] have been built on top of them. However, the semantics of the guarded actions often differs, depending on the preferred model of computation. As we consider the synchronous model of computation, the system's computation consists of micro and macro steps: Thus, we must analyse the causal order of the micro steps within one macro step, and variables have constant values during the analysis. Indeed, the major goal of causality analysis is to find a constructive schedule (dependent on the values of the input variables) to determine the successive values of the output variables.

The symbolic causality analysis of this synchronous system starts with known inputs x_i and unknown outputs y_i. For this reason, the initial environment $\mathcal{E}_0$ maps all input variables to known Boolean constants, while it maps the output variables to a third value $\bot$ that means 'still unknown' in the multi-valued causality analysis. Starting with the initial environment $\mathcal{E}_0$, the following iteration is performed in causality analysis to compute an environment $\mathcal{E}_{i+1}$ from a given environment $\mathcal{E}_i$. We first evaluate all guards of the actions and partition the actions into (1) must-actions that have a guard $\mathcal{E}_i(\gamma) =$ true and (2) can-actions that have a guard $\mathcal{E}_i(\gamma) \in \{\text{true}, \bot\}$, respectively. Then, we 'execute' the must-actions, so that the assigned variables' values are updated in the new environment $\mathcal{E}_{i+1}$. Moreover, if all actions of an output variable y are cannot-actions, we can assign a default value to y (which may be a fixed constant in case of event variables (transient) or the previous value in case of memorised variables (persistent)). In all other cases, we have to maintain the current value of y, which may still be unknown $\bot$. If a variable y should already have a known value, but a must-action assigns a different value, we update the value of y to $\top$ to indicate a write conflict.

$$
\begin{cases}
(x_0 \wedge y_5, y_0 = \text{true}), \\
(x_1 \vee y_0, y_1 = \text{true}), \\
(x_2 \wedge y_1, y_2 = \text{true}), \\
(x_0 \vee y_2, y_3 = \text{true}), \\
(x_1 \wedge y_3, y_4 = \text{true}), \\
(x_2 \vee y_4, y_5 = \text{true})
\end{cases}
\quad
\begin{cases}
y_0 = x_0 \wedge y_5 \\
y_1 = x_1 \vee y_0 \\
y_2 = x_2 \wedge y_1 \\
y_3 = x_0 \vee y_2 \\
y_4 = x_1 \wedge y_3 \\
y_5 = x_2 \vee y_4
\end{cases}
\quad
\begin{cases}
y_0 = x_0 \wedge (x_2 \vee x_1) \\
y_1 = x_1 \vee (x_0 \wedge x_2) \\
y_2 = x_2 \wedge (x_1 \vee x_0) \\
y_3 = x_0 \vee (x_2 \wedge x_1) \\
y_4 = x_1 \wedge (x_0 \vee x_2) \\
y_5 = x_2 \vee (x_1 \wedge x_0)
\end{cases}
$$

Fig. 1. An Example of a Cyclic Equation System due to [17]

In case of Boolean event variables, a set of guarded actions is equivalent to a (potentially cyclic) equation system, i.e., a combinational hardware circuit with potential feedback loops. As an example (due to Rivest [17]), consider the guarded actions given on the left hand side of Figure 1 for inputs x_0, x_1 and x_2 and outputs y_0, y_1, y_2, y_3, y_4 and y_5. This set of guarded actions is equivalent to the cyclic Boolean equation system shown in the middle of Figure 1. Using ternary simulation [13,22,4,19], it is possible to convert every causally correct set of guarded actions into an acyclic equation system. In case of our example of Figure 1, this equivalent acyclic equation system is given on the right of Figure 1.

While there are paper-and-pencil proofs for the equivalence of the causality analysis based on can-must analysis and the symbolic ternary simulation, there is no publication on a symbolic analysis for non-Boolean or non-event variables. In this paper, we present such a symbolic analysis as a generalisation of ternary simulation. To this end, we have to take into account the problem of write conflicts which does not appear for Boolean events (since disjunction is used as an implicit conflict resolution function). In addition, we prove the equivalence between the can-must causality analysis and our symbolic formulation based on extending each data type by the constants $\bot$ and $\top$.

To summarise, we present in this paper the formalisation of the traditional causality analysis for synchronous systems as well as the definition of a fully symbolic version of a general causality analysis. We use the HOL4 theorem prover to prove the equivalence of these two variants of the causality analysis, so that we can guarantee that what is checked by means of model checking in a compiler with the symbolic analysis exactly matches the definition of causality analysis as given by the can-must analysis.

Moreover, we extend the classical analysis by considering run-time errors like write conflicts, division by zero, or access to array elements outside the declared range. This is accomplished by generalising the traditional three-valued analysis to a four-valued setting using a constant $\top$ for 'run-time error' in addition to $\bot$ 'yet unknown'.

This paper is organised as follows: In Section 2 we present a formalisation of the traditional can-must analysis. Section 3 describes our symbolic approach to causality analysis, which is subsequently formalised and shown to be equivalent in Section 4. Finally, Section 5 draws some conclusions.

2 Formalisation of Traditional Can-Must Analysis

Before we formalise the traditional can-must analysis in the last part of this section, we first have to model the system description, which is based on guarded actions (Section 2.1) and the environment to allow the evaluation of terms (Section 2.2).

2.1 System Description

The main objective of our work is the reasoning about different procedures for causality analysis; we are less interested in causality analysis of a particular system. Therefore, we need a *deep embedding* of the system description to quantify over systems in the logic. Hence, our first step is the formalisation of the guarded actions.

A guarded action is a pair $(\gamma, x = \tau)$, where $\mathrm{grd}(\gamma, x = \tau) = \gamma$ is called the guard, $\mathrm{lhs}(\gamma, x = \tau) = x$ the left-hand side, and $\mathrm{rhs}(\gamma, x = \tau) = \tau$ the right-hand side of the action. The guard and the right-hand side are Boolean expressions, which are defined in HOL as follows:

$$\mathtt{BlExpr} \vdash_{\mathrm{def}} \mathit{false} \in \mathit{BlExpr} \mid \mathit{true} \in \mathit{BlExpr} \mid x, x \in \mathcal{V}$$
$$\mid \mathit{not}(e), e \in \mathit{BlExpr}$$
$$\mid \mathit{and}(e_1, e_2), e_1, e_2 \in \mathit{BlExpr} \mid \mathit{or}(e_1, e_2)e_1, e_2 \in \mathit{BlExpr}$$

2.2 Four-Valued Environment

The previous subsection only describes the syntax of synchronous systems in our HOL theory. To define their semantics, we have to formalise the environment of the system. In the traditional can-must analysis, the environment maps each variable to a three-valued truth value. We extend this definition here to integrate write conflicts and other run-time errors into the causality analysis. Thus, the environment maps each variable to one of the four truth values $\mathbb{F} = \{\bot, 0, 1, \top\}$.

$\ddot{\neg}$	
$\bot$	$\bot$
0	1
1	0
$\top$	$\top$

$\ddot{\wedge}$	$\bot$	0	1	$\top$
$\bot$	$\bot$	0	$\bot$	$\top$
0	0	0	0	$\top$
1	$\bot$	0	1	$\top$
$\top$	$\top$	$\top$	$\top$	$\top$

$\ddot{\vee}$	$\bot$	0	1	$\top$
$\bot$	$\bot$	$\bot$	1	$\top$
0	$\bot$	0	1	$\top$
1	1	1	1	$\top$
$\top$	$\top$	$\top$	$\top$	$\top$

$\sup_{\mathbb{F}}$	$\bot$	0	1	$\top$
$\bot$	$\bot$	0	1	$\top$
0	0	0	$\top$	$\top$
1	1	$\top$	1	$\top$
$\top$	$\top$	$\top$	$\top$	$\top$

Fig. 2. Definitions of Four-Valued Negation, Conjunction and Disjunction

In the rest of this section we implement a theory based on four-valued logic. To reason about monotonic functions, and the existence of fixpoints, we define a strict partial order $\stackrel{\cdot\cdot}{<}$ on the type $\mathbb{F}$ as follows: $x\stackrel{\cdot\cdot}{<}y :\Leftrightarrow x \neq y \wedge (x = \bot \vee y = \top)$. It is easily seen that $(\mathbb{F}, \stackrel{\cdot\cdot}{<})$ is a complete lattice, since two elements have a supremum and an infimum (see Figure 2).

Clearly, our theory contains four-valued generalisations for all Boolean operators that are defined according to the truth tables shown in Figure 2. As can be seen, all four-valued operations are maximal monotonic generalisations of the corresponding Boolean operations (with respect to our partial order). The reason for this choice will be discussed later in this paper, when we show the equivalence of this method to our model-checking approach.

The environment itself is defined as a finite map [7] from natural numbers, which represent variables in our case, to four-valued truth values. The actual implementation is hidden behind the functions $\mathcal{E}(x)$ and $\mathcal{E}_x^\tau$, which read and update the value of a given variable x in the environment $\mathcal{E}$, respectively. The evaluation $[\![\tau]\!]_\mathcal{E}^\mathbb{F}$ of an expression τ to a value in $\mathbb{F}$ with respect to the given environment $\mathcal{E}$ is defined recursively as follows:

blEval4_def $\vdash_{\mathrm{def}}$
$$([\![false]\!]_\mathcal{E}^\mathbb{F} = 0) \wedge ([\![true]\!]_\mathcal{E}^\mathbb{F} = 1) \wedge$$
$$([\![x]\!]_\mathcal{E}^\mathbb{F} = \mathcal{E}(x)) \wedge ([\![not(\tau)]\!]_\mathcal{E}^\mathbb{F} = \stackrel{\cdot\cdot}{\neg}[\![\tau]\!]_\mathcal{E}^\mathbb{F}) \wedge$$
$$([\![and(\tau_1, \tau_2)]\!]_\mathcal{E}^\mathbb{F} = [\![\tau_1]\!]_\mathcal{E}^\mathbb{F} \stackrel{\cdot\cdot}{\wedge} [\![\tau_2]\!]_\mathcal{E}^\mathbb{F}) \wedge ([\![or(\tau_1, \tau_2)]\!]_\mathcal{E}^\mathbb{F} = [\![\tau_1]\!]_\mathcal{E}^\mathbb{F} \stackrel{\cdot\cdot}{\vee} [\![\tau_2]\!]_\mathcal{E}^\mathbb{F})$$

2.3   Fixpoint Iteration

As already explained in the introduction, the can-must analysis iteratively computes values for all output variables of the environment $\mathcal{E}$ until a fixpoint is reached. To this end, we have to maintain two sets of guarded actions $\mathcal{A}$: The guard of the *must-actions* is true, while the guard of the *can-actions* is not false.[1] To this end, we provide the following definitions:

can4_def $\vdash_{\mathrm{def}}$
$$(\mathsf{can}(\bot) = \mathsf{true}) \wedge (\mathsf{can}(0) = \mathsf{false}) \wedge$$
$$(\mathsf{can}(1) = \mathsf{true}) \wedge (\mathsf{can}(\top) = \mathsf{false})$$
canActs_def $\vdash_{\mathrm{def}}$
$$\mathsf{canActs}_\mathcal{E}(\mathcal{A}) = \mathbf{filter}\ \lambda(\gamma, x = \tau).\ \mathsf{can}([\![\gamma]\!]_\mathcal{E}^\mathbb{F})\ \mathbf{from}\ \mathcal{A}$$
must4_def $\vdash_{\mathrm{def}}$
$$(\mathsf{must}(\bot) = \mathsf{false}) \wedge (\mathsf{must}(0) = \mathsf{false}) \wedge$$
$$(\mathsf{must}(1) = \mathsf{true}) \wedge (\mathsf{must}(\top) = \mathsf{true})$$
mustActs_def $\vdash_{\mathrm{def}}$
$$\mathsf{mustActs}_\mathcal{E}(\mathcal{A}) = \mathbf{filter}\ \lambda(\gamma, x = \tau).\ \mathsf{must}([\![\gamma]\!]_\mathcal{E}^\mathbb{F})\ \mathbf{from}\ \mathcal{A}$$

After the set of guarded actions is partitioned into must-, cannot-, and the remaining actions, their update of the environment is defined as follows: for must-actions, the variable on the left-hand side should be assigned the supremum of its old value and the right-hand side. This construction ensures two important properties. First, the order of

---

[1] For a more intuitive formalisation, we use *can-actions* and not the complement *cannot-actions*, which can be usually found in traditional causality.

assignments to the same variable is irrelevant, which is a result of the associativity and commutativity of the $\sup_{\mathbb{F}}$ operation. Thus, multiple actions can be executed sequentially without caring about the sequential order. Second, instead of changing a variable from one Boolean value to another one, $\top$ is assigned, which signals a write conflict. Hence, this definition exactly models the intended behaviour.

executeActionF_def $\vdash_{\text{def}}$
$$\text{execAct}_{\mathbb{F}}(\mathcal{E}, a) = \mathcal{E}_{\text{lhs}(a)}^{\sup_{\mathbb{F}}(\llbracket\text{lhs}(a)\rrbracket_{\mathcal{E}}^{\mathbb{F}}, \llbracket\text{rhs}(a)\rrbracket_{\mathcal{E}}^{\mathbb{F}})}$$
executeActionsF_def $\vdash_{\text{def}}$
$$(\text{execActs}_{\mathbb{F}}(\mathcal{E}, \langle\rangle) = \mathcal{E}) \wedge$$
$$(\text{execActs}_{\mathbb{F}}(\mathcal{E}, a :: \mathcal{A}) = \text{execAct}_{\mathbb{F}}(\text{execActs}_{\mathbb{F}}(\mathcal{E}, \mathcal{A}), a))$$

If no action is activated, the new value of a variable x is set to a default value, which is commonly referred to as *reaction to absence*. The function $\text{reactToAbsense}_{\mathbb{F}}(\mathcal{E}, x, \mathcal{A})$ checks for a given output variable x, whether there is a possibly enabled action in $\mathcal{A}$ for x. If this is not the case, x is assigned its default value in environment $\mathcal{E}$.

reactToAbsenseF_def $\vdash_{\text{def}}$
$$(\text{reactToAbsense}_{\mathbb{F}}(\mathcal{E}, x, \langle\rangle) = \mathcal{E}_x^{\text{default}(x)}) \wedge$$
$$(\text{reactToAbsense}_{\mathbb{F}}(\mathcal{E}, x, a :: \mathcal{A}) =$$
$$\quad \textbf{if } x = \text{lhs}(a) \textbf{ then } \mathcal{E} \textbf{ else reactToAbsense}_{\mathbb{F}}(\mathcal{E}, x, \mathcal{A}))$$
reactToAbsensesF_def $\vdash_{\text{def}}$
$$(\text{reactToAbsenses}_{\mathbb{F}}(\mathcal{E}, \langle\rangle, \mathcal{A}) = \mathcal{E}) \wedge$$
$$(\text{reactToAbsenses}_{\mathbb{F}}(\mathcal{E}, m :: \mathcal{V}, \mathcal{A}) =$$
$$\quad \text{reactToAbsense}_{\mathbb{F}}(\text{reactToAbsenses}_{\mathbb{F}}(\mathcal{E}, \mathcal{V}, \mathcal{A}), x, \mathcal{A}))$$

Each step of the causality analysis consists in executing all must-actions and reacting to absence for all variables that have no can-actions. Note that must-actions of the previous iteration step are executed again in the next iteration. This must be done, since the expression on the right-hand side might not have been known. So, its value can change during the iterations.

cmAnalysisStep_def $\vdash_{\text{def}}$ cmAnalysisStep$(\mathcal{A}, \mathcal{V}, \mathcal{E}) =$
$$\quad \text{reactToAbsenses}_{\mathbb{F}}(\text{execActs}_{\mathbb{F}}(\mathcal{E}, \text{mustActs}_{\mathcal{E}}(\mathcal{A})), \mathcal{V}, \text{canActs}_{\mathcal{E}}(\mathcal{A}))$$
cmAnalysis_dfn $\vdash_{\text{def}}$ cmAnalysis$(\mathcal{A}, \mathcal{V}, \mathcal{E}) =$
$$\quad \textbf{let } \mathcal{E}' = \text{cmAnalysisStep}(\mathcal{A}, \mathcal{V}, \mathcal{E}) \textbf{ in}$$
$$\quad \textbf{if } \mathcal{E} = \mathcal{E}' \textbf{ then } \mathcal{E} \textbf{ else cmAnalysis}(\mathcal{A}, \mathcal{V}, \mathcal{E}')$$

To prove the termination of the fixpoint iteration performed by the can-must analysis cmAnalysis$(\mathcal{A}, \mathcal{V}, \mathcal{E})$, we define a weight envWeight$(\mathcal{E}, \mathcal{V})$ for an environment $\mathcal{E}$ with output variables $\mathcal{V}$. The weight of a variable intuitively reflects the amount of knowledge it stores, and the weight of an environment is just the sum of the weights of all output variables. Since the set of output variables $\mathcal{V}$ is finite, it has the upper bound $2|\mathcal{V}|$.

weight4_def $\vdash_{\text{def}}$
$$(\text{wt}(\bot) = 0) \wedge (\text{wt}(0) = 1) \wedge (\text{wt}(1) = 1) \wedge (\text{wt}(\top) = 2)$$
envWeight_def $\vdash_{\text{def}}$
$$(\text{envWeight}(\mathcal{E}, \langle\rangle) = 0) \wedge$$
$$(\text{envWeight}(\mathcal{E}, v :: \mathcal{V}) = \text{envWeight}(\mathcal{E}, \mathcal{V}) + \text{wt}(\mathcal{E}(v)))$$

Since the termination is apparent for the case that no action is executed and the environment remains the same, it remains to prove that each call to cmAnalysisStep($\mathcal{A}, \mathcal{V}, \mathcal{E}$) is monotonic in that it increases the weight of the environment. The proof basically follows the argument that the executions of all actions and the reactions to absence are monotonic. A variable with a Boolean value is never reset to $\bot$, and a $\top$ is never replaced by any other value. The proof is done by induction on the variables, followed by an induction on the actions. However, some more subtle preconditions are required for a successful proof, which are due to the general assumptions of the theory:

- Since the list of variables $\mathcal{V}$ models rather a set than an actual list, all of its elements must be distinct, so that the weight can be computed correctly.
- varsDefined($\mathcal{E}, \mathcal{V}$) assures that values for all output variables $\mathcal{V}$ can be found in the environment $\mathcal{E}$. Otherwise, their weights would be undefined. Alternatively, the weight of undefined variables could be defined, which would not reflect the real situation.
- The actions $\mathcal{A}$ must only modify the variables given by output variables $\mathcal{V}$. Otherwise, some updates would not affect the weight computation. Note that the set of output variables cannot be determined from the set of actions, since there may be some variables whose behaviour is only given by the reaction to absence.
- Finally, as already noted, the default value default(x) should be independent of the current environment. Otherwise, the reaction to absence would be able to cause additional dependencies.

$$\texttt{varDefined\_def} \ \vdash_{\text{def}} \text{varDefined}(\mathcal{E}, v) = v \in \mathbf{domain}(\mathcal{E})$$
$$\texttt{varsDefined\_def} \ \vdash_{\text{def}}$$
$$(\text{varsDefined}(\mathcal{E}, \langle \rangle) = \text{true}) \wedge$$
$$(\text{varsDefined}(\mathcal{E}, v :: \mathcal{V}) = \text{varDefined}(\mathcal{E}, v) \wedge \text{varsDefined}(\mathcal{E}, \mathcal{V}))$$
$$\texttt{actionValid\_def} \ \vdash_{\text{def}} \text{actionValid}(\mathcal{V}, a) = \text{lhs}(a) \in \mathcal{V}$$
$$\texttt{actionsValid\_def} \ \vdash_{\text{def}}$$
$$(\text{actionsValid}(\mathcal{V}, \langle \rangle) = \text{true}) \wedge$$
$$(\text{actionsValid}(\mathcal{V}, a :: \mathcal{A}) = \text{actionValid}(\mathcal{V}, a) \wedge \text{actionsValid}(\mathcal{V}, \mathcal{A}))$$
$$\texttt{STEP\_MONOTONE} \ \vdash$$
$$\text{distinct}(\mathcal{V}) \rightarrow \text{varsDefined}(\mathcal{E}, \mathcal{V}) \rightarrow \text{actionsValid}(\mathcal{V}, \mathcal{A}) \rightarrow$$
$$(\forall \mathcal{E}_1 \, \mathcal{E}_2 \, x. \, (\llbracket \text{default}(x) \rrbracket_{\mathcal{E}_1}^{\mathbb{F}}) = (\llbracket \text{default}(x) \rrbracket_{\mathcal{E}_2}^{\mathbb{F}})) \rightarrow$$
$$(\text{envWeight}(\mathcal{V}, \mathcal{E}) \leq \text{envWeight}(\mathcal{V}, \text{cmAnalysisStep}(\mathcal{A}, \mathcal{V}, \mathcal{E})))$$

This concludes the formalisation of the can-must analysis, which serves as a specification for other algorithms for causality analysis.

3 Causality Analysis by Model Checking

In this section, we describe our symbolic approach to causality analysis, which is based on a symbolic description of a transition system, so that the causality analysis can be formulated as a model checking problem. In the following, we describe the transition system, before we explain the actual verification task in Subsection 3.2.

3.1 Modelling the Progress of Information

To encode the causality analysis as a model checking problem, we must create a transition system that explicitly models the progress of information and the occurrence of write conflicts. Therefore, in addition to the variables in the program, we introduce for every output variable x a Boolean-typed variable x^{kn} that holds iff the value of x is known. If this is the case, then the value of x is stored in the variable x, otherwise the value of x is not yet known and we have to ignore its content. Similarly, a second Boolean-typed variable x^{wc} is added that holds iff a write conflict occurred for variable x. If this bit is set, we can also ignore the content of the corresponding variable.

Using the variables x^{kn} and x^{wc}, we can explicitly model the progress of the information flow, which is obtained by evaluating the program expression step by step until either assignments can be executed that determine the current value of a variable or until it becomes clear that no assignment will modify the current value of a variable so that the reaction to absence will determine it.

As a first step, we define a function that maps a program expression σ to a Boolean formula $wc(\sigma)$ such that $wc(\sigma)$ holds iff the expression σ cannot be evaluated due to a write conflict. Basically, this is always the case if a subformula cannot be evaluated:

- For variables and constants, we define:
  - $wc(x) := \begin{cases} \text{false} & : \text{if } x \text{ is an input variable} \\ x^{wc} & : \text{otherwise} \end{cases}$
  - $wc(c) := \text{false}$

- For the Boolean operators, we define:
  - $wc(not(\varphi)) := wc(\varphi)$
  - $wc(and(\varphi, \psi)) := wc(\varphi) \vee wc(\psi)$
  - $wc(or(\varphi, \psi)) := wc(\varphi) \vee wc(\psi)$

Analogously, we formally define a function that maps a program expression σ to a Boolean formula $kn(\sigma)$ such that $kn(\sigma)$ holds if and only if the expression σ can be evaluated to a known value. The formula $kn(\sigma)$ encodes the lazy evaluation rules of the Boolean operators, which are shown in Figure 3.

- For variables and constants, we define:
  - $kn(x) := \begin{cases} \text{true} & : \text{if x is an input variable} \\ x^{kn} & : \text{otherwise} \end{cases}$
  - $kn(c) := \text{true}$

- For the Boolean operators, we define:
  - $kn(not(\tau_1)) := wc(\tau_1) \vee kn(\tau_1)$

$$\begin{array}{ll} x \wedge \text{false} = \text{false} & \text{false} \wedge x = \text{false} \\ x \vee \text{true} = \text{true} & \text{true} \vee x = \text{true} \\ \text{false} \rightarrow x = \text{true} & x \rightarrow \text{true} = \text{true} \end{array}$$

Fig. 3. Lazy Evaluation Rules

- $\mathsf{kn}(and(\tau_1, \tau_2)) := \mathsf{wc}(\tau_1) \vee \mathsf{wc}(\tau_2) \vee \mathsf{kn}(\tau_1) \wedge \mathsf{kn}(\tau_2) \vee$
  $\mathsf{kn}(\tau_1) \wedge (\tau_1 = \mathsf{true}) \vee \mathsf{kn}(\tau_2) \wedge (\tau_2 = \mathsf{true})$
- $\mathsf{kn}(or(\tau_1, \tau_2)) := \mathsf{wc}(\tau_1) \vee \mathsf{wc}(\tau_2) \vee \mathsf{kn}(\tau_1) \wedge \mathsf{kn}(\tau_2) \vee$
  $\mathsf{kn}(\tau_1) \wedge (\tau_1 = \mathsf{false}) \vee \mathsf{kn}(\tau_2) \wedge (\tau_2 = \mathsf{false})$

Clearly, in the actual behaviour of the program, we can only make use of an expression if we know its value. For this reason, the entire execution of the actions is controlled by the data flow: we start with unknown values for all output variables and try to determine their values with the micro steps that can occur in a macro step. The procedure is repeated for each reaction.

For this reason, we introduce a clock signal tick, which is true whenever all variables have become known values, and a new reaction may be started. If this happens, the system state may change according to the control flow, and the input variables are allowed to change their values in a non-deterministic way (which reflects the uncontrollable input of the environment).

Propagation of Knowledge:

$$\mathsf{KnownTrans}_x := \left(\mathsf{next}(x^{\mathsf{kn}}) :\Leftrightarrow \left(\begin{array}{l} \neg\mathsf{tick} \wedge \mathsf{kn}(x) \vee \\ \neg\mathsf{tick} \wedge \left(\bigvee_{j=1}^{p} \mathsf{kn}(\gamma_j) \wedge \gamma_j \wedge \mathsf{kn}(\tau_j) \right) \vee \\ \neg\mathsf{tick} \wedge \left(\bigwedge_{j=1}^{p} \mathsf{kn}(\gamma_j) \wedge \neg\gamma_j \right) \end{array} \right) \right)$$

Propagation of Write Conflicts:

$$\mathsf{WCTrans}_x := \left(\mathsf{next}(x^{\mathsf{wc}}) :\Leftrightarrow \left(\begin{array}{l} \neg\mathsf{tick} \wedge \mathsf{wc}(x) \vee \\ \neg\mathsf{tick} \wedge \mathsf{kn}(x) \wedge \\ \quad \left(\bigvee_{j=1}^{p} \mathsf{kn}(\gamma_j) \wedge \gamma_j \wedge \mathsf{kn}(\tau_j) \wedge (x \neq \tau_j) \right) \vee \\ \neg\mathsf{tick} \wedge \\ \quad \left(\begin{array}{l} \bigvee_{j=1}^{p-1} \bigvee_{k=j+1}^{p} \mathsf{kn}(\gamma_j) \wedge \gamma_j \wedge \mathsf{kn}(\tau_j) \wedge \\ \mathsf{kn}(\gamma_k) \wedge \gamma_k \wedge \mathsf{kn}(\tau_k) \wedge (\tau_j \neq \tau_k) \end{array} \right) \end{array} \right) \right)$$

Computation of Values:

$$\mathsf{ValTrans}_x := \left(\begin{array}{l} \left(\neg\mathsf{tick} \rightarrow \left(\bigwedge_{j=1}^{p} \neg\mathsf{kn}(x) \wedge \mathsf{kn}(\gamma_j) \wedge \gamma_j \wedge \mathsf{kn}(\tau_j) \rightarrow \mathsf{next}(x) = \tau_j \right) \right) \wedge \\ \left(\neg\mathsf{tick} \rightarrow \left(\mathsf{kn}(x) \vee \neg \left(\bigvee_{j=1}^{p} \mathsf{kn}(\gamma_j) \wedge \gamma_j \wedge \mathsf{kn}(\tau_j) \right) \rightarrow \mathsf{next}(x) = x \right) \right) \wedge \\ (\ \mathsf{tick} \rightarrow (\mathsf{next}(x) = \mathsf{default}(x))) \end{array} \right)$$

Fig. 4. Causality Transition Relation of Variable x

Macro steps are therefore separated by occurrences of the tick signal, and between two clock ticks, the micro steps of a macro step are executed: as long as tick is false, the immediate assignments to the variables are executed if the values of the guards and right hand side expressions are known, and the guard is true. We therefore distinguish between the *information flow* and the *data flow*. The information flow of a variable x is determined by the corresponding variables $\mathsf{kn}(x) = x^{\mathsf{kn}}$ and $\mathsf{wc}(x) = x^{\mathsf{wc}}$.

The transition relation of the information flow can be formulated as an equation system as shown in Figure 4 that contains the following cases for x^{kn}:

- If there is no clock tick, the value of x remains known if it was already known.
- If there is no clock tick, the value of x becomes known if a guarded action $(\gamma_j, x{=}\tau_j)$ with an assignment $x{=}\tau_j$ can be fired. This is the case if and only if the value of the guard γ_j is known to be true and if the value of the right hand side expression τ_j is known.
- If there is no clock tick, and all guards γ_j are known to be false, the reaction to absence determines the value of x: the formula of Figure 4 simply demands that $\mathsf{next}(x){=}x$ has to hold in this case, since x has been given the now desired value at the first step of the reaction as a preliminary value.
- Otherwise, the value of x is not known.

Write conflicts are propagated according to the following rules:

- If there is no clock tick, a write conflict for x persists.
- If there is no clock tick, a write conflict occurs, if a guarded action $(\gamma_j, x{=}\tau_j)$ is activated that assigns a different value to an already known variable.
- If there is no clock tick, a write conflict occurs, if there are two guarded actions that are activated and assign different values to the same variable.

The data flow of x is determined by the same cases as formalised in formula $\mathsf{ValTrans}_x$ given in Figure 4:

- If there is no clock tick and a guarded action $(\gamma_j, x{=}\tau_j)$ with an assignment $x{=}\tau_j$ can be fired, then x will receive the value of τ_j at the next point of time.
- If there is no clock tick and no guarded action can be fired, then x will keep its value. This covers two cases: first, if the reaction-to-absence should take place, since all guards γ_j are known to be false, then keeping the value of x is correct, since we already provided the desired value for x at the previous clock tick. Second, if a write conflict occurs, the value of x is irrelevant.
- If there is a clock tick, then the values of all variables are known, and therefore, we can execute all enabled delayed actions. If one of the delayed actions can be fired, then the value of x is known for the following macro step. Note again that there is no transition if several delayed guarded actions with different values π_i and π_j are fired.
- Finally, if there is a clock tick, x is initialised to its default value $\mathsf{default}(x)$.

The formulas given in Figure 4 describe the transition relation of a particular variable x. The complete transition relation is therefore the conjunction of the partial transition relations of all output variables. In addition to this, we also have to determine, when the causality analysis terminates. This is accomplished by the clock signal, which is defined as follows:

$$\mathsf{ClockTrans} :\equiv \left(\mathsf{tick} :\Leftrightarrow \bigwedge_x \mathsf{kn}(x) \wedge \neg\mathsf{wc}(x) \right)$$

The new clock tick can arrive as soon as all values of the local and output variables become known. New input variables can then be read for the next reaction.

3.2 Model Checking Tasks

It is easily seen that the micro step behaviour as formalised in the previous subsection describes the semantics of an *asynchronous circuit*. Obviously, a program is causally correct, if the values of all variables can be determined for all possible inputs. Due to the definition of ClockTrans, the states in which all variables are known can be identified by the tick variable, which also marks the beginning of a new reaction. Hence, a single reaction is modelled by a finite chain of transitions in this model leading from one state where tick holds to another state where tick holds. Causally incorrect transitions (which do not exist in the macro step model), are chains that do no lead to a valid new state, i.e. tick does not appear after the initial step. Such execution sequences end in a self-loop of a state without a tick. Thus, any system that is guaranteed to hit a *clock* state after the initialisation, is causally correct. This property can be simply denoted by XF tick in linear time temporal logic and can be checked by any state-of-the-art model checker that supports linear time logic (LTL).

4 Formalisation of Alternative Analysis

Traditional causality analysis differs from our symbolic procedure in two aspects: first, the data is represented in a different way. Instead of four-valued truth values, the symbolic formulation is based on the original data types plus two additional status bits (to allow the use of state-of-the-art model checkers). Second, the transition relation has a denotational style, while traditional causality analysis describes the fixpoint computation operationally. Our formalisation and the equivalence proof of the equivalence are therefore divided into two parts. The next two sections first link both data representations, while Section 4.3 bridges the second gap.

4.1 Two-Valued Environment

Similar to the definitions of the four-valued environment, we hide the functions to read and store values in the environment behind the notations $\mathcal{E}(x)$ and $\mathcal{E}_x^v$, respectively. Analogously, the additional status bits for knowledge and write conflicts can be accessed by $\mathcal{E}(x^{\mathsf{kn}})$, $\mathcal{E}_{x^{\mathsf{kn}}}^v$, $\mathcal{E}(x^{\mathsf{wc}})$ and $\mathcal{E}_{x^{\mathsf{wc}}}^v$, respectively.

The fundamental definition to link both variants of causality analysis, which is the basis for the equivalence proof, is given by $\mathsf{envEqual}(\mathcal{E}_1, \mathcal{E}_2)$. This relation takes two environments, a four-valued one and a two-valued one and defines whether they are equivalent or not. For this task, it makes a case distinction on the four possible truth values of the can-must analysis:

- If a four-valued variable is $\perp$, the two-valued counterpart should be marked as not known and without write conflicts.
- If the value of variable is 0 (or 1), it should be marked as known with no write conflict and its value is set to 0 (or 1, respectively).
- If the value of the variable is $\top$, it should be marked as known with a write conflict.
- The remaining case for the two-valued environment, i. e. a variable is not known and has a write conflict, is forbidden.

The following definition respects these considerations:

$$\text{envEqual\_def} \vdash_{\text{def}} \text{envEqual}(\mathcal{E}_1, \mathcal{E}_2) =$$
$$(\forall x.(\mathcal{E}_1(x) = \bot) = \neg\mathcal{E}_2(x^{\text{kn}}) \wedge \neg\mathcal{E}_2(x^{\text{wc}})) \wedge$$
$$(\forall x.(\mathcal{E}_1(x) = 0) = \mathcal{E}_2(x^{\text{kn}}) \wedge \neg\mathcal{E}_2(x^{\text{wc}}) \wedge (\mathcal{E}_2(x) = 0)) \wedge$$
$$(\forall x.(\mathcal{E}_1(x) = 1) = \mathcal{E}_2(x^{\text{kn}}) \wedge \neg\mathcal{E}_2(x^{\text{wc}}) \wedge (\mathcal{E}_2(x) = 1)) \wedge$$
$$(\forall x.(\mathcal{E}_1(x) = \top) = \mathcal{E}_2(x^{\text{kn}}) \wedge \mathcal{E}_2(x^{\text{wc}})) \wedge$$
$$(\forall x.\ \text{false} = \neg\mathcal{E}_2(x^{\text{kn}}) \wedge \mathcal{E}_2(x^{\text{wc}})$$

On this basis, the evaluation $[\![e]\!]_{\mathcal{E}}$ of the Boolean expressions is defined:

$$\text{blEval\_def} \vdash_{\text{def}}$$
$$([\![\textit{false}]\!]_{\mathcal{E}} = \text{false}) \wedge ([\![\textit{true}]\!]_{\mathcal{E}} = \text{true}) \wedge$$
$$([\![x]\!]_{\mathcal{E}} = \mathcal{E}(x)) \wedge$$
$$([\![\textit{not}(\tau)]\!]_{\mathcal{E}} = \neg[\![\tau]\!]_{\mathcal{E}}) \wedge$$
$$([\![\textit{and}(\tau_1, \tau_2)]\!]_{\mathcal{E}} = [\![\tau_1]\!]_{\mathcal{E}} \wedge [\![\tau_2]\!]_{\mathcal{E}}) \wedge$$
$$([\![\textit{or}(\tau_1, \tau_2)]\!]_{\mathcal{E}} = [\![\tau_1]\!]_{\mathcal{E}} \vee [\![\tau_2]\!]_{\mathcal{E}})$$

The definition for the write conflict status needs some considerations. Following the approach of the previous section, its determination should be strict. If a write conflict occurs in any subterm, it is propagated. This models the fact that a model that contains inconsistent variables is completely inconsistent.

$$\text{blWriteConflict\_def} \vdash_{\text{def}}$$
$$(\text{wc}_{\mathcal{E}}(\textit{false}) = \text{false}) \wedge (\text{wc}_{\mathcal{E}}(\textit{true}) = \text{false}) \wedge$$
$$(\text{wc}_{\mathcal{E}}(x) = \mathcal{E}(x^{\text{wc}})) \wedge$$
$$(\text{wc}_{\mathcal{E}}(\textit{not}(\tau)) = \text{wc}_{\mathcal{E}}(\tau)) \wedge$$
$$(\text{wc}_{\mathcal{E}}(\textit{and}(\tau_1, \tau_2)) = \text{wc}_{\mathcal{E}}(\tau_1) \vee \text{wc}_{\mathcal{E}}(\tau_2)) \wedge$$
$$(\text{wc}_{\mathcal{E}}(\textit{or}(\tau_1, \tau_2)) = \text{wc}_{\mathcal{E}}(\tau_1) \vee \text{wc}_{\mathcal{E}}(\tau_2))$$

In contrast to this, the known status makes use of lazy evaluation.

$$\text{blKnown\_def} \vdash_{\text{def}}$$
$$(\text{kn}_{\mathcal{E}}(\textit{false}) = \text{true}) \wedge (\text{kn}_{\mathcal{E}}(\textit{true}) = \text{true}) \wedge$$
$$(\text{kn}_{\mathcal{E}}(x) = \mathcal{E}(x^{\text{kn}})) \wedge$$
$$(\text{kn}_{\mathcal{E}}(\textit{not}(\tau)) = \text{wc}_{\mathcal{E}}(\tau) \vee \text{kn}_{\mathcal{E}}(\tau)) \wedge$$
$$(\text{kn}_{\mathcal{E}}(\textit{and}(\tau_1, \tau_2)) =$$
$$\quad \text{wc}_{\mathcal{E}}(\tau_1) \vee \text{wc}_{\mathcal{E}}(\tau_2) \vee (\text{kn}_{\mathcal{E}}(\tau_1) \wedge \text{kn}_{\mathcal{E}}(\tau_2)) \vee$$
$$\quad (\text{kn}_{\mathcal{E}}(\tau_1) \wedge ([\![\tau_1]\!]_{\mathcal{E}} = \text{false})) \vee (\text{kn}_{\mathcal{E}}(\tau_2) \wedge ([\![\tau_1]\!]_{\mathcal{E}} = \text{false}))) \wedge$$
$$(\text{kn}_{\mathcal{E}}(\textit{or}(\tau_1, \tau_2)) =$$
$$\quad \text{wc}_{\mathcal{E}}(\tau_1) \vee \text{wc}_{\mathcal{E}}(\tau_2) \vee (\text{kn}_{\mathcal{E}}(\tau_1) \wedge \text{kn}_{\mathcal{E}}(\tau_2)) \vee$$
$$\quad (\text{kn}_{\mathcal{E}}(\tau_1) \wedge ([\![\tau_1]\!]_{\mathcal{E}} = \text{true})) \vee (\text{kn}_{\mathcal{E}}(\tau_2) \wedge ([\![\tau_1]\!]_{\mathcal{E}} = \text{true})))$$

The first proof obligation is that the previous definitions comply with the four-valued ones. This corresponds to a lifting of the environments for the variables to the expressions, i. e. : provided that variables in the current environments are considered equal, all expressions can be considered to be equal, too. This goal can be proved with the calculation rules for the four-valued operations, the given definitions and a first-order tactic automatically, after initiating an induction on the structure of the expressions.

EXPR_EQUAL $\vdash$ envEqual$(\mathcal{E}_1, \mathcal{E}_2) \rightarrow$
$$(([\![\tau]\!]_{\mathcal{E}_1}^{\mathbb{F}} = \bot) = \neg \mathsf{kn}_{\mathcal{E}_2}(\tau) \wedge \neg \mathsf{wc}_{\mathcal{E}_2}(\tau)) \wedge$$
$$(([\![\tau]\!]_{\mathcal{E}_1}^{\mathbb{F}} = 0) = \mathsf{kn}_{\mathcal{E}_2}(\tau) \wedge \neg \mathsf{wc}_{\mathcal{E}_2}(\tau) \wedge ([\![\tau_1]\!]_{\mathcal{E}_2} = \mathsf{false})) \wedge$$
$$(([\![\tau]\!]_{\mathcal{E}_1}^{\mathbb{F}} = 1) = \mathsf{kn}_{\mathcal{E}_2}(\tau) \wedge \neg \mathsf{wc}_{\mathcal{E}_2}(\tau) \wedge ([\![\tau_1]\!]_{\mathcal{E}_2} = \mathsf{true})) \wedge$$
$$(([\![\tau]\!]_{\mathcal{E}_1}^{\mathbb{F}} = \top) = \mathsf{kn}_{\mathcal{E}_2}(\tau) \wedge \mathsf{wc}_{\mathcal{E}_2}(\tau)) \wedge$$
$$(\neg(\neg \mathsf{kn}_{\mathcal{E}_2}(\tau) \wedge \mathsf{wc}_{\mathcal{E}_2}(\tau)))$$

This proof is done by structural induction on expressions. Surprisingly, it revealed some glitches in the definitions in former versions of $\mathsf{kn}_{\mathcal{E}}(\tau)$, which did not respect some write conflicts.

4.2 Execution of Actions

The next step to define the execution of actions and to prove is that both variants perform exactly the same steps, each one in its representation. Provided that the environments have been equivalent before the execution of an action, they must be equivalent after the execution. This is assured by the following definitions and theorems:

executeGuardedAction4_def $\vdash_{\mathsf{def}}$ executeGuardedAction$_{\mathbb{F}}(a, \mathcal{E}) =$
  if must$([\![\mathsf{grd}(a)]\!]_{\mathcal{E}_{\mathbb{F}}})$ **then** execAct$_{\mathbb{F}}(\mathcal{E}_{\mathbb{F}}, a)$ **else** $\mathcal{E}_{\mathbb{F}}$
actionExecutable_def $\vdash_{\mathsf{def}}$ actionExecutable$(\mathcal{E}, a) =$
  $\mathsf{kn}_{\mathcal{E}}(\mathsf{grd}(a)) \wedge ([\![\mathsf{grd}(a)]\!]_{\mathcal{E}} = \mathsf{true}) \wedge \mathsf{kn}_{\mathcal{E}}(\mathsf{rhs}(a))$
actionConflictFree_def $\vdash_{\mathsf{def}}$ actionConflictFree$(\mathcal{E}, a) =$
  $\neg \mathsf{kn}_{\mathcal{E}}(\mathsf{lhs}(a)) \vee ([\![\mathsf{lhs}(a)]\!]_{\mathcal{E}} = [\![\mathsf{rhs}(a)]\!]_{\mathcal{E}})$
executeGuardedAction_def $\vdash_{\mathsf{def}}$ executeGuardedAction$(a, \mathcal{E}) =$
  if actionExecutable$(\mathcal{E}, a)$ **then**
      (**if** actionConflictFree$(\mathcal{E}, a)$ **then** execAct$_{\mathbb{F}}(\mathcal{E}, a)$ **else** $\mathcal{E}_{a^{\mathsf{wc}}}^1)_{a^{\mathsf{kn}}}^1$
  else $\mathcal{E}$
ACTION_EQUAL $\vdash$ envEqual$(\mathcal{E}_1, \mathcal{E}_2) \rightarrow$
  envEqual(executeGuardedAction$_{\mathbb{F}}(a, \mathcal{E}_1)$, executeGuardedAction$(a, \mathcal{E}_2))$

The proof of the last theorem basically makes a case distinction on the following situations given by the value and status of the expressions occurring in a guarded action a:

- The guard $\mathsf{grd}(a)$ is not known or it is false. In both cases, the value of the left-hand side is not changed by the execution of the guarded action, since the **else** branches in both representations are taken.
- The guard $\mathsf{grd}(a)$ is known and not false, and the right-hand side expression $\mathsf{rhs}(a)$ is unknown. The value of the left-hand side is not changed in both environment, due to the following reasons. In the environment $\mathcal{E}$, the whole action a is not executed. In the environment $\mathcal{E}_{\mathbb{F}}$, the action is executed, in principle. However, it does not have any effect, since the maximum of the old value and the value of the unknown right-hand side (which is $\bot$ by assumption) is always the old value.
- The guard of the guarded action is known and not false. The right-hand side expression is known. The environment is updated, where the update depends whether a write conflict is caused or not. If this is not the case, the value of the left-hand side is updated and set to be known. Otherwise, its write conflict status is set in both representations.

```
actionActive_def ⊢def
```
$$\mathsf{active}_{\mathcal{E}}(a) = \mathsf{kn}_{\mathcal{E}}(\mathsf{grd}(a)) \wedge [\![\mathsf{grd}(a)]\!]_{\mathcal{E}} \wedge \mathsf{kn}_{\mathcal{E}}(\mathsf{rhs}(a))$$
```
someActionActive_def ⊢def
```
$$(\mathsf{someActive}_{\mathcal{E}}(\langle\rangle) = \mathsf{false}) \wedge$$
$$(\mathsf{someActive}_{\mathcal{E}}(a :: \mathcal{A}) = \mathsf{active}_{\mathcal{E}}(a) \vee \mathsf{someActive}_{\mathcal{E}}(\mathcal{A}))$$
```
allActionsInactive_def ⊢def
```
$$(\mathsf{allInactive}_{\mathcal{E}}(\langle\rangle) = \mathsf{true}) \wedge$$
$$(\mathsf{allInactive}_{\mathcal{E}}(a :: \mathcal{A}) = \neg\mathsf{active}_{\mathcal{E}}(a) \wedge \mathsf{allInactive}_{\mathcal{E}}(\mathcal{A}))$$
```
conflictingActionActive_def ⊢def
```
$$\mathsf{conflActive}_{\mathcal{E}}(a) = \mathsf{active}_{\mathcal{E}}(a) \wedge \mathsf{kn}_{\mathcal{E}}(\mathsf{lhs}(a)) \wedge ([\![\mathsf{lhs}(a)]\!]_{\mathcal{E}} \neq [\![\mathsf{rhs}(a)]\!]_{\mathcal{E}})$$
```
conflictingActionsActive_def ⊢def
```
$$(\mathsf{conflsActive}_{\mathcal{E}}(\langle\rangle) = \mathsf{false}) \wedge$$
$$(\mathsf{conflsActive}_{\mathcal{E}}(a :: \mathcal{A}) = \mathsf{conflActive}_{\mathcal{E}}(a) \vee \mathsf{conflsActive}_{\mathcal{E}}(\mathcal{A}))$$
```
inconsistentActionActive_def ⊢def
```
$$(\mathsf{inconActive}_{\mathcal{E}}(a_0, \langle\rangle) = \mathsf{false})$$
$$(\mathsf{inconActive}_{\mathcal{E}}(a_0, a_1 :: \mathcal{A}) = \mathsf{inconActive}_{\mathcal{E}}(a_0, \mathcal{A}) \vee$$
$$\mathsf{active}_{\mathcal{E}}(a_0) \wedge \mathsf{active}_{\mathcal{E}}(a_1) \wedge ([\![\mathsf{rhs}(a_0)]\!]_{\mathcal{E}} \neq [\![\mathsf{rhs}(a_1)]\!]_{\mathcal{E}}))$$
```
inconsistentActionsActive_def ⊢def
```
$$(\mathsf{inconsActive}_{\mathcal{E}}(\langle\rangle) = \mathsf{false})$$
$$(\mathsf{inconsActive}_{\mathcal{E}}(a :: \mathcal{A}) = \mathsf{inconActive}_{\mathcal{E}}(a, \mathcal{A}) \vee \mathsf{inconsActive}_{\mathcal{E}}(\mathcal{A}))$$

```
trWCVar_def ⊢def
```
$$\mathsf{trWCVar}(\mathcal{A}, v, \mathcal{E}, \mathcal{E}') =$$
$$(\mathsf{wc}_{\mathcal{E}'}(v) = \mathsf{wc}_{\mathcal{E}}(v) \vee \mathsf{conflActive}_{\mathcal{E}}(\mathcal{A}_v) \vee \mathsf{inconsActive}_{\mathcal{E}}(\mathcal{A}_v)$$
```
trWC_def ⊢def
```
$$\mathsf{trWC}(\mathcal{A}, \mathcal{V}, \mathcal{E}, \mathcal{E}') = \bigwedge_{v \in \mathcal{V}} \mathsf{trWCVar}(\mathcal{A}, v, \mathcal{E}, \mathcal{E}')$$

```
trKnVar_def ⊢def
```
$$\mathsf{trKnVar}(\mathcal{A}, v, \mathcal{E}, \mathcal{E}') =$$
$$(\mathsf{kn}_{\mathcal{E}'}(v) = \mathsf{kn}_{\mathcal{E}}(v) \vee \mathsf{someActive}_{\mathcal{E}}(\mathcal{A}_v) \vee \mathsf{allInactive}_{\mathcal{E}}(\mathcal{A}_v))$$
```
trKn_def ⊢def
```
$$\mathsf{trKn}(\mathcal{A}, \mathcal{V}, \mathcal{E}, \mathcal{E}') = \bigwedge_{v \in \mathcal{V}} \mathsf{trKnVar}(\mathcal{A}, v, \mathcal{E}, \mathcal{E}'))$$

```
trValAct_def ⊢def
```
$$\mathsf{trValAct}(a, v, \mathcal{E}, \mathcal{E}') =$$
$$\neg\mathsf{kn}_{\mathcal{E}}(v) \wedge \mathsf{active}_{\mathcal{E}}(a) \rightarrow ([\![v]\!]_{\mathcal{E}'} = [\![\mathsf{rhs}(a)]\!]_{\mathcal{E}})$$
```
trValActs_def ⊢def
```
$$(\mathsf{trValActs}(\langle\rangle, v, \mathcal{E}, \mathcal{E}') = \mathsf{true}) \wedge$$
$$(\mathsf{trValActs}(a :: \mathcal{A}, v, \mathcal{E}, \mathcal{E}') = \mathsf{trValActs}(a, v, \mathcal{E}, \mathcal{E}') \wedge \mathsf{trValAct}(\mathcal{A}, v, \mathcal{E}, \mathcal{E}'))$$
```
trValVar_def ⊢def
```
$$\mathsf{trValVar}(\mathcal{A}, v, \mathcal{E}, \mathcal{E}') =$$
$$\mathsf{trValActs}(\mathcal{A}_v, v, \mathcal{E}, \mathcal{E}') \wedge (\mathsf{kn}_{\mathcal{E}}(v) \vee \mathsf{allInactive}_{\mathcal{E}}(\mathcal{A}_v) \rightarrow ([\![v]\!]_{\mathcal{E}} = [\![v]\!]_{\mathcal{E}'}))$$
```
trVal_def ⊢def
```
$$\mathsf{trVal}(\mathcal{A}, \mathcal{V}, \mathcal{E}, \mathcal{E}') = \bigwedge_{v \in \mathcal{V}} \mathsf{trValVar}(\mathcal{A}, v, \mathcal{E}, \mathcal{E}')$$

```
trAnalysisStep_def ⊢def
```
$$\mathsf{trAnalysisStep}(\mathcal{A}, \mathcal{V}, \mathcal{E}, \mathcal{E}') =$$
$$\mathsf{trKn}(\mathcal{A}, \mathcal{V}, \mathcal{E}, \mathcal{E}') \wedge \mathsf{trWC}(\mathcal{A}, \mathcal{V}, \mathcal{E}, \mathcal{E}') \wedge \mathsf{trVal}(\mathcal{A}, \mathcal{V}, \mathcal{E}, \mathcal{E}')$$

Fig. 5. Formalisation of the Transition Relation

4.3 Transition Relation

A last step to link both variants of causality analysis remains, the description style: the previous paragraph still formalises the analysis operationally, it describes how to move from one state to another. In contrast, the transition relation of the model-checking analysis has a different, rather denotational view. It describes the possible transitions and makes it possible that no behaviours or multiple behaviours exist (instead of a single one defined by the operational description). Hence, we have to prove that the transition relation exactly describes the steps that would be executed by an iteration in the operational description of the previous sections.

The complete formalisation of the transition relation is given in Figure 5. We closely follow the definitions for a single variable given in Section 3, but use some auxiliary definitions to keep the formalisation traceable and readable. Furthermore, we do not integrate the tick signal in our formalisation, but replace it by an initialisation of the output variables.

The following two theorems are the final step of our equivalence proof, which shows that a step in the transition relation corresponds to a step in the can-must analysis.

$$\text{TRANSREL\_CORRECTNESS} \vdash$$
$$\text{envEqual}(\mathcal{E}_1, \mathcal{E}_2) \to \text{envEqual}(\mathcal{E}'_1, \mathcal{E}'_2) \to$$
$$(\mathcal{E}'_1 = \text{cmAnalysisStep}(\mathcal{A}, \mathcal{V}, \mathcal{E}_1)) \to \text{trAnalysisStep}(\mathcal{A}, \mathcal{V}, \mathcal{E}_2, \mathcal{E}'_2)$$
$$\text{TRANSREL\_COMPLETENESS} \vdash$$
$$\text{envEqual}(\mathcal{E}_1, \mathcal{E}_2) \to \text{envEqual}(\mathcal{E}'_1, \mathcal{E}'_2) \to$$
$$\text{trAnalysisStep}(\mathcal{A}, \mathcal{V}, \mathcal{E}_2, \mathcal{E}'_2) \to (\mathcal{E}'_1 = \text{cmAnalysisStep}(\mathcal{A}, \mathcal{V}, \mathcal{E}_1))$$

Critical points are the variables that are updated multiple times in the course of a step, e. g. write conflicts are typical examples for this. There, we must abstract from the intermediate environments, which come from the sequential execution of actions within a single iteration in the traditional can-must analysis. The equivalence proof uses the fact that they can be reordered, i. e. the execution of actions has the Church-Rosser property. Hence, a cumulated action with the same effect can be defined, which is subsequently shown to be equivalent with the transition relation.

5   Conclusions

In this paper, we have presented a new symbolic causality analysis based on model checking, which supports arbitrary data types and run-time error checking. With the help of the HOL4 theorem prover, we formalised our new approach as well as the traditional can-must analysis and showed their equivalence. Thus, we gained a formally verified symbolic causality analysis, which can be used in particular by compilers of synchronous languages.

References

1. Andersen, F.: A Theorem Prover for UNITY in Higher Order Logic. PhD thesis, Horsholm, Denmark (March 1992)
2. Andersen, F., Petersen, K.D., Petterson, J.S.: Program verification using HOL-UNITY. In: Joyce, J.J., Seger, C.-J.H. (eds.) HUG 1993. LNCS, vol. 780, pp. 1–15. Springer, Heidelberg (1994)
3. Benveniste, A., Caspi, P., Edwards, S., Halbwachs, N., Le Guernic, P., de Simone, R.: The synchronous languages twelve years later. Proceedings of the IEEE 91(1), 64–83 (2003)

4. Berry, G.: The constructive semantics of pure Esterel (July 1999),
   http://www-sop.inria.fr/esterel.org/
5. Brzozowski, J.A., Seger, C.-J.: Asynchronous Circuits. Springer, Heidelberg (1995)
6. Chandy, K.M., Misra, J.: Parallel Program Design, May 1989. Addison Wesley, Austin, Texas (1989)
7. Collins, G., Syme, D.: A theory of finite maps. In: Schubert, E.T., Alves-Foss, J., Windley, P. (eds.) HUG 1995. LNCS, vol. 971, pp. 122–137. Springer, Heidelberg (1995)
8. Girault, A., Lee, B., Lee, E.: Hierarchical finite state machines with multiple concurrency models. IEEE Transactions on Computer Aided Design of Integrated Circuits and Systems 18(6), 742–760 (1999)
9. Huffman, D.: Combinational circuits with feedback. In: Mukhopadhyay, A. (ed.) Recent Developments in Switching Theory, pp. 27–55. Academic Press, London (1971)
10. Jantsch, A.: Modeling Embedded Systems and SoCs. Morgan Kaufmann, San Francisco (2004)
11. Kautz, W.: The necessity of closed circuit loops in minimal combinational circuits. IEEE Transactions on Computers C-19(2), 162–166 (1970)
12. Lee, E., Sangiovanni-Vincentelli, A.: A framework for comparing models of computation. IEEE Transactions on Computer-Aided Design of Integrated Circuits and Systems 17(12), 1217–1229 (1998)
13. Malik, S.: Analysis of cycle combinational circuits. IEEE Transactions on Computer Aided Design 13(7), 950–956 (1994)
14. Riedel, M.: Cyclic Combinational Circuits. PhD thesis, California Institute of Technology, Passadena, California (2004)
15. Riedel, M.D., Bruck, J.: Cyclic combinational circuits: Analysis for synthesis. In: International Workshop on Logic and Synthesis (IWLS), Laguna Beach, California (2003)
16. Riedel, M.D., Bruck, J.: The synthesis of cyclic combinational circuits. In: Design Automation Conference (DAC), Anaheim, California, USA, pp. 163–168. ACM Press, New York (2003)
17. Rivest, R.: The necessity of feedback in minimal monotone combinational circuits. IEEE Transactions on Computers C-26(6), 606–607 (1977)
18. Schneider, K.: The synchronous programming language Quartz. Internal Report, Department of Computer Science, University of Kaiserslautern (to appear, 2008)
19. Schneider, K., Brandt, J., Schuele, T.: Causality analysis of synchronous programs with delayed actions. In: Compilers, Architecture, and Synthesis for Embedded Systems (CASES), pp. 179–189. ACM Press, New York (2004)
20. Schneider, K., Brandt, J., Schuele, T., Tuerk, T.: Improving constructiveness in code generators. In: Synchronous Languages, Applications, and Programming (SLAP), Edinburgh, United Kingdom (2005)
21. Schneider, K., Brandt, J., Schuele, T., Tuerk, T.: Maximal causality analysis. In: Application of Concurrency to System Design (ACSD), St. Malo, France, pp. 106–115. IEEE Computer Society, Los Alamitos (2005)
22. Shiple, T.R., Berry, G., Touati, H.: Constructive analysis of cyclic circuits. In: European Design and Test Conference (EDTC), Paris, France. IEEE Computer Society Press, Los Alamitos (1996)

Imperative Functional Programming with Isabelle/HOL

Lukas Bulwahn[1], Alexander Krauss[1], Florian Haftmann[1,*],
Levent Erkök[2], and John Matthews[2]

[1] Technische Universität München,
Institut für Informatik, Boltzmannstraße 3, 85748 Garching, Germany
[2] Galois Inc., Beaverton, OR 97005, USA

Abstract. We introduce a lightweight approach for reasoning about programs involving imperative data structures using the proof assistant Isabelle/HOL. It is based on shallow embedding of programs, a polymorphic heap model using enumeration encodings and type classes, and a state-exception monad similar to known counterparts from Haskell. Existing proof automation tools are easily adapted to provide a verification environment. The framework immediately allows for correct code generation to ML and Haskell. Two case studies demonstrate our approach: An array-based checker for resolution proofs, and a more efficient bytecode verifier.

1 Introduction

A very common way of verifying programs in a HOL theorem prover is to use a shallow embedding and express the program as a set of recursive functions. Properties of the program can then be proved by induction. Despite some well-known limitations, shallow embeddings are widely used for verification. This success is due in part to the simplicity of the approach: A full-blown formal model of the operational or denotational semantics of the language is not required, and many technical difficulties (e.g. the representation of binders) are avoided altogether. Furthermore, the proof methods used are just standard induction principles and equational reasoning, and no specialized program logic is necessary. The specifactions may be turned into executable code directly by means of code generation.

Until recently, this approach has been used primarily for purely functional programs. As the notion of side-effect is alien to the world of HOL functions, programs with imperative updates of references or arrays cannot be expressed directly. However, there are many examples where for efficiency's sake imperative data structures are unavoidable to obtain practically usable executable programs.

We aim to permit Haskell's imperative specification style in Isabelle/HOL [11], where local state references and mutable arrays can be dynamically allocated without having to add their types to the enclosing function's type signature [6]. From such specifications we then generate efficient imperative functional code. Currently we need to restrict the contents of references and mutable arrays to first order values, but this is still sufficient for many applications.

---

* Supported by DFG project NI 491/10-1.

O. Ait Mohamed, C. Muñoz, and S. Tahar (Eds.): TPHOLs 2008, LNCS 5170, pp. 134–149, 2008.

Accordingly, the contributions of this paper are:

1. A purely definitional polymorphic *heap* allowing to encode dynamic allocation of polymorphic first-order references and mutable arrays (§2).
2. A Haskell-style *heap monad* encapsulating the primitive heap operations and supporting abnormal termination through exceptions (§3); an adaption for Isabelle's code generator allows to generate monadic Haskell and imperative ML[1] code (§4).
3. A set of proof rules that allows to reason about such monadic programs (§5).
4. Two case studies (§6): an imperative MiniSat proof replay oracle and an imperative Jinja bytecode verifier.

1.1 Related Work

Since the seminal paper by Peyton Jones and Wadler [13], the use of monads to incorporate effects in purely functional programs is standard. However, up to now, no practically usable verification framework for such monadic programs exists.

For imperative programs, there are such tools: The Why/Krakatoa/Caduceus toolset [3] works by translating the source language into an intermediate language and using a verification condition generator to generate proof obligations. Schirmer [14] proposes a similar method, which is closely integrated with Isabelle/HOL, and whose metatheory is formally verified. These approaches rely on Hoare logic and a verification condition generator. The actual reasoning then happens on the generated verification conditions and is often outsourced to automatic provers. The user must provide enough annotations in the source code that the verification conditions can be solved by the automated prover. In our approach, reasoning happens on the source code level, which we find better suited for interactive use. Proof principles are similar to those used for purely functional programs, i.e. induction and equational reasoning.

Probably closest to our work is the concept of *single-threaded objects* [1] in the ACL2 prover. By declaring an object as single-threaded (and obeying rigorous syntactic restrictions), one instructs the prover to replace non-destructive updates by destructive ones. The rules ensure that referential transparency is not violated, and thus the code can be treated as purely functional in the reasoning phase. Our approach is similar in the sense that our heap can be seen as a single-threaded object. However, we allow the dynamic allocation of arrays and references, whereas in ACL2 imperative fields must be statically declared in a single record.

Imperative language features have previously been embedded in higher order logic via a *state monad* [8, 15]. As long as the monad primitives do not duplicate the state, the resulting programs are single threaded and can be safely transformed to monadic Haskell or imperative ML code. However, just like single-threaded objects, the state monad approach requires the state record be statically declared as part of the monad type itself, either fixed or as an explicit type parameter. This makes it difficult to write specifications that dynamically allocate new references or mutable arrays, or to compose monadic specifications that work over different state types.

Our heap model has some similarities to the one used by Tuch, Klein, and Norrish [16], especially concerning the use of type classes and phantom types to manage

---

[1] In its two flavors SML and OCaml.

encodings. On the other hand, our model is slightly more abstract, since we are only dealing with functional languages instead of low level C code.

Nanevski et al. [10] describe how Hoare logic can be integrated in dependent type theory, yielding *Hoare Type Theory*, with a sophisticated type system and program logic. However, it seems that this requires significant modifications to the theorem prover in order to support such a system. In contrast, our approach was developed on top of standard Isabelle/HOL.

1.2 A Trivial Example: Array Reversal

To illustrate imperative functional programming in monadic HOL, we define two simple functions, one for swapping two elements in an array, and one for reversing an array:

$swap :: (\alpha::hrep)\ array \Rightarrow nat \Rightarrow nat \Rightarrow unit\ Heap$
$swap\ a\ i\ j = \textbf{do}\ x \leftarrow a[i];$
$\qquad\qquad\qquad y \leftarrow a[j];$
$\qquad\qquad\qquad a[i] := y;$
$\qquad\qquad\qquad a[j] := x;$
$\qquad\qquad\qquad \textbf{return}\ ()$

$rev :: (\alpha::hrep)\ array \Rightarrow nat \Rightarrow nat \Rightarrow unit\ Heap$
$rev\ a\ i\ j = (\textbf{if}\ i < j\ \textbf{then do}\ swap\ a\ i\ j;$
$\qquad\qquad\qquad\qquad\qquad rev\ a\ (i+1)\ (j-1)$
$\qquad\qquad\quad \textbf{else return}\ ())$

This idiom is well-known from Haskell: Manipulations of imperative arrays are monadic actions (of type $\alpha\ Heap$), and they can be composed into more complex actions using the sequencing operation provided by the monad. Other language constructs (conditionals, recursion, data types) are taken from the functional part of the language.

Let us prove a lemma that describes the behavior of swap:

$(h, h', r) \in [\![swap\ a\ i\ j]\!] \Longrightarrow$
$get\text{-}array\ a\ h'!\ k =$
$(\textbf{if}\ k = i\ \textbf{then}\ get\text{-}array\ a\ h\ !\ j$
$\ \textbf{else if}\ k = j\ \textbf{then}\ get\text{-}array\ a\ h\ !\ i\ \textbf{else}\ get\text{-}array\ a\ h\ !\ k)$

The notation $[\![c]\!]$ stands for the big-step semantics of a command c, which is a ternary relation: $(h, h', r) \in [\![c]\!]$ holds iff the computation c started on heap h does not generate an exception and yields the result r and the updated heap h'. The lemma expresses how the entries of the updated array are related to original entries. The function *get-array* returns the contents of an array on a given heap as a list, and infix ! denotes list indexing.

The lemma is proved by unfolding the call to swap and applying standard rules for the semantics of the monad operations and the basic commands, which can easily be automated using existing Isabelle technology. Now let us turn to the reversal function:

$(h, h', r) \in [\![rev\ a\ i\ j]\!] \Longrightarrow$
$get\text{-}array\ a\ h'!\ k =$
$(\textbf{if}\ k < i\ \textbf{then}\ get\text{-}array\ a\ h\ !\ k$
$\ \textbf{else if}\ j < k\ \textbf{then}\ get\text{-}array\ a\ h\ !\ k\ \textbf{else}\ get\text{-}array\ a\ h\ !\ (j - (k - i)))$

Since *rev* is defined recursively, we proceed by induction. The proof is as one would expect: In the step case, we distinguish the cases $k < i, k = i, i < k < j, k = j$ and $j < k$, and apply the induction hypothesis and the lemma about swap.

1.3 Dynamic Allocation: Linked Lists

The ability to explicitly allocate memory is another fundamental technique in imperative programming. To illustrate how this idiom can be coded in HOL, we will show how to build and traverse a dynamically allocated linked list.

Linked lists are represented by a recursive datatype, where the tail of the list is a mutable reference.

> **datatype** α *node* = *Empty* | *Node* α (α *node ref*)

To convert a HOL list of elements to a linked list, we simply recurse over each tail, allocating the nodes as we go along by calling the *ref* function:

> *make-llist* :: (α::*hrep*) *list* $\Rightarrow$ α *node Heap*
> *make-llist* [] = **return** *Empty*
> *make-llist* ($x \cdot xs$) = **do** *tl* $\leftarrow$ *make-llist xs*;
> $\qquad\qquad\qquad\qquad$ *next* $\leftarrow$ *ref tl*;
> $\qquad\qquad\qquad\qquad$ **return** (*Node x next*)

In the other direction, we can traverse a linked list as follows:

> *traverse* :: (α::*hrep*) *node* $\Rightarrow$ α *list Heap*
> *traverse Empty* = **return** []
> *traverse* (*Node x r*) = **do** *tl* $\leftarrow$!*r*;
> $\qquad\qquad\qquad\qquad$ *xs* $\leftarrow$ *traverse tl*;
> $\qquad\qquad\qquad\qquad$ **return** ($x \cdot xs$)

Note that the definitions of *make-llist* and *traverse* operationally mimic their equivalents in Haskell using the state monad, or in ML using imperative features.[2]

For reference, here is the *traverse* function as rendered by the code generator of our framework in ML:[3]

```
datatype 'a node = Node of 'a * 'a node ref | Empty;

fun traverse A_ (Node (x, r)) =
  (fn a as () =>
    let
      val tl = (fn () => ! r) ();
      val xs = traverse A_ tl ();
    in
      x :: xs
    end)
  | traverse A_ Empty = (fn () => []);
```

---

[2] Technical details on the definition of *traverse* can be found in §7.3.

[3] The A_ argument denotes the dictionary, which is not used in this particular example. See [4].

2   Modeling a Polymorphic Heap

In the following two sections we present our definitional model of a typed heap and the monad we are using. We present the theory in a bottom-up manner, and explain the most important design decisions.

Essentially, our heap will be a mapping $h :: \mathbb{N} \Rightarrow \mathcal{V}al$ from addresses to values. However, since values can generally have arbitrary types, this is difficult to model in a simply typed language. Since there is no HOL type that can contain all types, we are facing the problem what type to choose for $\mathcal{V}al$.

This problem could be solved using dependent types, but we want to stay in the simply typed framework, so we make a draconian restriction: We decide that functions cannot be stored on the heap, and use the natural numbers as value type, in which all first-order data objects will be encoded. We'll use phantom types (§2.2) to safely omit these encodings from the generated code.

Obviously this restrictive design decision excludes a fair number of relevant programs. But even then our model allows for interesting applications. Possibilities for lifting this restriction are discussed in §7.2.

2.1   Representable Types

Using encodings to circumvent restrictions in the type system seems very awkward at first, but we can make this transparent to the user by defining an axiomatic type class *countable*, with an axiom stating that the type can be encoded into the natural numbers:

> **axclass** *countable* $\subseteq$ *type*
> $\exists\,(enc :: \alpha \Rightarrow nat).\ inj\ enc$

Obviously, basic types like *nat*, *int* and all finite types are countable, and the well-known constructions can be used to show that if α and β are countable then so are $\alpha \times \beta$ and α *list*. In fact, such instance proofs are straightforward for first-order recursive data types and could be automated. The overloaded encoding and decoding functions are called *to-nat* and *from-nat*:

> *to-nat*   :: $(\alpha::countable) \Rightarrow nat$
> *from-nat* :: $nat \Rightarrow (\alpha::countable)$

2.2   Typed References

References are just addresses, i.e. natural numbers

> **datatype** $\alpha\ ref = Ref\,nat$

with the projection *addr-of* $(Ref\ n) = n$. Here, unlike above, the type system is again a useful tool instead of just a handicap: The *phantom type* α, which does not occur on the right hand side of the definition, allows us to view references as typed objects as we know them from ML, although the underlying representation is untyped.

Reference Equality. We will certainly need to reason about reference (in)equality. For example, we would expect the following simplification rule to hold, where r and s are references and h is a heap:

$$r \neq s \implies \textit{get-ref } r \;(\textit{set-ref } s \; x \; h) = \textit{get-ref } r \; h$$

Indeed, when we write down and prove this lemma, everything seems to work. Only if we look at the inferred types, there is an unpleasant surprise: Because of the equality in the premise, the references r and s have the same type, and we have thus only proved a special case. Of course we want to perform the same simplification if we have references of different types, and ideally, we want the condition $r \neq s$ to be immediate, when the references have different types.

The solution is to define a heterogeneous inequality relation for references, which just strips away the phantom type and compares the bare addresses:

$$r \ncong s \leftrightarrow (\textit{addr-of } r \neq \textit{addr-of } s) \; ^{4}$$

If r and s have the same type, the relation $\ncong$ coincides with $\neq$.

2.3 Type Reflection

Comparing references of different types is a little artificial. In a typed language, aliasing between e.g. an integer and a boolean reference is not possible, and the above rewrite rule should be applicable unconditionally. Our model will be built in such a way that we can automatically derive $r \ncong s$, whenever r and s have different types:

We define a type *typerep* and a type class *typeable* to reflect the syntax of (monomorphic) types back into the language of terms:

datatype $\textit{typerep} = \textit{Typerep string (typerep list)}$

class $\textit{typeable} = \textit{type} +$
  fixes $\textit{typerep} :: \alpha \; \textit{itself} \Rightarrow \textit{typerep}$

The predefined type α *itself* comes with a singleton term written $\textit{TYPE}(\alpha)$ which is used to embed types into terms. We write $\textit{RTYPE}(\alpha)$ for *typerep* ($\textit{TYPE}(\alpha)$). The overloaded function *typerep* constructs a concrete syntactic representation of a type name. Its definition for concrete types is completely schematic (and easily automated):

$$\textit{RTYPE}(\textit{nat}) = \textit{Typerep ''nat'' } []$$
$$\textit{RTYPE}(\textit{bool}) = \textit{Typerep ''bool'' } []$$
$$\textit{RTYPE}(\alpha \; \textit{list}) = \textit{Typerep ''list'' } [\textit{RTYPE}(\alpha)]$$

The result of this exercise (which is also common in the Haskell world) is that we can now compare types for equality explicitly. For example: $\textit{RTYPE}(\textit{nat}) \neq \textit{RTYPE}(\textit{bool})$ and $\textit{RTYPE}(\textit{char list}) = \textit{RTYPE}(\textit{string})$ are theorems[5], however $\textit{RTYPE}(\alpha) \neq \textit{RTYPE}(\beta)$ is not, since α and β could later be instantiated to the same type.

Now we can refine the definition of reference inequality as follows:

$$(r :: \alpha \; \textit{ref}) \ncong (s :: \beta \; \textit{ref}) \leftrightarrow \textit{RTYPE}(\alpha) \neq \textit{RTYPE}(\beta) \vee \textit{addr-of } r \neq \textit{addr-of } s$$

---

[4] The $\leftrightarrow$ denotes equality of *bool* values.

[5] Note that in Isabelle, *string* just abbreviates *char list* on the surface syntax level.

From this immediately follows that references of different types are always unequal. Hence, aliasing proof obligations like $r \ncong s$ can be solved automatically, if r and s are of different types.

2.4 The Heap

The heap is modelled as a mapping from type representations and addresses to *nat*-encoded values. We use two separate mappings for references and arrays (which are mapped to lists of encoded values). Additionally, we use a counter which bounds the currently used address space. It is incremented when new references are created.

> **record** *heap* =
>   *refs* :: *typerep* $\Rightarrow$ *addr* $\Rightarrow$ *nat*
>   *arrays* :: *typerep* $\Rightarrow$ *addr* $\Rightarrow$ *nat list*
>   *lim* :: *addr*

We can now define the basic heap operations, such as allocation, reading and writing of references. Note how the embeddings and projections are used here to convert the stored values into their respective encodings and back[6]:

> *get-ref* :: $(\alpha::hrep)$ *ref* $\Rightarrow$ *heap* $\Rightarrow$ α
> *get-ref* $(Ref\, r)$ $(h(\!|refs := f|\!)) = from\text{-}nat\, (f\, RTYPE(\alpha)\, r)$

> *set-ref* :: $(\alpha::hrep)$ *ref* $\Rightarrow$ α $\Rightarrow$ *heap* $\Rightarrow$ *heap*
> *set-ref* $(Ref\, r)\, x\, (h(\!|refs := f|\!)) =$
> $h(\!|refs := f(RTYPE(\alpha) := (f\, RTYPE(\alpha))(r := to\text{-}nat\, x))|\!)$

> *new-ref* :: *heap* $\Rightarrow$ $(\alpha::hrep)$ *ref* $\times$ *heap*
> *new-ref* $(h(\!|lim := l|\!)) = (Ref\, l, h(\!|lim := Suc\, l|\!))$

Operations for arrays are analogous. From these definitions, we can now easily prove the expected lemmas, expressing the interaction of the operations, e.g.:

> *get-ref* r $(set\text{-}ref\, r\, x\, h) = x$
> $r \ncong s \implies$ *get-ref* r $(set\text{-}ref\, s\, x\, h) = get\text{-}ref\, r\, h$

Since arrays and references occupy different heap areas, the corresponding heap operations always commute:

> *get-array* a $(set\text{-}ref\, r\, x\, h) = get\text{-}array\, a\, h$
> *get-ref* r $(set\text{-}array\, a\, xs\, h) = get\text{-}ref\, r\, h$

3   The Heap Monad

We now define a monad which characterizes computations affecting the heap. An imperative program with return type α will be logically represented as a value of type α *Heap*.

---

[6] The *hrep* type class just intersects the classes *countable* and *typeable*; also note that the $h(\!|\ldots := \ldots|\!)$ syntax denotes component assignment on records and may also be used for pattern matching.

Essentially, our monad is a state-exception monad, where the state is the heap from the previous section:

$$\textbf{datatype } \alpha \; Heap = Heap \; (heap \Rightarrow (\alpha + exception) \times heap)$$
$$heap \, f \qquad\quad = Heap \; (\lambda h. \; \textbf{let } (x, h') = f \, h \textbf{ in } (Inl \, x, h'))$$
$$execute \; (Heap \, f) \; = f$$

Exceptions are essentially strings generated by $error :: string \Rightarrow exception$ and are not caught inside the monad; they are a mere device to introduce a notion of abnormal termination. The monad operations $return$, ($\ggeq$) and $raise$ are defined as expected:

$$return \, x = heap \; (\lambda h. \; (x, h))$$
$$f \ggeq g \;\; = Heap$$
$$\qquad\qquad (\lambda h. \; \textbf{case } execute \, f \, h \textbf{ of } (Inl \, x, h') \Rightarrow execute \; (g \, x) \, h'$$
$$\qquad\qquad\qquad\qquad | \; (Inr \, e, h') \Rightarrow (Inr \, e, h'))$$
$$raise \, s \;\; = Heap \; (\lambda h. \; (Inr \; (error \, s), h))$$

Isabelle's syntax facilities allow for Haskell-style do-notation. Lifting the heap operations into the monad is straightforward:

$$ref \, x \;\; = heap \; (\lambda h. \; \textbf{let } (r, h') = new\text{-}ref \, h \textbf{ in } (r, set\text{-}ref \, r \, x \, h'))$$
$$!r \;\;\; = heap \; (\lambda h. \; (get\text{-}ref \, r \, h, h))$$
$$r := x = heap \; (\lambda h. \; ((), set\text{-}ref \, r \, x \, h))$$

$$array \, n \, x = heap \; (\lambda h. \; \textbf{let } (a, h') = new\text{-}array \, h \textbf{ in } (a, set\text{-}array \, a \; (replicate \, n \, x) \, h'))$$
$$length \, a \;\; = heap \; (\lambda h. \; (|get\text{-}array \, a \, h|, h))$$
$$a[i] \qquad = \textbf{do } len \leftarrow length \, a;$$
$$\qquad\qquad\quad (\textbf{if } i < len \textbf{ then } heap \; (\lambda h. \; (get\text{-}array \, a \, h \, ! \, i, h))$$
$$\qquad\qquad\quad \textbf{else raise } ''array \, lookup\text{: } index \, out \, of \, range'')$$
$$a[i] := x \; = \textbf{do } len \leftarrow length \, a;$$
$$\qquad\qquad\quad (\textbf{if } i < len \textbf{ then } heap \; (\lambda h. \; (a, set\text{-}array \, a \; (get\text{-}array \, a \, h[i := x]) \, h))$$
$$\qquad\qquad\quad \textbf{else raise } ''array \, update\text{: } index \, out \, of \, range'')$$

These are the necessary foundations to write stateful programs like in §1.2.

4   Execution

When we consider some parts of HOL as the shallow embedding of a programming language, then the inverse of that embedding is called *code generation*. Isabelle's code generator [4] can produce SML, OCaml and Haskell code from executable HOL specifications. In a first approximation, the executable fragment of HOL consists of datatype and function definitions, which are simply translated to their counterparts. This guarantees partial correctness: if $\langle s \rangle$ denotes the generated code from term s, then each abstract evaluation step from $\langle s \rangle$ to some t' in the target language corresponds to an equational rewrite step $s = t$ in HOL, such that $\langle t \rangle = t'$ (cf. Fig. 1(a)).

HOL	ML	Haskell
α *Heap*	*unit* $\Rightarrow \alpha$	*ST* ξ α
$t \gg= (\lambda x. f)$	$\lambda(). \ let \ x = t \ () \ in \ f \ () \ end$	$t \gg= (\lambda x. f)$
return t	$\lambda(). \ t$	*return* t
α *ref*	α *ref*	*STRef* ξ α
ref x	$\lambda(). \ ref \ x$	*newSTRef* x
$!r$	$\lambda(). \ !r$	*readSTRef* r
$r := t$	$\lambda(). \ r := x$	*writeSTRef* r x
α *array*	α *array*	*STArray* ξ *Integer* α
array n x	$\lambda(). \ Array.array \ (n, x)$	*newArray* $(0, n)$ x
$a[i]$	$\lambda(). \ Array.sub \ (a, i)$	*readArray* a i
$a[i] := x$	$\lambda(). \ Array.update \ (a, i, x)$	*writeArray* a i x
length a	$\lambda(). \ Array.length \ a$	*liftM snd* (*getBounds* a)
raise s	$\lambda(). \ raise \ Fail \ s$	*error* s

(a) Partial correctness

$s \xrightarrow{\text{code gen.}} \langle s \rangle$

equational rewriting

evaluation

$t \xrightarrow[\text{code gen.}]{} \langle t \rangle = t'$

(b) Translating monadic constructs to ML and Haskell

Fig. 1. Code generation

The reference and array operations are mapped to the target language as given in Fig. 1(b). Since ML expressions may already contain side effects, the monad vanishes and is just replaced by a unit abstraction to ensure the correct evaluation order.

For Haskell we use the built-in *ST* state monad, together with the corresponding *STRef* and *STArray* types. Recall that our HOL programs only raise exceptions but never handle them – instead of dealing with them inside the monad, we treat them as partiality, using the *error* primitive.

Note that the extended executable fragment of HOL does not include the constructions that were used to define the heap monad: If we break the monad abstraction (e.g. by writing *heap* $(\lambda h. \ (h, h)))$, the results are no longer executable and trying to generate code for them causes an error, just like for other purely logical notions like quantifiers.

5 Verification

Having defined the model of execution for our stateful programs, we need verification tools which can be used to prove an individual program correct. Our model does not force us to use a particular technique: We can choose any calculus (like e.g. Hoare logic) that is sound with respect to the semantics we have defined. After a bit of experimenting, we opted for a very simple method, which seems to fit well with the structured proof language Isabelle/Isar. A deeper comparison of this relational style with Hoare logic is beyond the scope of this paper.

We use the relational description of the big-step semantics we have already seen in §1.2. The relation is defined by:

$$(h, h', r) \in [\![c]\!] = ((Inl \ r, h') = execute \ c \ h)$$

We can prove rules which connect this relation to the different basic commands. Here is the rule for the bind operation.

$$(h, h'', r') \in [\![f \gg= g]\!] \Longrightarrow$$
$$(\bigwedge h' r. (h, h', r) \in [\![f]\!] \Longrightarrow (h', h'', r') \in [\![g\, r]\!] \Longrightarrow P) \Longrightarrow P$$

Note the elimination rule format. Since the $(\dots) \in [\![\dots]\!]$ relation usually lives in the premise of a statement, we use elimination rules to manipulate it: if our goal has a premise of the form $(h, h'', r) \in [\![f \gg= g]\!]$, we can obtain the intermediate heap h' and the new assumptions $(h, h', a) \in [\![f]\!]$ and $(h', h'', r) \in [\![g\, a]\!]$. These elimination rules allow us to systematically decompose compound statements into primitive steps. Here are some other rules:

$$(h, h', r) \in [\![\textit{return}\, x]\!] \Longrightarrow (r = x \Longrightarrow h = h' \Longrightarrow P) \Longrightarrow P$$

$$(h, h', r) \in [\![a[i]]\!] \Longrightarrow$$
$$(r = \textit{get-array}\, a\, h\, !\, i \Longrightarrow h = h' \Longrightarrow i < \textit{length-array}\, a\, h \Longrightarrow P) \Longrightarrow P$$

$$(h, h', r) \in [\![a[i] := v]\!] \Longrightarrow (r = a \Longrightarrow h' = \textit{Heap.upd}\, a\, i\, v\, h \Longrightarrow P) \Longrightarrow P$$

By feeding these rules into Isabelle's *auto* method, we obtain a reasonable *ad-hoc* automation, which makes proofs quite short.

6   Case Studies

6.1   A SAT Checker

Our first case study is motivated by the wish to integrate SAT solvers into Isabelle in a scalable way, such that they can be used to solve large propositional proof obligations.

We aim at a compromise between performing a full replay of the proof within Isabelle and trusting the SAT solver completely. The first approach was taken by Weber and Amjad [17] and gives the usual high assurance of the LCF principle, but is computationally expensive. On the other end of the spectrum, trusting the external tool is obviously cheap but unsatisfactory.

A reasonable compromise is to run the external proof (a standard propositional resolution proof) through a checker, which is itself proved correct in Isabelle. This gives a good balance between assurance and cost, since unlike the SAT solver, the checker is formally verified, and checking a proof is about an order of magnitude faster than replaying the inferences in Isabelle.

Usually, for such a reflective approach, the checker would need to be purely functional. Using our framework, we can implement a checker that uses destructive arrays instead, which gives us another 30% speedup over a purely functional implementation with balanced trees.

The core of our checker operates on a table that stores the clauses that have already been derived. Clauses are modeled (purely functionally) as sorted lists of integers, where a negative number signifies a negated variable:

types		**datatype** *ProofStep =*
idx	*= nat*	*Root idx clause*
lit	*= int*	*| Resolve idx resolvants*
clause	*= lit list*	*| Delete idx*
resolvants = idx × (lit × idx) list		

A proof step can either (a) add a new so-called *root clause* to the array, (b) derive a new clause from existing clauses and store it in the array, or (c) delete a clause from the array, to free some memory.

The root clauses are the initial clauses from which a contradiction is derived. It is a specialty of the MiniSAT [2] proof format that root clauses may be added any time during the proof, hence our checker must accumulate all root clauses it encounters in a list. Then, if the checker succeeds in deriving the empty clause, the root clauses it has collected must be inconsistent.

A *Resolve* step derives a new clause in a series of resolutions: *Resolve i (j, rs)* starts with clause no. *j* and resolves it with the clause/variable pairs in *rs*. In the end, the result is stored at position *i*. The *Delete* proof step removes a clause from the array. This weakening step is simply an optimization to reduce memory usage of the checker by removing clauses that are no longer needed.

With clauses modeled as sorted lists, resolution is essentially a merge operation and can be done in just one traversal. However, the operation may fail if the literal does not occur in the clause. It is convenient to let the monad deal with such failures, even if no heap access is required. Hence our *resolve* operation has the following type (for brevity, we omit the implementation, which does not contain surprises):

$$resolve :: lit \Rightarrow clause \Rightarrow clause \Rightarrow clause \; Heap$$

The function *get-clause* retrieves a clause from the array. It fails if it sees a *None*:

$$get\text{-}clause :: clause \; option \; array \Rightarrow idx \Rightarrow clause \; Heap$$

The heart of our checker is the function *step*, which processes a single proof step, collecting root clauses in the accumulator list *rcs*:

$$step :: clause \; option \; array \Rightarrow ProofStep \Rightarrow clause \; list \Rightarrow clause \; list \; Heap$$

$step \; a \; (Root \; cid \; clause) \; rcs = $ **do** $a[cid] := Some \; (remdups \; (sort \; clause));$
    return $(clause \cdot rcs)$
$step \; a \; (Resolve \; saveTo \; (i, rs)) \; rcs = $
do $cli \leftarrow get\text{-}clause \; a \; i;$
    $result \leftarrow foldM \; (\lambda(l, j) \; c. \; get\text{-}clause \; a \; j \gg= resolve \; l \; c) \; rs \; cli;$
    $a[saveTo] := Some \; result;$
    return rcs
$step \; a \; (Delete \; cid) \; rcs = $ **do** $a[cid] := None;$
    return rcs

Finally, a wrapper function *checker* just allocates an array of a given size, folds the step function over a list of proof steps, and finally checks for the empty clause at some given position. Our main result is the following partial correctness theorem:

$$(h, h', cs) \in [\![\; checker \; n \; p \; i \;]\!] \Longrightarrow inconsistent \; cs$$

Integration. Since we have verified our checker, we may now choose to use it to import proofs into Isabelle. This can be done using a generic *monadic evaluation oracle*, which implements the following inference rule:

$$\frac{\bigwedge h\, h'.\ (h, h', r) \in [\![\, c\,]\!] \Longrightarrow P\, r}{P\, r} \quad \text{(if } c, \text{ when executed in ML, evaluates to } r)$$

Thus we can discharge the premise of a partial correctness theorem by just running the generated code in ML.

The soundness of this rule relies on the assumption that the semantics of ML is compatible with our model of monadic programs. At the moment, we have no proof of this assumption.

However, such a generic reflection mechanism, which provides a clearly defined way to extend the theorem prover by reflected imperative proof tools, still provides higher assurance than an ad-hoc extension, since the monadic code is verified, and no additional "glue code" is required for the integration.

In particular, nothing in our particular development of the SAT checker needs to be trusted.

6.2 A Jinja Bytecode Verifier

Our second case study is a modification of the Jinja bytecode verifier. Jinja [7] is a complete formal model of a Java-like language, which includes a formal semantics, type system, virtual machine model, compiler, and bytecode verifier.

Essentially, the bytecode verifier performs an abstract interpretation of the bytecode instructions, keeping track of the abstract state, that is, the types of values in registers and on the stack. The central data structure is a mapping that assigns such an abstract state to every bytecode instruction. Then, this information is propagated to the successors of the instruction until a fixed point is reached.

In the existing implementation, this mapping is represented by a list of fixed length. In our modification, we use an imperative array instead, with the obvious advantages: constant-time access and no garbage.

Fortunately, the bytecode verifier is modeled in a very abstract framework using a semilattice (type σ), which hides all the technical details of the virtual machine. Later, the "real thing" can be obtained by instantiation. A bytecode method is modelled by a function $step :: nat \Rightarrow \sigma \Rightarrow (nat \times \sigma)\ list$ that maps a given program position and an abstract machine state to a list of possible successor positions and states. Additional requirements for the $step$ function (e.g. monotonicity) are detailed in [7].

Figure 2 shows the pure version of the bytecode verifier together with its monadic counterpart. This side-by-side comparison shows that the differences between the two versions are small. Consequently, proving partial correctness of *kildallM* wrt. *kildall* is straightforward:

$$\tau s \in list\ n\ A \Longrightarrow (h, h', \tau s') \in [\![\, kildallM\ \tau s\,]\!] \Longrightarrow \tau s' = kildall\ \tau s$$

This shows that it is relatively easy to move from a purely functional specification to a monadic one, which can then be executed efficiently.

$propa\ []\ \tau s\ w = (\tau s, w)$
$propa\ ((q, \tau)\cdot qs)\ \tau s\ w =$
$(let\ u = \tau \sqcup \tau s\ !\ q;$
$\quad w' = if\ u = \tau s\ !\ q\ then\ w\ else\ \{q\} \cup w$
$in\ propa\ qs\ (\tau s[q := u])\ w')$

$propaM\ []\ \tau s\ w = \textit{return}\ w$
$propaM\ ((q, \tau)\cdot qs)\ \tau s\ w =$
$\textit{do}\ \tau' \leftarrow \tau s[q];$
$\quad let\ u = \tau \sqcup \tau';$
$\quad let\ w' = (if\ u = \tau'\ then\ w\ else\ \{q\} \cup w);$
$\quad \tau s[q] := u;$
$\quad propaM\ qs\ \tau s\ w'$

$iter\ (\tau s, w) =$
$(if\ w = \emptyset\ then\ \tau s$
$\quad else\ let\ p = SOME\ p.\ p \in w$
$\quad\quad in\ iter$
$\quad\quad\quad (propa\ (step\ p\ (\tau s\ !\ p))\ \tau s\ (w -$
$\{p\})))$

$iterM\ \tau s\ w =$
$(if\ w = \emptyset\ then\ freeze\ \tau s$
$\quad else\ let\ p = SOME\ p.\ p \in w$
$\quad\quad in\ \textit{do}\ v \leftarrow \tau s[p];$
$\quad\quad\quad w' \leftarrow propaM\ (step\ p\ v)\ \tau s$
$\quad\quad\quad\quad (w - \{p\});$
$\quad\quad\quad iterM\ \tau s\ w')$

$kildall\ \tau s = iter\ (\tau s, unstables\ \tau s)$

$kildallM\ \tau s =$
$\textit{do}\ a \leftarrow of\text{-}list\ \tau s;$
$\quad iterM\ a\ (unstables\ \tau s)$

Fig. 2. Pure vs. monadic versions of the bytecode verifier

7 Problems and Limitations

7.1 No Monad Polymorphism

Of course, one would like to specify a monad as a constructor class, and see our heap monad just as a particular instance of the general concept. However, for this we would need type constructor polymorphism, which is not supported in HOL. We must be satisfied with the possibility of defining concrete instances of monads.

Huffman, Matthews, and White [5] describe how to simulate constructor classes in an extension of HOL, but their embedding does not seem practical for our application.

7.2 Heap Model

Our simple heap model prohibits storing any kind of function value in mutable references. Although many applications can live with this limitation, it may be painful in other situations. One can think of different ways to improve this situation:

Encoding types of order n. Just like we now encode all first-order values in $\mathbb{N}$, one could also encode all functions on such values by $\mathbb{N} \Rightarrow \mathbb{N}$, and all functions that take such arguments by $(\mathbb{N} \Rightarrow \mathbb{N}) \Rightarrow \mathbb{N}$, and so on. For any given order, we can encode all "smaller" types in a single type. Again, this can be made transparent using type classes. Probably, order 3 or 4 would be enough for most practical purposes.

Dependent types. In a dependently typed system, one could do without explicit encodings, and represent heap values as a dependent pair of a type and a value. In such a system, the type $heap$ would live in some higher universe than the types used in a program.

ZF extension. HOLZF [12] is a consistent extension of HOL which declares a set-theoretic universe Z, in which all HOL types can be embedded. In such a system we could store the full tower of (pure, monomorphic) higher order functions over the naturals, since our heap function could take values in Z.

However, even these extensions will not allow us to store *monadic* functions in the heap. The collection of heap monad functions has at least the cardinality of $heap \Rightarrow heap$, which is strictly larger than $heap$ itself in classical HOL.

One avenue of escape would be to limit ourselves to the constructive portion of HOL and build some kind of impredicative datatype facility to represent the heap. A more pragmatic option is to store only a representable subset of the full function space in the heap, for example just the continuous functions as is done in Isabelle/HOLCF[9]. We would retain the full power of classical HOL while still allowing to store all (partially) executable functions, which are the only ones we are really interested in.

7.3   Recursive Functions

Monadic functions can be defined recursively just like any other function by using the available packages in Isabelle. However, proving termination of the functions can sometimes be tricky, as the following example demonstrates:

$$f :: nat\ ref \Rightarrow nat \Rightarrow nat\ Heap$$
$$f\ r\ n = \mathbf{do}\ x \leftarrow !r;$$
$$(\mathbf{if}\ x = 0\ \mathbf{then}\ return\ n\ \mathbf{else}\ \mathbf{do}\ r := x - 1;$$
$$f\ r\ (Suc\ n))$$

Since there is no wellfounded order for which $(r, Suc\ n) \prec (r, n)$ holds, we cannot hope to define f by wellfounded recursion on its arguments. Instead, the recursion happens on the heap itself, which is not an explicit argument of the function. To define f, we must first break the monad abstraction and define a function $f' :: nat\ ref \Rightarrow nat \Rightarrow heap \Rightarrow (nat + exception) \times heap$, which explicitly recurses over the heap. Then f can be defined in terms of f', deriving the above recursion equation.

Another issue is that when building pointer structures on the heap, many functions are actually partial, since the structures can become cyclic. The attentive reader may have noticed that the *traverse* function in §1.3 is in fact such an example.

However, it turns out that even nonterminating recursive functions are definable if the recursion happens within the heap monad, since such definitions always have a total model. This argument is similar to the observation that tail-recursive functions can always be defined in HOL (e.g. using a *while* combinator). The details are beyond the scope of this paper, so we just mention that *traverse* was defined using a monadic recursion combinator *MREC*, which satisfies the following recursion equation:

$$MREC :: (\alpha \Rightarrow (\beta + \alpha)\ Heap) \Rightarrow (\alpha \Rightarrow \alpha \Rightarrow \beta \Rightarrow \beta\ Heap) \Rightarrow \alpha \Rightarrow \beta\ Heap$$
$$MREC\ f\ g\ x =$$
$$\mathbf{do}\ y \leftarrow f\ x;$$
$$(\mathbf{case}\ y\ \mathbf{of}\ Inl\ r \Rightarrow return\ r \mid Inr\ s \Rightarrow \mathbf{do}\ z \leftarrow MREC\ f\ g\ s;$$
$$g\ x\ s\ z)$$

In the future, we plan to provide automation for defining such recursive functions.

7.4　External I/O

Another practical limitation is that our heap monad does not support any kind of interaction with the outside world. This means, for example, that the full sequence of MiniSat proof steps needs to be passed into our SAT checker from the start. This becomes a problem for long-running proofs where the number of steps may exceed the total size of Isabelle's memory.

However, if our monad supported IO actions then we could incrementally ask MiniSat to supply us just the next portion of the proof to check, and never have to represent the entire proof at once. Supporting I/O would require us to extend our heap model to include relevant aspects of the outside system, plus some kind of nondeterminism for I/O actions, to take into account that we can never model the world in its entirety.

8　Conclusion

We presented a lightweight approach to reuse our favorite theorem prover for verifying monadic programs that manipulate a state. Our shallow embedding of imperative constructs in HOL is a continuation of the traditional way of modeling programs and systems by recursive functions, which can be translated to "real" programs by a code generator. Although there are still some limitations (see §7), our case studies show that it is already quite useful in its current form. Equipped with that, we want to tackle specification, verification and prototypic code generation for compute-intensive applications like e.g. microprocessor models. Another important application is the extension of the Isabelle system itself by means of verified monadic proof procedures as we have sketched it in §6.1.

Future work will also focus on alleviating current limitations, most notably to allow a broader range of heap-representable types, monadic I/O, and more automation for defining monadic recursive functions.

Acknowledgments

We would like to thank David Hardin, Joe Hurd, Matt Kaufmann, Dylan McNamee, Tobias Nipkow, Konrad Slind, and Tjark Weber for their useful discussions, encouragement and feedback on our work.

References

[1] Boyer, R.S., Moore, J.S.: Single-threaded objects in ACL2. In: Krishnamurthi, S., Ramakrishnan, C.R. (eds.) PADL 2002. LNCS, vol. 2257, pp. 9–27. Springer, Heidelberg (2002)

[2] Een, N., Sörensson, N.: An extensible sat-solver. In: Goos, G., Hartmanis, J., van Leeuwen, J. (eds.) SAT 2003. LNCS, vol. 2919, p. 502. Springer, Heidelberg (2004)

[3] Filliâtre, J.-C., Marché, C.: The Why/Krakatoa/Caduceus platform for deductive program verification. In: Damm, W., Hermanns, H. (eds.) CAV 2007. LNCS, vol. 4590. Springer, Heidelberg (2007)

[4] Haftmann, F., Nipkow, T.: A code generator framework for Isabelle/HOL. Technical Report 364/07, Department of Computer Science, University of Kaiserslautern (August 2007)

[5] Huffman, B., Matthews, J., White, P.: Axiomatic constructor classes in Isabelle/HOLCF. In: Hurd, J., Melham, T. (eds.) TPHOLs 2005. LNCS, vol. 3603, pp. 147–162. Springer, Heidelberg (2005)

[6] Jones, S.P., Launchbury, J.: Lazy functional state threads. In: SIGPLAN Conference on Programming Language Design and Implementation, pp. 24–35 (1994)

[7] Klein, G., Nipkow, T.: A machine-checked model for a Java-like language, virtual machine and compiler. ACM Trans. Program. Lang. Syst. 28(4), 619–695 (2006)

[8] Krstić, S., Matthews, J.: Verifying BDD algorithms through monadic interpretation. In: Cortesi, A. (ed.) VMCAI 2002. LNCS, vol. 2294, pp. 182–195. Springer, Heidelberg (2002)

[9] Müller, O., Nipkow, T., Oheimb, D.V., Slotosch, O.: HOLCF = HOL + LCF. Journal of Functional Programming 9, 191–223 (1999)

[10] Nanevski, A., Morrisett, G., Birkedal, L.: Polymorphism and separation in hoare type theory. In: ICFP 2006: Proceedings of the eleventh ACM SIGPLAN international conference on Functional programming, pp. 62–73. ACM Press, New York (2006)

[11] Nipkow, T., Paulson, L.C., Wenzel, M.T.: Isabelle/HOL. LNCS, vol. 2283. Springer, Heidelberg (2002)

[12] Obua, S.: Partizan games in Isabelle/HOLZF. In: Barkaoui, K., Cavalcanti, A., Cerone, A. (eds.) ICTAC 2006. LNCS, vol. 4281, pp. 272–286. Springer, Heidelberg (2006)

[13] Jones, S.P., Wadler, P.: Imperative functional programming. In: Proc. 20th ACM SIGPLAN-SIGACT Symposium on Principles of Programming Languages (POPL 1993), pp. 71–84 (1993)

[14] Schirmer, N.: A verification environment for sequential imperative programs in Isabelle/HOL. In: Baader, F., Voronkov, A. (eds.) Logic for Programming, Artificial Intelligence, and Reasoning, vol. 3452, pp. 398–414. Springer, Heidelberg (2005)

[15] Sprenger, C., Basin, D.A.: A monad-based modeling and verification toolbox with application to security protocols. In: Schneider, K., Brandt, J. (eds.) TPHOLs 2007. LNCS, vol. 4732, pp. 302–318. Springer, Heidelberg (2007)

[16] Tuch, H., Klein, G., Norrish, M.: Types, bytes, and separation logic. In: Hofmann, M., Felleisen, M. (eds.) Proc. 34th ACM SIGPLAN-SIGACT Symposium on Principles of Programming Languages (POPL 2007), Nice, France, January 2007, pp. 97–108 (2007)

[17] Weber, T., Amjad, H.: Efficiently checking propositional refutations in HOL theorem provers. Journal of Applied Logic (to appear, 2007)

HOL-Boogie — An Interactive Prover for the Boogie Program-Verifier[*]

Sascha Böhme[1], K. Rustan M. Leino[2], and Burkhart Wolff[3]

[1] Technische Universität München
boehmes@in.tum.de
[2] Microsoft Research, Redmond
leino@microsoft.com
[3] Universität Saarbrücken
bwolff@wjpserver.cs.uni-sb.de

Abstract. *Boogie* is a program verification condition generator for an imperative core language. It has front-ends for the programming languages C# and C enriched by annotations in first-order logic.

Its verification conditions — constructed via a *wp* calculus from these annotations — are usually transferred to automated theorem provers such as *Simplify* or *Z3*. In this paper, however, we present a proof-environment, HOL-Boogie, that combines Boogie with the interactive theorem prover Isabelle/HOL. In particular, we present specific techniques combining automated and interactive proof methods for code-verification.

We will exploit our proof-environment in two ways: First, we present scenarios to "debug" annotations (in particular: invariants) by interactive proofs. Second, we use our environment also to verify "background theories", i.e. theories for data-types used in annotations as well as memory and machine models underlying the verification method for C.

1 Introduction

Verifying properties of programs at their source code level has attracted substantial interest recently. While not too long ago, "real programming languages" like Java or C have been considered as too complex to be tackled formally, there are meanwhile verification systems like ESC/Java [11], Why/Krakatoa/-Caduceus [10], and Boogie used both for Spec# [3,1] and C [21]. The latter system is also used in a substantial verification effort for the Microsoft Hypervisor as part of the Verisoft XT project ([5], see also http://www.verisoft.de/).

Combining Boogie with an interactive prover has a number of incentives:

- verification attempts can be debugged by interactive proofs,
- background theories can be proven consistent,
- existing front-end compilers for Spec# and C to the Boogie-Core-Language represent an alternative to a logical embedding of these languages.

Debugging Verification Attempts. Starting to annotate a given program can lead to situations where the automated prover fails and can neither find a

---

[*] Supported by BMBF under grant 01IS07008.

O. Ait Mohamed, C. Muñoz, and S. Tahar (Eds.): TPHOLs 2008, LNCS 5170, pp. 150–166, 2008.
© Springer-Verlag Berlin Heidelberg 2008

proof nor a counterexample. All existing systems report of a degree of automation approaching 100%, causing wide-spread and understandable enthusiasm. However, there is also a slight tendency to overlook that the remaining few percent are usually the critical ones, related to the underlying theory of the algorithm rather than implementation issues like memory and sharing. Moreover, these figures tend to hide the substantial effort that may have been spent to end up with a formulation that can be finally proven automatically; there is even some empirical evidence that in the difficult cases, the labor to massage the specification can be comparable to the effort of an interactive proof [4].

The reason for a prover failure might be:
- specification-related (i.e., annotations and "background theories" (see below) are inconsistent, incomplete, or specify unintended behavior),
- program-related, e. g. a program simply does not behave as intended, or
- it can be a problem of the prover, by just using a wrong heuristics for the concrete task, or even by bad luck (e.g., Z3 [6] uses random-based heuristics).

An interactive proof, suitably adapted to the problems arising from automated formula generation, decomposes the verification conditions along the program structure and finally the logical structure of the annotations and can thus lead to insufficient preconditions or invariants systematically.

Consistency of Background Theories. Conceptually, the Boogie-Core-Language (called *BoogiePL*) allows for specifying transition systems; these transitions are described in terms of first-order logic over a model comprising user-defined types, constants, axioms as well as program variables. The signature and axiom set is called the "background theory" of a program. A background theory can be just program specific or programming-language specific. In the case of the *Verified C Compiler* (VCC, e.g., Boogie with C front-end, the configuration of Boogie mostly studied in this paper), the operations to be axiomatized consist of elementary operations of a machine model (called *C Virtual Machine* (CVM)), allowing reading and storing byte-wise, word-wise, and double-word-wise, doing signed and unsigned operations in bitvector arithmetic, etc. This machine model presents a (slight) abstraction over an x86 processor architecture, taking into account the processor intricacies of little-endianness, bit-padding, etc., but abstracting from registers and jumps (which are represented by **goto**'s in BoogiePL). A crucial part of the model is concerned with the representation of memory, memory regions etc. in order to formulate frame conditions. VCC compiles an ANSI-C program into a BoogiePL program based on the CVM.

Getting an axiomatization of this size consistent is a non-trivial task, and for several automated and interactive provers to work together, one has to make sure that all provers agree on this axiomatization.

An Alternative to Embeddings. Compiling ANSI-C to a transition system described in the fairly small and logically clean BoogiePL represents an alternative to a *logical embedding* into HOL (such as, for example, [12,20] describing a small-step transition semantics for the C fragment C0 comprising only side-effect free expressions). While we still consider the logical embedding method as "near perfection" with respect to logical foundations, it is obvious that an embedding of a more substantial C fragment is an enormous effort with questionable value. Given that ANSI-C language semantics is heavily under-specified, given that optimizing compilers tend to make their own story over the variety of "semantic deviation points", and given that a realistic concurrency model depends on the granularity of the atomic operations defined by the target assembler language, it is debatable if we should care about the sterile myth of some general C semantics or rather concentrate our efforts on a specific compiler and target assembler language. For a given compiler, one can exchange the code-generator against a BoogiePL translator, and validate compiled assembler sequences against the abstract model traces in the CVM by tests (what has been done intensively) or by simulation proofs if needed. There is meanwhile sufficient empirical evidence that a carefully constructed and tested C front-end to a verification condition generator such as Boogie can achieve a reasonable degree of trustworthiness.

Outline of the Paper. We will address the first two issues. After presenting the background of this work, namely Isabelle/HOL, Boogie, and the HOL-Boogie architecture, we present three scenarios of using HOL-Boogie and will explain the underlying machinery at need: In the first scenario, we use HOL-Boogie to verify Dijkstra's Shortest Path Algorithm given as BoogiePL program (only a high-level memory model involved). In the second scenario, we verify a C program, converted into BoogiePL, i.e. a program over the CVM. In the third scenario, we show how CVM axiomatizations can be proven consistent with HOL-Boogie, enabling a "safe mode" of C program verification.

2   Background

2.1   Isabelle/HOL and the Isar Framework

Isabelle is a generic theorem prover [17], i.e. new object logics can be introduced by specifying their syntax and inference rules. Isabelle/HOL is an instance of Isabelle with Church's higher-order logic (HOL), a classical logic with equality. Substantial libraries for sets, lists, maps, have been developed for Isabelle/HOL, based on definitional techniques, allowing the use of Isabelle/HOL as a "functional language with quantifiers".

Isabelle is based on the so-called "LCF-style architecture" which allows one to extend a small trusted logical kernel by user-programmed procedures in a logically safe way. Moreover, on top of the kernel, there is a generic system framework Isabelle/Isar [22] that can be compared in a rough analogy to the Eclipse programming system framework. It provides (1) a hierarchical organization of theory documents, (2) incremental document processing for interactive theory and proof development (with unlimited undo) and an Emacs-based GUI, and (3) extensible syntax for top-level commands, embedded methods and attributes, and the inner term language. HOL-Boogie is yet another instance of the Isabelle/Isar framework. It comes with a loader of the verification conditions generated by Boogie, a proof-obligation management and specific tactic support

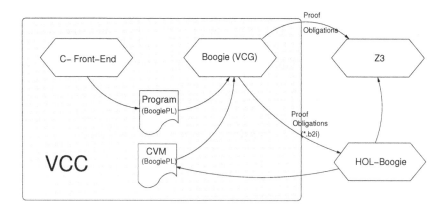

Fig. 1. VCC and the HOL-Boogie back-end

for the formulas arising in this scenario as well as interactions with external provers such as Z3 which have been integrated via the Isabelle oracle mechanism.

2.2 The VCC System Architecture

The *Verified C Compiler*(VCC) evolved from the Spec# project (see `http://research.microsoft.com/specsharp/`, [3]). It comprises a C front-end supporting ANSI-C and — geared towards verification of programs close to the hardware-level — bitwise representation of e. g. integers, structs, and unions in memory. The core component of VCC is Boogie, a verification condition generator. Its input language BoogiePL provides constants and functions and first-order axioms, as well as a small imperative language with assignments, first-order assertions, unstructured **goto** and structured control constructs (**if**, **while**, **break**). From these annotated imperative programs, Boogie computes (optimized) verification conditions over the program and the axiomatization of a *background theory*. In the case of VCC, an abstract machine model is given in the background theory, describing linear memory (a map from references to bitvectors), allocation operations, little-endian word-wise load- and store operations, and a family of word-wise operations abstracting the x64 processor architecture.

Boogie also provides a framework into which converters to external prover formats may be "plugged in". Our HOL-Boogie integration is based on such a plug-in that we implemented for interactive back-ends. We also coupled HOL-Boogie to the default target prover *Z3* in such a way that formulas constructed in the former can be discharged by the latter.

3   Foundations of Boogie and HOL-Boogie

3.1   Introduction to BoogiePL

BoogiePL is a many-sorted logical specification language extended by an imperative language with variables, contracts, and procedures.

The type system of BoogiePL has several built-in as well as user-defined types. The former cover basic types like **bool** and **int**, as well as one- and two-dimensional arrays which can be indexed by any valid type.

BoogiePL includes the following kinds of top-level declarations:

– user-defined types:

```
type Vertex;
```

– symbolic *constants* having a fixed but possibly unknown value:

```
const Infinity: int;
```

– *uninterpreted functions*:

```
function Distance(from: Vertex, to: Vertex) returns (result: int);
```

– *axioms* constraining symbolic constants and functions:

```
axiom 0 < Infinity;
```

– global variables:

```
var ShortestPath: [Vertex] int;
```

– *procedure contracts*, i.e. signatures with pre- and postconditions, and
– *implementations* of procedures.

An implementation begins with local-variable declarations which are followed by a sequence of basic blocks. We will only consider the latter in more detail here, and we omit the structured control structures, which can be desugared into the statements and **goto**'s shown here. Each basic block has a name, a body, and a possibly empty set of successors. Expressions are first-order logic formulas with equality and integer operations.

Semantically, each block corresponds to a transition relation over the variables of a program; **goto** statements correspond to a composition with the intersection of the successor transition relations, loops to fixpoints: Boogie represents a partial correctness framework. The basic assertion **assert** constrains the subsequent transition, while **assume** weakens it. Pragmatically, **assert** produces obligations for the programmer, while **assume** leaves him "off-the-hook", see, e.g., [15,16].

$$
\begin{array}{rcl}
\textit{BlockSeq} & ::= & \textit{Block}^+ \\
\textit{Block} & ::= & \textit{BlockId} : [\ \textit{Statement}\ ;] \ \textit{Goto}\ ; \\
\textit{Statement} & ::= & \textit{Var} := \textit{Expression} \\
& | & \textbf{havoc}\ \textit{VarId} \\
& | & \textbf{assert}\ \textit{Expression} \\
& | & \textbf{assume}\ \textit{Expression} \\
& | & \textbf{call}\ [\ \textit{Var}^+ :=]\ \textit{ProcId}\ (\ \textit{Expression}^*\ ) \\
& | & \textit{Statement}\ ;\ \textit{Statement} \\
\textit{Goto} & ::= & \textbf{goto}\ \textit{BlockId}^*\ |\ \textbf{return}
\end{array}
$$

Fig. 2. Schematic syntax of blocks in BoogiePL

An assignment statement $x := E$ updates the program state by setting the variable x to the value of the expression E. The statement **havoc** x sets the the variable x to an arbitrary value. The statement S ; T corresponds to the relation composition. The procedure call statement, i.e. **call**, is just a short-hand for suitable **assert**, **havoc** and **assume** statements, encoding the callee's pre- and postconditions [14]. The **return** command is a short-hand for the procedure's postconditions and a **goto** with no successors.

BoogiePL also comes with a structured syntax with which one can express loops (**while**) and branches (**if**) directly. These can be defined as a notation for certain basic blocks; for example, the following schematic **while** loop:

```
while (G) invariant P; { B }
```

is encoded by the following basic blocks [1]:

$$LoopHead : \textbf{assert } P \text{ ; } \textbf{goto } LoopBody, LoopDone;$$
$$LoopBody : \textbf{assume } G \text{ ; } B \text{ ; } \textbf{goto } LoopHead;$$
$$LoopDone : \textbf{assume } \neg G \text{ ; } \ldots$$

More details of BoogiePL can be found in [1,7].

3.2   Generating Verification Conditions

Verification condition generation proceeds in the following steps: First, the expansion of syntactic sugar and (safe) cutting of loops result in an acyclic control-flow graph. Second, a single-assignment transformation is applied. Third, the result is turned into a passive program by changing assignment statements into **assume** statements. Finally, a verification condition of the unstructured, acyclic, passive procedure is generated by means of weakest preconditions.

We will present only the final step here, the reader interested in the first three is referred to [2]. Each basic block in a preprocessed program consists only of a sequence of **assert** and **assume** statements, followed by a final **goto** command.

For any statement S and predicate Q on the post-state of S, the weakest precondition of S with respect to Q, written $wp(S, Q)$, is a predicate that characterizes all pre-states of S whose reachable successor states satisfy Q. The computation of weakest preconditions follows the following well-known rules:

$$wp(\textbf{assert } P, \ Q) \ = \ P \wedge Q$$
$$wp(\textbf{assume } P, \ Q) \ = \ P \implies Q$$
$$wp(S \text{ ; } T, \ Q) \ = \ wp(S, wp(T, Q))$$

For every block

$$A : S \text{ ; } \textbf{goto } B_1, \ldots, B_n;$$

an auxiliary variable $A_{correct}$ is introduced, the intuition being that $A_{correct}$ is true if the program is in a state from which all executions beginning from block A are correct. Formally, there is the following block equation:

$$A_{correct} \quad \equiv \quad wp(S, \bigwedge_{B \in \{B_1,...,B_n\}} B_{correct})$$

Each block contributes one block equation, and from their conjunction, call it R, the procedure's verification condition is

$$R \implies Start_{correct}$$

where *Start* is the name of the first block of the procedure. Note that the verification condition generated this way is linear in the size of the procedure.

3.3   Labeling in Boogie

Boogie is able to output source code locations of errors and also execution traces leading to these errors. The underlying basic idea is to enrich formulas by labels, i.e. uninterpreted predicate symbols intended to occur in counterexamples of verification conditions. In verification conditions generated by Boogie, labels are either positive (**lblpos** L: P) or negative (**lblneg** L: P). Logically, these formulas are equivalent to P; the labels occur in counterexamples if P has the indicated sense (i.e. P or $\neg P$). Their formal definition is as follows:

$$(\textbf{lblneg } L\colon P) \ = \ P \vee L$$
$$(\textbf{lblpos } L\colon P) \ = \ P \wedge \neg L$$

Negative labels tag formulas of assertions (including invariants and postconditions) with their location in the source program. If an assertion cannot be proven, the accompanying label allows Boogie to emit an error location identifying which program check failed. Positive labels tag the beginning of a block by an additional assertion which is always true. This way, execution traces contain information reflecting the order in which basic blocks were processed. If execution terminates in an error, the positive labels represent an error trace.

A more detailed description of this use of labels is found in [13].

3.4   Attribution in BoogiePL

We implemented a new feature in Boogie: The top-level declarations for types, constants, functions, axioms, and global variables, can be tagged by *attributes*; previously, Boogie allowed such attributes only on quantifier expressions. For example, an attributed axiom looks as follows: **axiom** {attr1} ... {attrN} P. These attributes are opaque for Boogie; they may carry information for external provers and may influence Boogie's back-ends. In the case of Z3, for example, attributes are used to tag some axioms as built-in to Z3.

The attribution mechanism provided by Boogie is flexible enough to add new attributes for any prover back-end.

4   Scenario I: Interactive Verification of Algorithms

4.1   Dijkstra's Shortest Path Algorithm

Widely known and yet fairly complex, this algorithm already poses a reasonable challenge for verification efforts. The following code, written by Itay Neeman,

```
type Vertex;
const Graph: [Vertex, Vertex] int;
axiom (∀ x: Vertex, y: Vertex :: x ≠ y ⟹ 0 < Graph[x,y]);
axiom (∀ x: Vertex, y: Vertex :: x == y ⟹ Graph[x,y] == 0);

const Infinity: int;
axiom 0 < Infinity;

const Source: Vertex;
var SP: [Vertex] int; // shortest paths from Source

procedure Dijkstra();
  modifies SP;
  ensures SP[Source] == 0;
  ensures (∀ z: Vertex, y: Vertex ::
    SP[y] < Infinity ∧ Graph[y,z] < Infinity ⟹ SP[z] ≤ SP[y] + Graph[y,z]);
  ensures (∀ z: Vertex :: z ≠ Source ∧ SP[z] < Infinity ⟹
    (∃ y: Vertex :: y ≠ z ∧ SP[z] == SP[y] + Graph[y,z]));

implementation Dijkstra()
{
  var v: Vertex;
  var Visited: [Vertex] bool;
  var oldSP: [Vertex] int;

  havoc SP;
  assume (∀ x: Vertex :: x == Source ⟹ SP[x] == 0);
  assume (∀ x: Vertex :: x ≠ Source ⟹ SP[x] == Infinity);

  havoc Visited;
  assume (∀ x: Vertex :: ¬Visited[x]);

  while ((∃ x: Vertex :: ¬Visited[x] ∧ SP[x] < Infinity))
    invariant SP[Source] == 0;
    invariant (∀ y: Vertex, z: Vertex ::
      ¬Visited[z] ∧ Visited[y] ⟹ SP[y] ≤ SP[z]);
    invariant (∀ z: Vertex, y: Vertex ::
      Visited[y] ∧ Graph[y,z] < Infinity ⟹ SP[z] ≤ SP[y] + Graph[y,z]);
    invariant (∀ z: Vertex :: z ≠ Source ∧ SP[z] < Infinity ⟹
      (∃ y: Vertex :: y ≠ z ∧ Visited[y] ∧ SP[z] == SP[y] + Graph[y,z]));
  {
    havoc v;
    assume ¬Visited[v];
    assume SP[v] < Infinity;
    assume (∀ x: Vertex :: ¬Visited[x] ⟹ SP[v] ≤ SP[x]);
    Visited[v] := true;
    oldSP := SP;
    havoc SP;
    assume (∀ u: Vertex ::
      Graph[v,u] < Infinity ∧ oldSP[v] + Graph[v,u] < oldSP[u] ⟹
        SP[u] == oldSP[v] + Graph[v,u]);
    assume (∀ u: Vertex ::
      ¬(Graph[v,u] < Infinity ∧ oldSP[v] + Graph[v,u] < oldSP[u]) ⟹
        SP[u] == oldSP[u]);
  }
}
```

presents a high-level implementation of Dijkstra's algorithm, abstracting from any memory model and even shortening several initializations and assignments by logical expressions.

While developing algorithms and their specifications like the one given here, it commonly happens that, even if a program behaves as intended, its specification is incomplete or inconsistent. Indeed, when letting Boogie check the given program, it reports the following error message:

```
Spec# Program Verifier Version 0.88, Copyright (c) 2003-2007, Microsoft.
dijkstra.bpl(34,5): Error BP5005: This loop invariant might not be
   maintained by the loop.
Execution trace:
   dijkstra.bpl(26,3): anon0
   dijkstra.bpl(33,3): anon2_LoopHead
   dijkstra.bpl(42,5): anon2_LoopBody
Spec# Program Verifier finished with 0 verified, 1 error
```

Using HOL-Boogie we can navigate to the cause for this error and inspect it. The underlying techniques, described later in more detail, split the verification condition into altogether 11 subgoals and pass each of them to Z3, which can discharge all of them except one. The remaining subgoal, without its premises, reads as follows in HOL-Boogie :

$$\textit{assert-at Line-34-Column-5 (SP-2 [Source] = 0)}$$

This formula corresponds to a negatively labeled formula in the verification condition generated by Boogie. Note that *SP-2* is an inflection of the program variable SP holding the computed shortest paths after arbitrary runs of the **while** loop.

The subgoal found by HOL-Boogie is exactly the cause of the error reported by Boogie, as the position label indicates. The associated premises represent the complete execution trace until the point where the above invariant is checked. Among those premises, only two express properties of *SP-2*, while a third one states something similar to the subgoal above:

$$\bigwedge u. \; G[v\text{-}1,\, u] < \textit{Infinity} \wedge \textit{SP-1}[v\text{-}1] + G[v\text{-}1,\, u] < \textit{SP-1}[u]$$
$$\implies \textit{SP-2}[u] = \textit{SP-1}[v\text{-}1] + G[v\text{-}1,\, u]$$

$$\bigwedge u. \; \neg(G[v\text{-}1,\, u] < \textit{Infinity} \wedge \textit{SP-1}[v\text{-}1] + G[v\text{-}1,\, u] < \textit{SP-1}[u])$$
$$\implies \textit{SP-2}[u] = \textit{SP-1}[u]$$

$$\textit{SP-1}[Source] = 0$$

Based on those three properties, we attempt to prove the subgoal. Consider the following Isar extract:

proof (*ib-split try-z3*)
  case *goal1*
    note *H1* = ⟨⋀u. G[v-1, u] < Infinity ∧ SP-1[v-1] + G[v-1, u] < SP-1[u]
            ⟹ SP-2[u] = SP-1[v-1] + G[v-1, u]⟩
    note *H2* = ⟨⋀u. ¬(G[v-1, u] < Infinity ∧ SP-1[v-1] + G[v-1, u] < SP-1[u])
            ⟹ SP-2[u] = SP-1[u]⟩
    note *H3* = ⟨SP-1[Source] = 0⟩

show *?case*
proof *ib-assert*
  show *SP-2[Source] = 0*
  proof (*cases*
              G[v-1, Source] < Infinity ∧
              SP-1[v-1] + G[v-1, Source] < SP-1[Source])
    case *True*
    moreover with *H3* **have** *SP-1[v-1] + G[v-1, Source] < 0* **by** *simp*
    ultimately have *SP-2[Source] < 0* **using** *H1* **by** *simp*
oops

Here, it becomes obvious what exactly caused the error in Boogie/Z3 before. Besides the contradiction in the proof attempt, computed shortest paths are always non-negative in Dijkstra's algorithm. From this observation, we can infer an additional invariant for the **while** loop of the implementation:

```
invariant (∀ x: Vertex :: SP[x] >= 0);
```

This addition suffices to correct the specification; the program can now be verified automatically by Boogie and Z3.

4.2   Tracking Program Positions

Relating formulas to locations in the original program is one of the key aspects of HOL-Boogie; this feature results from exploiting the labeling mechanism of Boogie. Since assertions, subsuming also invariants and postconditions, form the crucial parts of verification conditions, they are tagged by labels holding their program position. After producing a verification condition and loading it in HOL-Boogie, the labels then occur at the formulas to be proven, in the way shown along the example of Dijkstra's algorithm before.

4.3   Specific Tactic Support

HOL-Boogie comes with a set of specific tactics to manipulate verification conditions. They allow the user to navigate to assertions, to prune some of them by applying Z3, and to restrict the list of premises associated with assertions. Some of these tactics are already shown in the verification of Dijkstra's algorithm.

Based on the structure of verification conditions generated by Boogie, the central tactic of HOL-Boogie, *ib-split*, extracts all assertions and associates them with their execution trace, expressed as a list of premises. Each assertion then forms a subgoal for the proof of the original verification condition.

After splitting a verification condition, each subgoal is passed to Z3 if the argument *try-z3* is given to the tactic *ib-split*. This essentially gives the "debugging flavor" to HOL-Boogie, since Z3 usually discharges all subgoals except those that are incorrect or inconsistent. The method is based on an oracle calling Z3; the communication uses the SMT-LIB format [19]. Due to this standardized format, it is possible to replace Z3 with other SMT solvers, or combine them for better results.

The list of premises of a subgoal can be pruned by the tactic *ib-filter-prems*. It selects all premises potentially necessary to solve the current subgoal, while cutting off all other premises. Note, however, that this tactic, due to its heuristics, may remove too many premises. Therefore, its purpose is only to assist in finding a draft of a proof for a subgoal, especially in the case of a long list of premises.

Finally, the tactic *ib-assert* serves to unwrap a formula of an assertion by cutting off the label.

4.4   Structured Proofs and Isabelle Proof Support

Without using Z3 from inside HOL-Boogie, many subgoals of a verification condition can already by proven by tools included in Isabelle. In simple academic experiments, the built-in simplifier is already able to solve some subgoals. A more substantial help, however, comes from *sledgehammer*. When applied to a subgoal, it uses external first-order provers to identify necessary facts which are then combined into a proof, usually by passing the facts to a resolution-based built-in prover. Since the amount of facts given as axioms in BoogiePL as well as the number of premises for an assertion can easily grow to an unmanageable size, sledgehammer is of an invaluable help. Usually, around 50% of all subgoals generated from a verification condition can be shown by this method.

5   Scenario II: Interactive Verification of C-Programs

Verifying C programs in HOL-Boogie seems to be a straightforward extension to the previous section. The C front-end of VCC compiles a C program like the following example (computing a maximal unsigned byte for an array whose size is bounded by 2^{40}):

```
#include "vcc.h"
. . .
static UINT8 maximum(__inout_ecount(len) UINT8 arr[], UINT64 len)
  requires(0 < len ∧ len < (1UI64 << 40))
  ensures (∀(UINT64 i; i<len ⟹ arr[i]≤result))
{
  UINT8 max;
  UINT64 p;

  max = 0;
  for (p = 0; p < len; p++)
    invariant(p ≤ len)
    invariant(∀(UINT64 i; i < p ⟹ arr[i] ≤ max))
  {
    if (arr[p] > max) { max = arr[p]; }
  }
  assert(p == len);
  return max;
}
```

into a BoogiePL-program. This BoogiePL program is significantly larger (about 2400 lines), since it contains the axiomatization of the CVM. In order to give an impression of its abstraction level, we show some code resulting from the **invariant**'s:

```
invariant $cle.u8(p, len);
invariant (∀ i: bv64 :: $_inrange.u8(i) ⟹ $clt.u8(i, p) ⟹
  $cle.u4($ld.u1($mem, $add.ptr(arr, i, 1bv64)), max));
free invariant $only_region_changed_or_new(
  old($region(arr, $mul.u8(len, 1bv64))),
  old($gmem), $gmem, old($mem), $mem);
invariant $alloc_grows(#temp10, $gmem);
```

The primitives of the CVM provide operations for:

1. **dereference, load and store in memory**: $ld.u1, $ld.u2, $ld.u4, $ld.u8,
   $st.u1, $st.u2, $st.u4, $st.u8,... The index indicates the length of the bitvector
   in bytes. These operations take the padding conventions of the little-endian
   x86 architecture into account.
2. **bitvector computations**: e.g. cle.u8, clt.u8, mul.u8, etc, ...
3. **pointer arithmetic**: e.g. $add.ptr, $sub.ptr, $base, $offset, ...
4. **memory regions (= pointer sets)**: e.g. $region, $contains, $overlap, ...
5. **memory operations**: e.g. malloc, free, memcopy, ...
6. **framing conditions**: $only_region_changed_or_new(X, mem, mem') expresses
   that memory mem in the state and memory mem' in its successor state remain
   unchanged for all pointers not in X, ...
7. **typed ghost memory**: $gmem and its infrastructure.

Ghost memory is a separate memory, which is updated in a way that does
not affect the program control flow, where syntactic restrictions guarantee that
information never flows from ghost states to concrete program states. Thus, ghost
state and any code using it can be eliminated when the program is compiled. It
is used in particular to specify the concept of a $valid reference or the $size of an
array into which all references are $valid memory. Conceptually, it is a map from
references to records with arbitrarily many fields with possibly different types.

When compiling the axioms referring to ghost state, a problem arises: while
the typing discipline of BoogiePL is simply many-sorted first-order in most cases,
there is a non-standard built-in type construct <x>T (used here for a type name
that stands for *field names*) that requires special treatment. There are several
axioms that quantify over ghost memory which has the BoogiePL array type
[$gid,<x>name]x. We interpret this type by functions of type $gid ⟹ α$ *name* $⟹ α$
(where *gid* is the type of ghost references for which an injection from standard
memory references exist). For each ghost field, such as $size, the axiomatization
also defines a *field tag constant*:

```
const unique $_size : <bv64>name;
```

which we convert into a constant declaration $\$\_size :: bv64$ *name*. Thus, so far,
this concept can be safely embedded into Hindley-Milner style polymorphism.
However, there are axioms with quantifications over *name* (intended to mean:
"over all fields") such as in:

```
axiom(∀
 r:$region ,n:name,oldgmem:[$gid,<x>name]x,newgmem:[$gid,<x>name]x,
        oldmem:  $memory,  newmem:$memory  ::
           . . . n . . .
```

We interpret a BoogiePL axiom of this form as an axiom scheme and create for each field tag constant an instance for it.

The compiled BoogiePL code of the above C program can still be loaded within 36 seconds (on standard hardware) into HOL-Boogie. Its proof is fairly straightforward but profits substantially from the tactic firing Z3.

6   Scenario III: Verification of Background Theories

At present, the axiomatization of the CVM —called Prelude Version 7.0— consists of about 750 axioms (where a certain number of axioms were not made explicit since they are "built-in" into the target prover; for example, reflexivity of equality or the laws of arithmetic). There had been a number of errors in the current and similar formalizations of background theories; and consistency is even a greater issue if Boogie is used with *different* memory/machine models. Since the abstraction level of a machine model is tantamount for deduction efficiency, more refined models should be used only when inherently needed. This is the case if, for example, the allocation function itself must be verified, which is atomic in a more abstract model, or when inherently untyped memory is required such as in unions, where everything is translated into bitvectors [5].

From the perspective of a HOL system, proving the consistency of a complex first-order system is not exactly an easy task, but at least routine: Just build up a theory by conservative extensions, i.e. constant or type definitions, and derive all the "axioms" from it. In the sequel, we report on a verification of a previous version of the CVM model (Prelude Version 3.0). Since the CVM model is rapidly changing at present, we plan to repeat this effort at a later stage.

The conservative theory for Prelude 3.0 is constructed as follows: First, a simple bitvector library is built; bitvectors were represented as lists of boolean, and operations like *length*, *extract*, and *concat* were defined as usual. Since the CVM operations work only in byte and word formats, the necessary side-conditions referring to *length* can be omitted if these formats were already expressed at the type level, for example:

typedef $bv32 = \{x :: bool\ list.\ length\ x = 32\ \}$

Arithmetic operations for signed and unsigned integers were defined over $bv32$, as well as bitwise conjunction or disjunction. For example, consider the definition:

constdefs $shr\text{-}i4 :: [bv32,\ bv32] \Rightarrow bv32$
$shr\text{-}i4\ v\ w \equiv Abs\text{-}bv32\ (bv\text{-}shr\ (Rep\text{-}bv32\ v)\ (Rep\text{-}bv32\ w))$

where $bv\text{-}shr$ (omitted here) is defined on bitvectors directly representing the usual intuition "division by two". Moreover, following the conventions on signs of the x86 architecture, it is enforced that the most significant bit is replicated and the size of the bitvector remains identical.

Similarly, the type of pointers *ptr* is introduced as a pair of unsigned 64 bit integers (references called *ref*) and an integer; the former is called the *base* and the latter the *offset*. On *ptr*'s, pointer arithmetic operations are defined allowing byte-wise addressing of memory. The core of the memory model is:

> **typedef** *memory* = {*x* :: *ref* ⇒ *Bitvector*. *True*}
> **types** *state* = (*ref* ⇒ *bool*) × *memory*

The pivotal concept of a valid reference, for example, is defined as:

> **constdefs** *valid* :: [*state*, *ptr*, *int*] ⇒ *bool*
> *valid σ p l* ≡ (*fst σ*) (*base p*) ∧ *0* ≤ (*offset p*) + *1* ∧
> *offset p * 8* < *length* (*lkup* (*snd σ*) (*base p*))

Definitions for *malloc* and *free* are straightforward.

We implemented a little compiler that takes a *Boogie-Configuration* — i.e. a list of theorems, their names, and attributes — and compiles this information into a BoogiePL background-theory file. In particular, attributes are generated that correspond to Isabelle's internal naming in the theory, for example:

```
axiom {:isabelle "id_prelude.basics_axms_1"}(∀ x:int:: exp(x, 0) == (1));
```

Since Boogie re-feeds attributes to its target provers, HOL-Boogie can check that every axiom in the background theory of a verification condition indeed exists (and by construction is derived) in its own logical environment; thus, a *strict checking* mode can be implemented that makes sure that a verification in an external prover is based only on a consistent axiomatization.

7   Conclusion

We have presented a novel HOL-based proof environment, called HOL-Boogie, that is integrated into a verification tool chain for imperative programs, in particular C (and C#; not supported yet). Key issues of the integration are:

1. the support of labels and positions at the proof level, which enables tracking back missing properties to assertions in the source,
2. specific tactic support for decomposition of verification conditions in a way stable under certain changes of the source,
3. a mechanism to generate background theories from consistent, conservative models in HOL,
4. the integration of the target prover in order to discharge as many subgoals as possible, and
5. mechanisms to track attributes in order to exchange meta-information between tools.

7.1 Related Work

As such, combining an interactive prover with a Boogie-like VCG is not a new idea. In Figure 1, just replace C-Front-End by *Caduceus* [9], Boogie by *Why* [8], and Z3 by the default prover ERGO, and one gets (nearly) the architecture of the Why/Caduceus system [10]. However, its interactive prover-back-end cannot be used to decompose verification conditions and send the "splinters" to the target prover (ERGO is not integrated into Coq), it offers no mechanism for tracking back unsatisfiable subgoals to the source, and it offers little specific tactic support for this application scenario. With respect to the C front-end and the underlying CVM, VCC is more detailed since it leverages features such as byte-wise access into unions. Moreover, support for fine-grain concurrency is actively under development [5].

There is a quite substantial body on programming language embeddings into HOL, be it *shallow* [20,18] or *deep* [12]. In particular, Leinenbach [12] provides a small-step semantics for a language C0, which has been used for system level verification, and Schirmer [20] derives a (shallow-ish) Hoare-Logic from this semantics and formally developed a verification condition generator. C0 assumes a typed memory model (although bitvectors and conversions to standard data-types could easily be integrated). However, the size of the supported language fragment, many complications in the semantic representation, and the degree of automatic proof support have limited its use in case studies substantially. In contrast to VCC and the Hypervisor Verification Project [5], the idea is to adapt the code to be verified instead of trying to live with the existing code and adapt the tool chain.

7.2 Future Work

We see the following directions for future work:

1. **More Stable Proof Formats.** In our scenario, where the specification of a program is essentially re-constructed post-hoc, it is the annotations that change constantly under development. This means that positions of assertions change easily, which can (but must not) have influence on proofs resulting from previous proof attempts. A proof style using control-flow labels (as generated by Boogie) would be more stable under changes of the source in this scenario.
2. **Verified Current CVM Model.** The verified C background theory containing the memory and machine axiomatization is currently rapidly evolving; at a later stage, we would like to verify the consistency on a more recent model. From our experience, this is a substantial task (several man-months), but routine.
3. **More Automated Proof Support in Isabelle.** Currently, there is not enough automated proof support for bitvectors and for the logical reasoning required to discharge formulas related to memory-framing and updates in the C model.

References

1. Barnett, M., Chang, B.-Y.E., DeLine, R., Jacobs, B., Leino, K.R.M.: Boogie: A modular reusable verifier for object-oriented programs. In: de Boer, F.S., Bonsangue, M.M., Graf, S., de Roever, W.-P. (eds.) FMCO 2005. LNCS, vol. 4111, pp. 364–387. Springer, Heidelberg (2006)
2. Barnett, M., Leino, K.R.M.: Weakest-precondition of unstructured programs. In: PASTE 2005, pp. 82–87. ACM Press, New York (2005)
3. Barnett, M., Leino, K.R.M., Schulte, W.: The Spec# Programming System: An Overview. In: Barthe, G., Burdy, L., Huisman, M., Lanet, J.-L., Muntean, T. (eds.) CASSIS 2004. LNCS, vol. 3362, pp. 49–69. Springer, Heidelberg (2005)
4. Basin, D., Kuruma, H., Miyazaki, K., Takaragi, K., Wolff, B.: Verifying a signature architecture: A comparative case study. Formal Aspects of Computing 19(1), 63–91 (2007)
5. Cohen, E., Hillebrand, M., Leinenbach, D., der Rieden, T.I., Moskal, M., Paul, W., Santen, T., Schirmer, N., Schulte, W., Tobies, S., Wolff, B.: The Microsoft Hypervisor Verification Project (manuscript in preparation) (2008)
6. de Moura, L., Bjørner, N.: Z3: An efficient SMT solver. In: TACAS 2008. LNCS, vol. 4963, pp. 337–340. Springer, Heidelberg (2008)
7. DeLine, R., Leino, K.R.M.: BoogiePL: A typed procedural language for checking object-oriented programs. Tech. Rep. 2005-70, Microsoft Research (2005)
8. Filliâtre, J.-C.: Why: A multi-language multi-prover verification condition generator. Tech. Rep. 1366, LRI, Université Paris Sud (2003)
9. Filliâtre, J.-C., Marché, C.: Multi-prover verification of C programs. In: Davies, J., Schulte, W., Barnett, M. (eds.) ICFEM 2004. LNCS, vol. 3308, pp. 15–29. Springer, Heidelberg (2004)
10. Filliâtre, J.-C., Marché, C.: The Why/Krakatoa/Caduceus platform for deductive program verification. In: Damm, W., Hermanns, H. (eds.) CAV 2007. LNCS, vol. 4590, pp. 173–177. Springer, Heidelberg (2007)
11. Flanagan, C., Leino, K.R.M., Lillibridge, M., Nelson, G., Saxe, J.B., Stata, R.: Extended static checking for Java. In: PLDI 2002, pp. 234–245. ACM Press, New York (2002)
12. Leinenbach, D., Paul, W., Petrova, E.: Towards the formal verification of a C0 compiler: Code generation and implementation correctness. In: SEFM 2005, pp. 2–12. IEEE Computer Society Press, Los Alamitos (2005)
13. Leino, K.R.M., Millstein, T., Saxe, J.B.: Generating error traces from verification-condition counterexamples. Science of Computer Programming 55(1-3), 209–226 (2005)
14. Leino, K.R.M., Saxe, J.B., Stata, R.: Checking Java programs via guarded commands. In: FTfJP 1999, Tech. Rep. 251. Fernuniversität Hagen (1999)
15. Morgan, C.: The specification statement. ACM TOPLAS 10(3), 403–419 (1988)
16. Nelson, G.: A generalization of Dijkstra's calculus. ACM TOPLAS 11(4), 517–561 (1989)
17. Nipkow, T., Paulson, L.C., Wenzel, M.T.: Isabelle/HOL. LNCS, vol. 2283. Springer, Heidelberg (2002)
18. Norrish, M.: C formalised in HOL. Ph.D. thesis, Computer Laboratory, University of Cambridge (1998)
19. Ranise, S., Tinelli, C.: The SMT-LIB standard: Version 1.2. Tech. rep., Dept. of Comp. Sci., The University of Iowa (2006), http://www.smt-lib.org

20. Schirmer, N.: Verification of Sequential Imperative Programs in Isabelle/HOL. Ph.D. thesis, Technische Universität München (2006)
21. Schulte, W., Xia, S., Smans, J., Piessens, F.: A glimpse of a verifying C compiler (extended abstract). In: C/C++ Verification Workshop (2007)
22. Wenzel, M., Wolff, B.: Building Formal Method Tools in the Isabelle/Isar Framework. In: Schneider, K., Brandt, J. (eds.) TPHOLs 2007. LNCS, vol. 4732, pp. 351–366. Springer, Heidelberg (2007)

Secure Microkernels, State Monads and Scalable Refinement

David Cock[1], Gerwin Klein[1,2], and Thomas Sewell[1]

[1] Sydney Research Lab., NICTA[*], Australia
[2] School of Computer Science and Engineering, UNSW, Sydney, Australia
{david.cock,gerwin.klein,thomas.sewell}@nicta.com.au

Abstract. We present a scalable, practical Hoare Logic and refinement calculus for the nondeterministic state monad with exceptions and failure in Isabelle/HOL. The emphasis of this formalisation is on large-scale verification of imperative-style functional programs, rather than expressing monad calculi in full generality. We achieve scalability in two dimensions. The method scales to multiple team members working productively and largely independently on a single proof and also to large programs with large and complex properties.

We report on our experience in applying the techniques in an extensive (100,000 lines of proof) case study—the formal verification of an executable model of the seL4 operating system microkernel.

1   Introduction

This paper touches on three main topics: the verification of a secure operating system microkernel, the state monad as used in Haskell programs, and formal refinement as the verification technique in the correctness proof.

The main motivation for our work is the first of these three. In the larger context, we are aiming to design and fully formally verify the seL4 microkernel down to the level of its ARM11 C implementation. The seL4 microkernel [3,6] is an evolution of the L4 family [15] for secure, embedded devices. As described elsewhere [5], the design of seL4 involved building a binary compatible prototype of the kernel in the programming language Haskell which subsequently was automatically translated into Isabelle/HOL to arrive at a very detailed, executable formal model of the kernel. This operational model is inherently state based, and the corresponding Haskell program makes extensive use of the state monad to express the corresponding state transformations. The model is low level, using data types such as 32 bit wide finite machine words, modelling the heap memory of the eventual C program explicitly as part of its state, and mutating typical pointer data structures such as doubly linked lists on that heap.

Complementing this executable model is a still operational, but more abstract specification of the functional behaviour of seL4. This more abstract model uses

---

[*] NICTA is funded by the Australian Government as represented by the Department of Broadband, Communications and the Digital Economy and the Australian Research Council.

O. Ait Mohamed, C. Muñoz, and S. Tahar (Eds.): TPHOLs 2008, LNCS 5170, pp. 167–182, 2008.

nondeterminism to leave details unspecified and uses, for instance, abstract functions instead of explicit pointer representations (although it still makes use of *references* on many occasions, to model the user-visible sharing behaviour of particular data structures).

This paper presents the main techniques we used in verifying that the executable model correctly implements its abstract specification. It should be noted explicitly that we did not aim for maximum generality and theoretical depth in either the formalisations or the techniques. Instead, we focused on simplicity, easy applicability, and most importantly scalability of the methods. As a microkernel, seL4 is neither nicely modular, nor does it implement a nicely self contained abstract algorithm. Compared to other verifications, the main challenge was to deal with a highly complex, intermingled set of low-level data structures with high reliance on global invariants exploited in various optimisations. The size of the specifications, with about 3K lines of Isabelle definitions on the abstract and 7K lines on the concrete side, implies a massive proof effort which we aimed to spread over multiple people working concurrently, with as little need for interaction and coordination as possible.

In summary, this paper can be seen as a study on how far you can get with the simplest possible methods. It is our hypothesis that it was precisely this simplicity that enabled us to achieve this large-scale verification.

The contributions of this paper are as follows.

- We formalise the nondeterministic state monad with exceptions and failure. This subsumes the state monad with exceptions that is commonly used in Haskell. The formalisation is a shallow embedding into the logic of Isabelle/HOL.
- We present a Hoare Logic and refinement calculus on the above, both simple yet scalable and practical.
- We report on our experience in applying the above to a binary compatible, executable model of seL4 microkernel translated from Haskell.

The following sections provide detail on each of these in turn.

2   State Monads

A state monad allows a pure functional model of computation with side effects. For result type $'a$ and state type $'s$, the associated monad type (abbreviated $('s, 'a)$ *state-monad*) is $'s \Rightarrow 'a \times 's$. That is, a function from previous state to next state together with a computation result. A pure state transformer is typically denoted by the one-valued return type *unit* i.e. $'s \Rightarrow unit \times 's$.

All monads define two constructors, here called return and bind. For the state monad they are defined as follows:

return   :: $'a \Rightarrow ('s, 'a)$ *state-monad*
return $a \equiv \lambda s.\ (a,\ s)$

bind   :: $('s, 'a)$ *state-monad* $\Rightarrow ('a \Rightarrow ('s, 'b)$ *state-monad*$) \Rightarrow ('s, 'b)$ *state-monad*
bind $f\ g \equiv \lambda s.$ let $(v,\ s') = f\ s$ in $g\ v\ s'$

Note that as Isabelle/HOL is simply typed, it is not possible to straightforwardly define the monad type constructor as in Haskell, defining return, bind and associated syntax once, and thereby proving results generically about the class of monadic types. The solution that we adopt is to instantiate the class for specific monads and monad constructors, e.g., return :: $'a \Rightarrow ('s, 'a)$ *state-monad*.

The constructor return simply injects the value a into the monad type, passing the state unchanged, whilst bind sequentially composes a computation f, and a computation g (a function from the return type of f). The expression bind $f\, g$ is abbreviated as $f >>= g$. To allow concise description of longer computations, we define a do syntax in a similar fashion to Haskell:

$$\text{do } x \leftarrow f;\ g\, x \text{ od} \equiv f >>= g$$

A state monad also defines two additional constructors: get and put, the primitive state transformers (here () is the sole element of type *unit*):

get :: $('s, 's)$ *state-monad*   put :: $'s \Rightarrow (unit, 's)$ *state-monad*

get $\equiv \lambda s.\ (s, s)$ put $s \equiv \lambda\text{-}.\ ((), s)$

The constructors of all monads must obey the following three laws, which we have instantiated and proved for each monad instance:

$$
\begin{aligned}
\text{return } x >>= f &= f\, x & \text{RETURN\_BIND}\\
m >>= \text{return} &= m & \text{BIND\_RETURN}\\
(m >>= f) >>= g &= m >>= (\lambda x.\ f\, x >>= g) & \text{BIND\_ASSOC}
\end{aligned}
$$

The simple state monad is able to model sequential computations with side-effects, but does not provide good notation for non-local flow control (e.g. exceptions). A straightforward way to model try-catch-style exceptions is to instantiate the state monad return type with the sum type $'e + 'a$ (for result type $'a$ and exception type $'e$) in place of the simple result type. Every component in the monad now returns either Inr a in case of success, or Inl e in case of failure with exception e. To complete the model we require a new bind constructor, bindE which propagates exceptions, and the catch constructor to embed the error monad into the non-error state monad.

lift $f\, v$ $\equiv$ case v of Inl $e \Rightarrow$ return v | Inr $a \Rightarrow f\, a$

bindE $f\, g$ $\equiv f >>= $ lift g

catch f *handler* $\equiv$ do $x \leftarrow f$;
                case x of Inl $e \Rightarrow$ *handler* e | Inr $b \Rightarrow$ return b
       od

In formulating an abstract behavioural model, it is convenient to express computation nondeterministically. This is readily modelled as an extension of the state monad by allowing each computation to return a (possibly empty) set of value-state pairs: $'s \Rightarrow ('a \times 's)$ set, and redefining bind as $\lambda s.\ \bigcup\{g\, a\, s'$ | $(a,s') \in f\, s\}$. This formulation has a drawback however: we wish to model catastrophic failure e.g. kernel panic, and show its absence; the obvious definition fail $\equiv \lambda s.\ \{\}$, admits only the existential statement: *For all states s, not all paths fail*. What we desire, however, is the universal statement: *For all states s, no path fails*, which cannot be expressed as a simple predicate on the state set, as the failure case ($\{\}$) is dominated in the union by the non-failure case. Our solution is to append a failure flag which is propagated separately, and which dominates non-failure in bind. This leads us to the following definitions (n.b. f 'A is the image of A under f):

$$\text{return } a \equiv \lambda s.\ (\{(a, s)\}, \textsf{False})$$
$$\text{bind } f\ g \equiv \lambda s.\ (\bigcup \textsf{fst}\ `\ (\lambda(x, y).\ g\ x\ y)\ `\ \textsf{fst}\ (f\ s),$$
$$\textsf{True} \in \textsf{snd}\ `\ (\lambda(x, y).\ g\ x\ y)\ `\ \textsf{fst}\ (f\ s) \vee \textsf{snd}\ (f\ s))$$

In addition to the state monad constructors $\textsf{get}$ and $\textsf{put}$, we define $\textsf{select}$, a nondeterministic $\textsf{return}$, which takes a set of values. We also define $\textsf{fail}$ as indicated above:

get	::	$('s, 's)\ nd\text{-}monad$	put	::	$'s \Rightarrow (unit, 's)\ nd\text{-}monad$
get	$\equiv$	$\lambda s.\ (\{(s, s)\}, \textsf{False})$	put s	$\equiv$	$\lambda\text{-}.\ (\{((), s)\}, \textsf{False})$
select	::	$'a\ set \Rightarrow ('s, 'a)\ nd\text{-}monad$	fail	::	$('s, 'a)\ nd\text{-}monad$
select A	$\equiv$	$\lambda s.\ (A \times \{s\}, \textsf{False})$	fail	$\equiv$	$\lambda s.\ (\{\}, \textsf{True})$

The nondeterminism inherent in the model allows us to model input conveniently: do $x \leftarrow \textsf{select}\ InputActions$; $f\ x$ od.

3   Hoare Logic on State Monads

The Hoare triple $\lbrace P \rbrace\ f\ \lbrace Q \rbrace$ is a predicate on the computation f, stating that if the precondition P holds before execution, then the postcondition Q will hold afterwards. Since HOL is a logic of total functions, and f is just a HOL function that must terminate, our Hoare triples express total correcntess. For the nondeterministic state monad, the basic Hoare triple also needs to take into account the return value and is defined as follows:

$$\lbrace P \rbrace\ f\ \lbrace Q \rbrace \equiv \forall s.\ P\ s \longrightarrow (\forall (r, s') \in \textsf{fst}\ (f\ s).\ Q\ r\ s')$$

Note that the postcondition Q is a binary predicate while P is unary. For the state monad with exceptions, we define:

$$\lbrace P \rbrace\ f\ \lbrace Q \rbrace,\ \lbrace R \rbrace \equiv \lbrace P \rbrace\ f\ \lbrace \lambda r\ s.\ \textsf{case } r \textsf{ of } \textsf{Inl}\ a \Rightarrow R\ a\ s \mid \textsf{Inr}\ b \Rightarrow Q\ b\ s \rbrace$$

This specifies a separate postcondition for the exception and non-exception cases. All of the following rules have a natural expression for the state-exception monad in terms of this augmented Hoare triple.

To build a calculus for reasoning about monadic computations, we first state and prove axiomatic rules for the basic constructors:

$$\lbrace \lambda s.\ P\ ()\ x \rbrace\ \textsf{put}\ x\ \lbrace P \rbrace\ \textsc{put-wp} \qquad \lbrace \lambda s.\ P\ s\ s \rbrace\ \textsf{get}\ \lbrace P \rbrace\ \textsc{get-wp}$$
$$\lbrace P\ x \rbrace\ \textsf{return}\ x\ \lbrace P \rbrace\ \textsc{return-wp}$$

Constructor bind requires a more complicated rule, to capture the interaction of the pre- and post-conditions of composed computations:

$$\frac{\forall x.\ \lbrace B\ x \rbrace\ g\ x\ \lbrace C \rbrace \qquad \lbrace A \rbrace\ f\ \lbrace B \rbrace}{\lbrace A \rbrace\ f >>= g\ \lbrace C \rbrace}\ \textsc{seq}$$

Note that the premises of the SEQ rule are reversed with respect to the program order, this is done to ease the repeated application of the rule by the VCG.

Finally, to complete the basic calculus we introduce the WEAKEN rule, to substitute arbitrary preconditions. An analogous rule (STRENGTHEN) exists to substitute postconditions.

$$\frac{\{Q\}\,f\,\{R\} \qquad \forall\, s.\ P\, s \longrightarrow Q\, s}{\{P\}\,f\,\{R\}} \ \text{WEAKEN}$$

An equivalent set of rules exists to reason about the presence or absence of failure.

4 Verification Condition Generator

As usual in Hoare Logic, reasoning within this calculus can be substantially automated by the use of a verification condition generator (VCG) if we phrase our structural Hoare rules in weakest-precondition (WP) form. The rules given for put, get and return in Sect. 3 are weakest-precondition rules. As an example, consider the following definition of the modify constructor, and the proof of its associated weakest-precondition rule:

modify $f \equiv$ do $s \leftarrow$ get; put $(f\, s)$ od

We wish to show $\{\lambda s.\ P\ ()\ (f\, s)\}$ modify f $\{P\}$. Before invoking the VCG, we unfold definitions until the goal is phrased in terms of known operations. The VCG then produces the following proof steps automatically.

It starts by applying the WEAKEN rule to replace the concrete precondition with a schematic[1] precondition, $?Q$, and an implication. We get two new goals:

1. $\{\lambda s.\ ?Q\}$ do $s \leftarrow$ get; put $(f\, s)$ $\{P\}$
2. $\forall\, s.\ P\ ()\ (f\, s) \longrightarrow ?Q$

The VCG now repeatedly tries to apply one of its set of WP rules. The rule SEQ will match the current goal, as its postcondition is fully general (matching the concrete P) and the precondition, which is concrete in the rule, matches the schematic precondition $?Q$ that we have just created. If the WP set is constructed correctly, the goals will always remain in this form, with concrete postcondition and schematic precondition, and for every top-level operator there will be one rule that matches. In our example, the VCG would apply SEQ, PUT-WP, and GET-WP in turn, leaving the user with only the implication introduced at the first step. This is a HOL formula, free of both monad and Hoare syntax:

1. $\forall\, s.\ P\ ()\ (f\, s) \longrightarrow P\ ()\ (f\, s)$

Here, the goal is trivial, as the precondition we set out to prove was in fact the weakest. We could now add this new WP rule to the set available to the VCG, to avoid having to unfold the definition of modify in future. In this manner we progressively build the calculus towards a higher and higher level of abstraction.

Note that, if we add rules that are not strictly weakest preconditions, we do not affect the soundness of the VCG, we simply take the risk that the implication goal produced may be too weak to be provable, indicating that our rules need to be strengthened.

---

[1] Schematic variables in Isabelle stand for terms that can be syntactically instantiated (as opposed to free variables that need to remain fixed in proofs).

The weakest-precondition rules mentioned so far all apply to an arbitrary postcondition. For elementary functions like put, get and modify, rules of this form are easily stated. In principle such a rule can be stated for any of the monadic functions we use. In practice, however, we find that the preconditions in these rules are typically of exponential term size with respect to the complexity of the operator. The tractable solution we have found is to supply the VCG instead with Hoare triples that have specific postconditions and manually simplified preconditions. In principle these can still be weakest precondition rules, however this is not normally the case. An example is SET-EP-VALID-OBJS:

$$\{\lambda s.\ \text{valid-objs}\ s \wedge \text{valid-ep}\ v\ s\}\ \text{set-endpoint}\ ep\ v\ \{\lambda rv\ s.\ \text{valid-objs}\ s\}$$

The set-endpoint function models pointer update for the communication endpoint type. Like other models of pointer update, it simply replaces the contents of the heap at the given address with the new value. The valid-objs predicate in the postcondition is one of our global invariants, and establishes that all objects satisfy certain validity criteria. Clearly for set-endpoint to preserve valid-objs the new endpoint value must satisfy the appropriate validity predicate, valid-ep. This is not the weakest possible precondition, as it globally asserts in valid-objs that the value about to be replaced is valid, which is unnecessary. The precise weakest precondition would be tedious to define, and the precondition given, although not weakest, is always true in practice.

Hoare triples with specific postconditions complicate the VCG, which must labour to connect the postconditions available to the one that is needed. To illustrate this problem, consider the scenario in which we wish to establish valid-objs after a pair of endpoint updates.

$$\{\lambda s.\ \text{valid-objs}\ s \wedge \text{valid-ep}\ v\ s \wedge \text{valid-ep}\ v'\ s\}$$
do set-endpoint $p\ v$; set-endpoint $p'\ v'$ od
$$\{\lambda rv\ s.\ \text{valid-objs}\ s\}$$

The VCG can divide the problem using SEQ and apply SET-EP-VALID-OBJS to the second problem. The postcondition for the first update will then be $\lambda rv\ s.$ valid-objs $s \wedge$ valid-ep $v'\ s$. To apply SET-EP-VALID-OBJS again, the VCG must first use the conjunction lifting rule.

$$\frac{\{P\}\ f\ \{Q\} \qquad \{P'\}\ f\ \{Q'\}}{\{\lambda s.\ P\ s \wedge P'\ s\}\ f\ \{\lambda rv\ s.\ Q\ rv\ s \wedge Q'\ rv\ s\}}\ \text{CONJ-LIFT}$$

The conjunction operator is one of a family of first-order logic operators that have a VCG lifting rule. Conjunction, disjunction, and the universal and existential quantifiers have lifting rules, but the negation operator does not. Implication is dealt with by reducing to a disjunction and negation, after which the negation must be dealt with explicitly.

The only such lifting rule that the VCG will use by default is CONJ-LIFT, and it will only be used conservatively, that is, when one of the subproblems created can be immediately solved using another rule. The VCG can also be configured to use any lifting rule aggressively, that is, whenever possible.

The VCG could apply all lifting rules by default. However, should one or more of the created subgoals be unresolvable, the resulting proof state may be difficult to understand or work with. It is thus pragmatically useful for the VCG to fail early, returning an interactive state that is amenable to further manual progress. Conjunction occurs in our postconditions so frequently that the VCG must handle it explicitly, but other operators are rare enough to be handled manually.

The VCG is not limited to Hoare triples. Rules for absence of failure, as mentioned in Sect. 3, can be similarly automated by the same tool.

5   Refinement Calculus

The ultimate objective of our effort is to prove *refinement* [2] between an abstract and a concrete process. We define a process as a triple containing an initialisation function, which creates the process state with reference to some external state, a step function which reacts to an event, transforming the state, and a finalisation function which reconstructs the external state.

$$\textbf{record } process = \textsf{Init} :: \textit{'external} \Rightarrow \textit{'state set}$$
$$\textsf{Step} :: \textit{'event} \Rightarrow (\textit{'state} \times \textit{'state}) \textsf{ set}$$
$$\textsf{Fin} :: \textit{'state} \Rightarrow \textit{'external}$$

The execution of a process, starting from a initial external state, via a sequence of input reactions results in a *set* of external states: (n.b. R `` S is the image of the set S under the relation R)

$$\textsf{steps } \delta \textit{ s events} \quad \equiv \textsf{foldl } (\lambda \textit{states event. } (\delta \textit{ event}) \textit{ `` states}) \textit{ s events}$$
$$\textsf{execution } A \textit{ s events} \equiv (\textsf{Fin } A) \textit{ ` } (\textsf{steps } (\textsf{Step } A) (\textsf{Init } A \textit{ s}) \textit{ events})$$

Process A is refined by C, if with the same initial state and input events, execution of C yields a subset of the external states yielded by executing A:

$$A \sqsubseteq C \equiv \forall \textit{s events. } \textsf{execution } C \textit{ s events} \subseteq \textsf{execution } A \textit{ s events}$$

Refinement is commonly proven by establishing forward simulation [2], of which it is a consequence. To demonstrate forward simulation we define a relation, SR, between states of the two processes. We must show that the relation is established by Init, is maintained if we advance the systems in parallel, and implies equality of the final external states: (n.b. S ;; T is the composition of relations S and T.)

$$\textsf{fw-sim } SR \textit{ C A} \equiv (\forall \textit{s. } \textsf{Init } C \textit{ s} \subseteq SR \textit{ `` } \textsf{Init } A \textit{ s})$$
$$\wedge \ (\forall \textit{ event. } SR \textit{ ;; } \textsf{Step } C \textit{ event} \subseteq \textsf{Step } A \textit{ event ;; } SR)$$
$$\wedge \ (\forall \textit{s s'. } (s, s') \in SR \longrightarrow \textsf{Fin } C \textit{ s'} = \textsf{Fin } A \textit{ s})$$

To address our scalability concerns, we wish to decompose the refinement problem into smaller subproblems and translate the statement to the state monad. The simplest way to do this is to break the forward simulation problem down to component functions. The corres predicate captures forward simulation between a single concrete monadic computation, C, and its abstract counterpart, A, with SR instantiated to our standard state relation, state-relation. It takes three additional parameters: R relates abstract and concrete return values, and the preconditions P and P' restrict the input states, allowing use of information such as global invariants:

corres R P P' A C $\equiv$ $\forall (s,\, s') \in$ state-relation. $P\, s \wedge P'\, s' \longrightarrow$
$(\forall (r',\, t') \in$ fst $(C\, s')$. $\exists (r,\, t) \in$ fst $(A\, s)$. $(t,\, t') \in$ state-relation $\wedge R\, r\, r')$
$\wedge$ (snd $(C\, s') \longrightarrow$ snd $(A\, s))$

Note that the outcome of the monadic computation is a pair of result and failure flag. The last conjunct of the corres statement is stronger than strictly necessary for refinement. It states that failure on the concrete m' implies failure on the abstract m. This means we only have to show absence of failure on the most abstract level to get absence of failure on all concrete levels by refinement.

The key property of corres is that it decomposes over the bind constructor through the CORRES-SPLIT rule.

CORRES-SPLIT:

$$\frac{\text{corres } R'\, P\, P'\, A\, C \qquad \forall r\, r'.\ R'\, r\, r' \longrightarrow \text{corres } R\, (S\, r)\, (S'\, r')\, (B\, r)\, (D\, r')}{\text{corres } R\, (P \text{ and } Q)\, (P' \text{ and } Q')\, (A >>= B)\, (C >>= D)}$$

$$\{Q\}\, A\, \{S\} \qquad \{Q'\}\, C\, \{S'\}$$

Similar splitting rules exist for other common monadic constructs including bindE, catch and conditional expressions. There are terminating rules for the elementary monadic functions, for example:

CORRES-RETURN:

$$\frac{R\ a\ b}{\text{corres } R\ (\lambda s.\ \text{True})\ (\lambda s.\ \text{True})\ (\text{return } a)\ (\text{return } b)}$$

The corres predicate also has a weakening rule, similar to the Hoare Logic.

CORRES-PRECOND-WEAKEN:

$$\frac{\text{corres } R\, Q\, Q'\, A\, C \qquad \forall s.\ P\, s \longrightarrow Q\, s \qquad \forall s.\ P'\, s \longrightarrow Q'\, s}{\text{corres } R\, P\, P'\, A\, C}$$

Proofs of the corres property take a common form: first the definitions of the terms under analysis are unfolded and the CORRES-PRECOND-WEAKEN rule is applied. As with the VCG, this allows the syntactic construction of a precondition to suit the proof. The various splitting rules are used to decompose the problem; in some cases with carefully chosen return value relations. Existing results are then used to solve the component corres problems. Some of these existing results, such as CORRES-RETURN, require compatibility properties on their parameters. These are typically established using information from previous return value relations. The VCG eliminates the Hoare triples, bringing preconditions assumed in corres properties at later points back to preconditions on the starting states. Finally, as in Dijkstra's postcondition propagation [4], the precondition used must be proved to be a consequence of the one that was originally assumed.

6   Case Study – The seL4 Microkernel

In this section, we give an overview of the seL4 microkernel, its two formalisations in Isabelle/HOL, some of the properties we have proved on them, and our experience in this verification. With about 10,000 lines of C code; 7,500 lines of executable model and 3,000 lines of abstract Isabelle/HOL specification, the kernel is too large for us to provide any kind of useful detail in a conference paper, or even a comprehensive overview of its formalisation. We do not attempt to do so; instead we provide a very high-level view of its functionality, and show bits and pieces of the formalisation to give an impression of the general flavour.

6.1   Overview

As mentioned in the introduction, seL4 is an evolution of the L4 microkernel family. The main difference to L4 is that it is entirely capability based, unifying all resource accounting and access control into a single mechanism.

All kernel abstractions and system calls are provided via named, first-class kernel objects. Authorised users obtain kernel services by invoking operations on kernel objects. Authority over these objects is conferred via capabilities only. System call arguments can be either data, or other capabilities. Similarly to L4, seL4 provides three basic abstractions: threads, address spaces and inter-process communication (IPC). In addition, seL4 introduces an abstraction, *untyped memory* (UM), which represents a region of currently unused physical memory.

An important part of the seL4 design is that all memory, used directly by an application (e.g. memory frames) or indirectly through the kernel (e.g. page tables), is fully accounted for by capabilities. A *parent* capability to untyped memory can be refined into *child* capabilities to smaller untyped memory blocks or other kernel objects via the *retype* operation. The creator can then delegate all or part of the its authority over the object to one or more of its clients. Untyped capabilities can be *revoked*: this removes all corresponding child capabilities from clients and prepares the memory spanned by that capability for retyping.

These mechanisms make seL4 a highly flexible microkernel, supporting a number of practical application scenarios. A simple example is running a full legacy guest OS (e.g. Linux) next to a critical, trusted communications stack; another is to provide full separation between components at multiple security levels with strict controls on explicit information channels between them.

6.2   Formalisation

We now give a very brief introduction to the formalisation of seL4. We begin with the state space of the abstract model.

This state is embedded into a process, modelling machine execution, of which we make only the kernel execution precise. User-level execution is assumed free to mutate any user-accessible part of the state. The transitions for kernel execution are defined by a nondeterministic monadic function, in the manner described above. The events triggering these transitions are: timer interrupts, kernel trap instructions (user level kernel calls), page faults, and user-level faults. We collect all of these in the data structure *event*, shared with the executable level:

datatype *syscall* = *Send* | *Wait* | *SendWait* | *Identify* | *Yield*

datatype *event* = *SyscallEvent syscall* | *UnknownSyscall nat* |
            UserLevelFault nat | *TimerInterrupt* | *VMFaultEvent vptr bool*

The type *syscall* models user level calls (sending/replying to IPC, identifying capabilities, yielding the current time slice). The other events are machine generated. Arguments to system calls are read from machine registers in binary form and decoded for further processing. This decoding phase is fully precise in the abstract specification, and therefore very similar on the executable and the abstract level. It is a major part of the programmer-visible API specification; in fact, typical kernel reference manuals describe almost exclusively this syntactic part, and only sketch the semantics of the system; the latter is the bulk of the specification in our case.

The abstract state space of seL4 is a record with the following components:

record *abstract-state* = *pspace* :: *obj-ref* ⟶ *kernel-object*
$\qquad\qquad\qquad$ *cdt* :: *cte-ptr* ⟶ *cte-ptr*
$\qquad\qquad\qquad$ *cdt-revokable* :: *cte-ptr* ⟹ *bool*
$\qquad\qquad\qquad$ *cur-thread* :: *obj-ref*
$\qquad\qquad\qquad$ *machine-state* :: *machine-state*

datatype *kernel-object* = *CapTable cap-ref* ⟶ *cap* | *TCB tcb* |
$\qquad\qquad$ *Endpoint endpoint* | *AsyncEndpoint async-ep* | *Frame*

The whole state space declaration[2] comprises approximately 200 lines of Isabelle definitions, we mention only the salient points. The *pspace* component models the kernel-accessible part of memory. In this abstract view, it is a partial function from object references (machine words) to kernel objects. The capability derivation tree (CDT) is a data structure that keeps track of the parent/child relationship between capabilities; it is realised as a partial function from child *capability table entry* (CTE) locations to parent CTE locations, i.e., a tree of CTE locations. A CTE location is fully determined by the location of the kernel object (an *obj-ref*) and a position within that kernel object (a *cap-ref*). As mentioned, *CapTable* objects store capabilities; *TCB* objects implement the kernel accounting for threads; *Endpoint* and *AsyncEndpoint* objects implement IPC, and *Frame* objects stand for user data frames. The remaining two components of the global state are a pointer to the TCB of the current thread and the machine state (e.g. register state). Currently, we do not model the machine state in detail, but instead use a set of axiomatised functions such as loadRegister/storeRegister on type *machine-state*.

Since the machine context is the main part of the shared outside-observable part of the two models, we have proved during refinement that the observable effect of reads, writes, cache flushes, TLB flushes, etc. is the same on both levels. In the next step of refinement, to C, we plan to eliminate these remaining straightforward axioms and provide a direct model for the machine context.

In the concrete model, our abstract views of the CDT, and kernel object states vanish, and are replaced with much more detailed alternatives. The CDT, for instance, becomes a doubly linked list together with a number of flags for level information, stored in machine words within CTEs. In addition, we gain

---

[2] We present a slightly simpler, earlier version of the model here. The current version also contains interrupt tables and page table data structures.

a number of state components implementing data structures that were not necessary on the abstract level. These are: a table of ready-queues for scheduling (indexed by a priority byte), and a scheduler action which effectively points to the next thread's TCB.

> **record** *concrete-state* = *ksPSpace* :: *pspace*
>   *ksReadyQueues* :: *8 word* ⇒ *ready-queue*
>   *ksCurThread* :: *32 word*
>   *ksSchedulerAction* :: *scheduler-action*
>   *ksMachineState* :: *machine-state*

The *ksPSpace* component corresponds to the C heap, and *ksMachineState* to the machine context, as on the abstract side. The rest are global pointer variables. In this way, the executable model is close to the final implementation.

For refinement, we need to define the process types of the models. The executable model has a single entry point, callKernelC, which handles the *event* type defined above. It is natural then to define the Step component of the process datatype as the outcome of this nondeterministic monadic operator. Likewise, the Init component resets the state to the default newKernelStateC and then calls initKernelC. The Fin component is simply the projection, *ksMachineState*. The abstract process is defined similarly.

The refinement property can then be proven using corres properties and Hoare triples. First, we establish that our abstract and concrete global invariant collections, (invsA and invsC), are invariants of the respective processes.

$\{\!|\lambda s.\ s =$ newKernelStateA$|\!\}$ initKernelA *entry frames offset kFrames* $\{\!|\lambda r.$ invsA$|\!\}$
$\{\!|$invsA$|\!\}$ callKernelA e $\{\!|\lambda r.$ invsA$|\!\}$, $\{\!|\lambda r.$ invsA$|\!\}$

$\{\!|\lambda s.\ s =$ newKernelStateC$|\!\}$ initKernelC *entry frames offset kFrames* $\{\!|\lambda r.$ invsC$|\!\}$
$\{\!|$invsC$|\!\}$ callKernelC e $\{\!|\lambda r.$ invsC$|\!\}$, $\{\!|\lambda r.$ invsC$|\!\}$

Secondly, we establish that all elements of the Init sets are related.

(newKernelStateA, newKernelStateC) ∈ state-relation

corres dc $(\lambda s.\ s =$ newKernelStateA$)$ $(\lambda s.\ s =$ newKernelStateC$)$
        (initKernelA *entry frames offset kFrames*)
        (initKernelC *entry frames offset kFrames*)

Finally we establish that the main execution steps correspond.

corres (intr ⊕ dc) invsA invsC (callKernelA *event*) (callKernelC *event*)

From these we establish forward simulation, which implies refinement. The statements above are slightly simplified versions of our theorems which involve more preconditions on machine behaviour.

6.3   Properties

We now describe some of the properties and invariants we proved on these two formalisations, in addition to the main refinement theorem that states that the concrete model is a correct implementation of the abstract specification.

One of the first properties proved on both levels was that all system calls terminate. Since HOL is a logic of total functions, this is a necessary condition to formalise the kernel behaviour. The proof for most of the kernel was straight-forward; we only had one complex, mutually recursive case that models a nested loop in the C code: the *delete* operation that removes capabilities.

The main invariant of the kernel is simple: *all references in the kernel—be it in capabilities, kernel objects or other data structures—always point to an object of the expected type.* This is a dynamic property as memory can be re-typed at runtime. Despite its simplicity, it is the major driver for almost all other kernel invariants. Exceptions are low-level invariants like *address 0 is never inhabited by any object*, and *objects are always aligned to their size*.

The main validity predicates (including valid-objs and valid-ep mentioned pre-viously) are liftings of the well-typedness criterion above to the entire heap, thread states, scheduler queues and other state components. An example of a more complex invariant, needed to prove that well-typedness is preserved is: *A kernel object k_1 contains a reference to kernel object k_2 if and only if there exists a (possibly transitive) reference from k_2 back to k_1.* This symmetry condition can be used to conclude that if an object contains no references itself, there will be no dangling references to it in the rest of the kernel. It would therefore be safe to remove such an object once capability references are checked. To avoid inef-ficient object state checks, we additionally observe: *If an object is live (contains references to other objects), there exists a capability to it.*

Testing for capabilities is much easier, because they are tracked explicitly in the CDT. CDT-related properties include:

Linked List. The doubly-linked list structure is consistent (back/forward point-ers are implemented correctly), the lists always terminate in NULL, and the list together with the additional tags correctly implements a tree. This is a basic shape property.

Chunks. If two CTEs point to the same memory location, they have a common ancestor and all entries between them in the CDT point to this same memory location. This ensures various tests in the kernel can be implemented locally.

Cap Ancestry. If an untyped capability c_1 covers a sub-region of another ca-pability c_2, then c_1 must be a descendant of c_2 according to the CDT.

Object Ancestry. If a capability c_1 points to a kernel object whose memory is covered by an untyped capability c_2, then c_1 must be a descendant of c_2.

All of these together ensure that memory can be retyped safely and with minimal local checks; if an untyped capability has no children, then all kernel objects in its region must be non-live (otherwise there would be capabilities to them, which in turn would have to be children of the untyped capability). If the objects are not live and no capabilities to them exist, there is no further reference in the whole system that could be made unsafe by the type change.

This example is the most complex chain of invariants we had to create for a single operation. Other operations, such as IPC and scheduling have their own requirements.

6.4   Experience and Lessons Learned

The total effort for the refinement proof described here was 100,000 lines of Isabelle/HOL and 5 person years. The proof lead to over 100 changes in each of the two specifications. The majority of the changes were for ease of proof: slight rearrangement of code, avoidance of unnecessary optimisations, local tests instead of global assumptions. The majority of actual bugs were typographical and copy & paste errors that slipped through prior testing. Unsurprisingly, there were far more of these simple mistakes on the abstract level than on the executable one. The abstract level was only type checked, never run, since it is not executable. We found on the order of 10 conceptual problems and oversights which would have lead to crashes or security exploits—as would have most of the typos. These were mainly missing checks on user input, subtle side effects in the middle of operations, or (rarely) too-strong assumptions on what is invariant during execution. Security attacks became apparent via invariant violations.

We found that the kernel programming team usually knew the invariants precisely and used them for optimisations. In fact, the developers were often able to articulate clearly why a certain property should hold or why a certain test was unnecessary. A number of the security breaches mentioned above were discovered during these discussions with the developers. On the other hand, it was the formal proof that forced us to have this discussion in the first place.

In terms of lessons learned, we confirm the usual observation that the more abstract the easier, and the less assumptions on global state the easier. In this light, it was unsurprising that the low-level CDT and large, concrete initialisation phase of the kernel were unpleasant parts of the proof.

After an initial full proof of refinement was achieved, we found that new features could be added with reasonable effort. This depends on how independent the new feature is from the rest of the system. If it uses its own data structure that is not accessed anywhere else, the effort is largely proportional to the size of the feature. For instance, adding multiple capability arguments to system calls (as opposed to only one) was easy, with about 2 person weeks of effort, although it concerned changes fairly deep in IPC message decoding and transfer. If, on the other hand, the feature is highly intermingled with the rest of the system, a factor of the size of the kernel times the size of the feature is to be expected. We hypothesise that the effort for the whole verification so far was quadratic in the size of the kernel. Since in a microkernel almost every basic feature relies on properties of almost all others (IPC, TCBs, CTEs, CDT are all highly connected), proving preservation of a new invariant on one feature will involve significant work not only on this, but on all other features in the kernel as well. The refinement proof itself remains linear in the size of the kernel. With more modular code, one would expect independent data structures and therefore invariant proofs of a size proportional to the code.

Another observation concerns invariant discovery. We began with a simple invariant that was needed for the refinement proof (well-typedness) and let that drive the invariant discovery process. In hindsight it would, after a short initial phase, probably have been more effective to simply use the strongest invariant that we suspected we could prove. At several points we hoped to get away with a simpler formulation, but were then caught out halfway through the proof by a particular operation after several thousand lines of proof. We then ended up with the complex, precise form anyway. The lesson is: in such a complex system, the simple formulation is unlikely to succeed. Take the precise formulation instead, even if it looks like more work initially. As mentioned above, a good source of invariants is the development team.

In terms of proof engineering and the methods presented in this paper, we believe we have achieved our goal of scalability. Up to four people worked on this proof concurrently and independently without much conflict. We estimate that for code of this size (10K lines of C code) a team of more than five or six persons would need a more serious effort in planning and synchronisation.

Once the framework and invariants are established, and more importantly once the kernel is well understood, the proof is not too hard and should be readily repeatable. We see potential for more automation in the refinement proof, and in exploratory automatic invariant proofs. Simple invariants can often be stated and proved automatically for many functions at a time. This could automatically be tried for a set of basic properties before manual proof starts. We have developed first steps in this direction, but have not made it the focus so far.

In conclusion, code verification at this size and level of detail is entirely feasible with current theorem proving technology.

7   Related Work

Earlier work on OS verification includes PSOS [7] and UCLA Secure Unix [20]. Later, KIT [1] describes verification of process isolation properties down to object code level, but for an idealised kernel far simpler than modern microkernels. The Verisoft project [9] is attempting to verify a whole system stack, including hardware, compiler, applications, and a simplified microkernel VAMOS. The VFiasco [13] project is attempting to verify the Fiasco kernel, another variant of L4 directly on the C++ level. The Coyotos [17] kernel is being designed for verification, but it is unclear how much progress has been made.

The House and Osker kernels [11] (in Haskell) and the Hello kernel [8] (in Standard ML) demonstrated that modern functional languages can be used to develop bare metal implementations of operating systems. In contrast, we see our Haskell implementation of the kernel as a prototype only.

There are other approaches to translating Haskell into Isabelle [10,12,14]. Since none of these approaches were able to parse our code base, we use our own translator; for the work presented here, we need to assume its correctness. In the longer term however, this is unnecessary, because the final theorem will be a refinement theorem between the abstract Isabelle model and the C program. We have already invested significant effort into modelling C precisely [19].

Our treatment of Hoare Logic on monads is much less general than that of Mossakowski et al. [16]. We do not make assertions part of the program, which in our setting would provide barriers to splitting up the same proof among multiple persons. As mentioned before, we trade generality for simplicity, and for lightweight infrastructure with an emphasis on scalability.

8 Conclusion

We have presented simple, but effective techniques for reasoning about state-based functional programs and for proving formal refinement on them. Although we have not aimed at full generality, we are convinced that the combination of basic monads we used covers a wide range of practical programs in languages such as Haskell and ML. Our case study has shown that it is practical to fully formally verify programs of thousands of lines of code at this level.

The salient point of our Hoare Logic is that it is simple enough to be automated effectively; but despite its simplicity, expressive enough to be easily applicable. Our extension of classic refinement to the nondeterministic state monad is formally largely straightforward, and the calculus presented is not complete. Again, the main point is that it is engineered such that a large-scale proof can be effectively divided up into mostly independent parts. Classical step-wise refinement calculi do not necessarily work well within this paradigm, and often require window reasoning and other complex context tracking [18].

The case study we report on constitutes the formal, fully machine-checked verification of a binary-compatible executable model of seL4. Binary compatible meaning that the corresponding Haskell program, together with a hardware simulator, can execute normal, compiled user-level ARM11 binaries that would run unchanged on bare hardware. This includes low-level hardware feedback: cache flushes, TLB loads, etc. To our knowledge, this is the first such verification of an OS microkernel of this size and complexity.

Although the verification reported on here reaches a level of detail far greater than that usually present when a software system is claimed to be verified, we refrain from calling seL4 itself "fully formally verified" yet. Our goal is to take the verification from the executable model down to the level of C code, compiled and running on hardware.

Acknowledgements. We thank the other current and former members of the L4. verified and seL4 teams: Michael Norrish, Jia Meng, Catherine Menon, Jeremy Dawson, Simon Winwood, Harvey Tuch, Rafal Kolanski, David Tsai, Andrew Boyton, Kai Engelhardt, Kevin Elphinstone, Philip Derrin and Dhammika Elkaduwe for their help and support.

References

1. Bevier, W.R.: Kit: A study in operating system verification. IEEE Transactions on Software Engineering 15(11), 1382–1396 (1989)
2. de Roever, W.-P., Engelhardt, K.: Data Refinement: Model-Oriented Proof Methods and their Comparison. Cambridge Tracts in Theoretical Computer Science, vol. 47. Cambridge University Press, Cambridge (1998)

3. Derrin, P., Elphinstone, K., Klein, G., Cock, D., Chakravarty, M.M.T.: Running the manual: An approach to high-assurance microkernel development. In: Proc. ACM SIGPLAN Haskell Workshop, Portland, OR, USA (September 2006)
4. Dijkstra, E.W.: Guarded commands, nondeterminacy and formal derivation of programs. Commun. ACM 18(8), 453–457 (1975)
5. Elkaduwe, D., Derrin, P., Elphinstone, K.: A memory allocation model for an embedded microkernel. In: Proc. 1st MIKES, Sydney, Australia, pp. 28–34 (2007)
6. Elphinstone, K., Klein, G., Derrin, P., Roscoe, T., Heiser, G.: Towards a practical, verified kernel. In: Proc. 11th Workshop on Hot Topics in Operating Systems, San Diego, CA, USA (May 2007)
7. Feiertag, R.J., Neumann, P.G.: The foundations of a provably secure operating system (PSOS). In: AFIPS Conf. Proc., 1979 National Comp. Conf., New York, NY, USA, June 1979, pp. 329–334 (1979)
8. Fu, G.: Design and implementation of an operating system in Standard ML. Master's thesis, Dept.of Information and Computer Sciences, Univ.Hawaii at Manoa (1999)
9. Gargano, M., Hillebrand, M., Leinenbach, D., Paul, W.: On the correctness of operating system kernels. In: Hurd, J., Melham, T. (eds.) TPHOLs 2005. LNCS, vol. 3603, pp. 1–16. Springer, Heidelberg (2005)
10. Hallgren, T., Hook, J., Jones, M.P., Kieburtz, R.B.: An overview of the Programatica Tool Set. In: High Confidence Software and Systems Conference (2004)
11. Hallgren, T., Jones, M.P., Leslie, R., Tolmach, A.: A principled approach to operating system construction in Haskell. In: Proc. ICFP 2005, pp. 116–128. ACM Press, New York (2005)
12. Harrison, W.L., Kieburtz, R.B.: The logic of demand in Haskell. Journal of Functional Programming 15(6), 837–891 (2005)
13. Hohmuth, M., Tews, H.: The VFiasco approach for a verified operating system. In: Proc. 2nd ECOOP-PLOS Workshop, Glasgow, UK (October 2005)
14. Huffman, B., Matthews, J., White, P.: Axiomatic constructor classes in Isabelle/HOLCF. In: Hurd, J., Melham, T. (eds.) TPHOLs 2005. LNCS, vol. 3603, pp. 147–162. Springer, Heidelberg (2005)
15. Liedtke, J.: On μ-kernel construction. In: 15th ACM Symposium on Operating System Principles (SOSP) (December 1995)
16. Mossakowski, T., Schröder, L., Goncharov, S.: A generic complete dynamic logic for reasoning about purity and effects. In: Fiadeiro, J., Inverardi, P. (eds.) Proc. FASE 2008. LNCS, vol. 4961, pp. 199–214. Springer, Heidelberg (2008)
17. Shapiro, J.: Coyotos (2006), http://www.coyotos.org
18. Staples, M.: A Mechanised Theory of Refinement. PhD thesis, University of Cambridge (1999)
19. Tuch, H., Klein, G., Norrish, M.: Types, bytes, and separation logic. In: Hofmann, M., Felleisen, M. (eds.) Proc. 34th ACM SIGPLAN-SIGACT Symposium on Principles of Programming Languages, Nice, France, pp. 97–108. ACM Press, New York (2007)
20. Walker, B., Kemmerer, R., Popek, G.: Specification and verification of the UCLA Unix security kernel. Commun. ACM 23(2), 118–131 (1980)

Certifying a Termination Criterion Based on Graphs, without Graphs*

Pierre Courtieu[1], Julien Forest[2], and Xavier Urbain[2]

[1] CÉDRIC – CNAM, Paris, France
[2] CÉDRIC – ENSIIE, Évry, France

Abstract. Although graphs are very common in computer science, they are still very difficult to handle for proof assistants as proving properties of graphs may require heavy computations. This is a problem when it comes to issues such as the certification of a proof of well-foundedness, since premises of generic theorems involving graph properties may be at least as difficult to prove as their conclusion. We define a framework and propose an original approach based on both shallow and deep embeddings for the mechanical certification of these kinds of proofs without the help of any graph library. This framework actually avoids concrete models of graphs and handles those implicitly. We illustrate this approach on a powerful refinement of the dependency pairs approach for proving termination. This refinement makes heavy use of graph analysis and our technique is powerful enough to deal efficiently –and with full automation– with graphs containing thousands of arcs, as they may occur in practice.

1 Introduction

The halting problem is a well-known undecidable problem and its related property, *termination*, plays a fundamental role at several levels in many proofs and definitions. For instance the termination of a relation $\rightarrow$, i.e. the *well-foundedness* of its inverse, is crucial for induction proofs with reference to $\leftarrow$; functions which are *total*, i.e. functions whose computation always terminates, are often compulsory in some proofs assistants; some properties like the confluence of a relation become decidable as soon as the relation is proven to be terminating, etc. Termination is also compulsory when proving *total correctness* of programs.

Discovering a termination proof is often very tricky. The past decade has been rich in developments of automated tools dedicated to termination proofs [14, 10, 12, 18, 17], in particular in the context of first order term/string rewriting systems which we address here. However, skeptical proof assistants [21, 20] cannot take for granted the answers of these tools: they need a formal proof of handled properties. Hence, two of the main concerns are: 1) developing powerful techniques to prove termination of more and more relations with full automation, and 2) obtaining formal (mechanical) certificates of well-foundedness for these relations, in order to enable their definition and use in skeptical proof assistants. We will address here point 2).

Regarding termination criteria, most tools now use the Dependency Pairs (DP) approach, introduced in 1997 by Arts & Giesl [1, 2]. Contrary to the historical Manna and

---

* Work partially supported by A3PAT project of the French ANR (ANR-05-BLAN-0146-01).

O. Ait Mohamed, C. Muñoz, and S. Tahar (Eds.): TPHOLs 2008, LNCS 5170, pp. 183–198, 2008.
© Springer-Verlag Berlin Heidelberg 2008

Ness criterion, the main idea of dependency pairs does not consist in discovering a well-founded, monotonic (i.e., closed by context) and stable (w.r.t. instantiation) ordering for which *each rule* of the system decreases strictly. Roughly speaking, it focuses instead on the possible inner recursive calls of rules (the so-called dependency pairs). This leads to constraints on suitable orderings that are much easier to fulfil. For details, see [2]. This approach has been made even more powerful by use of multiple refinements: in particular it can benefit greatly from the analysis of a *dependency graph*, especially when different orderings can be used [13].

Regarding certification of automated proofs, the project A3PAT[1] aims at improving cooperation between automated provers and skeptical proof assistants [9]. The idea is to get some proof trace from an (efficient) automated prover and translate it into a proof script which can be certified by a targeted proof assistant (which often lacks automation), possibly with the help of dedicated libraries. Eventually we obtain *automatically* a proof of the wanted theorem. One original point of our approach is that it mixes shallow and deep embeddings, this may ease the work of the proof assistant significantly. This work takes place in project A3PAT and focuses on proofs involving *graph analysis*.

Regarding termination in particular, a few libraries model the base theory of dependency pairs [9,8,7]. However, regarding the use of graphs, some properties may require heavy computations and particularly involved algorithmics, and may be very difficult to overcome for a proof assistant (even with the help of dedicated libraries). For instance the strong version of the enhanced dependency graph theorem [13] states that one has to find a suitable ordering "for *each* cycle of the graph", that is: the property has to be shown for each cycle separately, and moreover one has to prove that *all* cycles have been considered. Applying directly such a theorem is currently out of reach regarding the size of graphs that occur in the practice of termination proof.

We propose here a method which allows certification of (termination) proofs based on a graph analysis. Our technique can manage efficiently graphs containing thousands of arcs; it is implemented in the prototype developed by the A3PAT project [9]. Our prototype instantiates our approach and generates termination proof scripts that just have to be compiled and type-checked by the COQ proof assistant[2], possibly with the help of our library COCCINELLE [8].

It is to date the only one able to use the power of the enhanced graph criterion in its strongest version (yet without the so-called usable rules). We consider enhanced graph criteria [13] only as they are much more powerful than original ones [2] which they subsume; for the sake of readability, we will simply write "graph criteria".

Since our approach uses shallow embedding, we describe an instantiation of it, on termination proofs. Thus, in Preliminaries, we will recall some notions and results about graphs, termination of term rewriting, termination proofs and the DP approach with graphs refinements. As noted in [9] and pictured as *processors* in [15] criteria can be expressed in a uniform setting. We will see in Section 3 that we can even write them as formal inference rules. For each of the considered graph criteria, we will give the corresponding rule. Then, in Section 4, we will focus on how we model dependency graphs

---

[1] http://a3pat.ensiie.fr
[2] http://coq.inria.fr

so as to certify termination proof using a proof assistant like Coq. We will eventually provide some experiments to illustrate the efficiency of our approach.

2 Preliminaries

We assume the reader to be familiar with basic concepts of graphs [6], and of term rewriting [11,5] and termination, in particular with the Dependency Pairs approach [2]. We recall the usual notions, and give our notations.

2.1 Graphs

Definition 1 (Graph, Path). *A graph $\mathcal{G}$ is a pair (N, A) where N is a set of* vertices, *and $A \subset N^2$ is a set of* arcs. *We say that an arc $a = (n_1, n_2)$ goes from vertex n_1 to vertex n_2 (noted $n_1 \mapsto_\mathcal{G} n_2$).*
A finite path from a vertex u to a vertex v is a sequence of arcs $(u \mapsto n_0, \dots, m_i \mapsto n_i, \dots, m_{k-1} \mapsto v)$ such that $\forall i, 0 < i < k, n_{i-1} = m_i$. In our case a path is completely determined by the sequence of vertices that it encounters.

Definition 2 (Subgraph). *let $\mathcal{G} = (\mathcal{D}, A)$ be a graph, the* subgraph *of $\mathcal{G}$ generated by $\mathcal{D}' \subset \mathcal{D}$ is the graph $\mathcal{G}_{\mathcal{D}'} = (\mathcal{D}', A')$ such that $A \supseteq A' = \{n \mapsto n' \mid n, n' \in \mathcal{D}'\}$. We also note $\mathcal{G}\backslash d$ the subgraph $\mathcal{G}_{\mathcal{D}\backslash d}$.*

Definition 3 (Strongly connected parts of a graph). *Let $\mathcal{G} = (N, A)$ be a graph. A* strongly connected part *(scp) of $\mathcal{G}$ is a subset C of N such that for all $a, b \in C$ there is a path from a to b in subgraph $\mathcal{G}_C$. We denote $\mathcal{SCP}(\mathcal{G})$ the set of all scp of $\mathcal{G}$.*
A strongly connected component *(scc) of a graph $\mathcal{G}$ is a maximal strongly connected part of $\mathcal{G}$. We denote $\mathcal{SCC}(\mathcal{G})$ the set of all scc of $\mathcal{G}$.*

In the example below, we see that a scc contains a set of scp:

$$\mathcal{SCP}\left(\text{①} \rightleftarrows \text{②} \circlearrowright \right) = \left\{ \text{①} \rightleftarrows \text{②} \; ; \; \text{②} \circlearrowright \; ; \; \text{①} \rightleftarrows \text{②} \circlearrowright \right\}$$

Definition 4 (directed acyclic graph of connected components). *Let $\mathcal{G}$ be a graph. The directed acyclic graph (DAG) of strongly connected components of G is the graph:*
$$\mathcal{G}^{SCC} = (\mathcal{SCC}(\mathcal{G}), \{c_1 \mapsto c_2 | c_1, c_2 \in \mathcal{SCC}(\mathcal{G}) \text{ and } \exists (n_1, n_2) \in c_1 \times c_2, n_1 \mapsto_\mathcal{G} n_2\}).$$

It is easy to see that $\mathcal{G}^{SCC}$ is the directed acyclic graph of strongly connected components of $\mathcal{G}$ since components are *maximal* strongly connected parts. All graphs in this work are finite, in particular $\mapsto_{\mathcal{G}^{SCC}}^+$ is a finite (well-founded) ordering.

2.2 Rewriting

A *signature* $\mathcal{F}$ is a finite set of *symbols* with arities. Let X be a countable set of *variables*; $T(\mathcal{F}, X)$ denotes the set of finite *terms* on $\mathcal{F}$ and X. $\Lambda(t)$ is the symbol at root position in term t. We write $t|_p$ for the subterm of t at position p and $t[u]_p$ for term t

where $t|_p$ has been replaced by u. *Substitutions* are mappings from variables to terms and $t\sigma$ denotes the application of a substitution σ to a term t.

A *term rewriting system* (TRS for short) over a signature $\mathcal{F}$ is a set R of *rewrite rules* $l \to r$ with $l, r \in T(\mathcal{F}, X)$. In this work we restrict to *finite* systems. A TRS R defines a monotonic relation $\to_R$ closed under substitution (aka a *rewrite relation*) in the following way: $s \to_R t$ (s *reduces to* t) if there is a position p such that $s|_p = l\sigma$ and $t = s[r\sigma]_p$ for a rule $l \to r \in R$ and a substitution σ. We shall omit systems and positions that are clear from the context. We denote the reflexive-transitive closure of a relation $\to$ by $\to^\star$. Symbols occurring at root position in the left-hand sides of rules in R are said to be *defined*, the others are said to be *constructors*.

A term is *R-strongly normalizable* (R-SN) if it cannot reduce infinitely many times for $\to_R$. A rewrite relation $\to_R$ *terminates* if any term is R-SN, which we denote $\mathsf{SN}(\to_R)$. In such case we may say that R terminates. This is equivalent to $\leftarrow_R$ is well-founded that is, every term is *accessible* for $\leftarrow_R$.

Dependency pairs. In this section, we briefly recall main definitions and results about dependency pairs, dependency chains and dependency graphs. We introduce some of our notations.

Definition 5 (Dependency pairs, Dependency chain [2]). *The set of* un-marked *dependency pairs of a TRS R, denoted $\mathsf{DP}(R)$ is defined as $\{\langle u, v \rangle \mid u \to t \in R \text{ and } t|_p = v \text{ and } \Lambda(v) \text{ is defined}\}$. Let $\mathcal{D}$ be a set of dependency pairs, a dependency chain in $\mathcal{D}$ is a sequence of dependency pairs $\langle u_i, v_i \rangle$ with a substitution σ such that*

$$\forall i, \ v_i\sigma \xrightarrow[R]{\neq \Lambda \ \star} u_{i+1}\sigma$$

It is worth noticing that distinguishing root symbols of dependency pairs (by means of marks, or 'tuple-symbols') enhances significantly this technique. Marking or not dependency pairs does *not* interfere with our approach, thus for readability's sake, we will restrict to unmarked pairs. Further note that our approach and prototype handle marks without any problem [9].

Definition 6. *Given a TRS R, we note $s \to_{\mathsf{DPR}(\mathcal{D})} t$ iff $s \xrightarrow[R]{\neq \Lambda \ \star} u\sigma \xrightarrow[\langle u,v \rangle \in \mathcal{D}]{\Lambda} v\sigma \equiv t$.*

The main theorem of dependency pairs of [2] can be rephrased using the DPR relation:

Theorem 1 (Dependency Pairs Criterion). *Let R be a TRS, $\to_{\mathsf{DPR}(\mathsf{DP}(R))}$ terminates if and only if $\to_R$ terminates.*

Termination of relation $\to_{\mathsf{DPR}(\mathcal{D})}$ may be directly proved by mean of an ordering pair, a very general notion of which may be found in [19]. Due to our definition of the DPR relation, we use a slightly restricted definition of those but it does not interfere with the topic of this work.

Definition 7 (Ordering pair). *An ordering pair is a pair $(\succeq, >)$ of relations over $T(\mathcal{F}, X)$ such that: 1) $\succeq$ is a quasi-ordering, i.e. reflexive and transitive, 2) $>$ is a strict ordering, i.e. irreflexive and transitive, and 3) $\succeq \cdot > = >$.*
An ordering pair $(\succeq, >)$ is well-founded if there is no infinite strictly decreasing sequence $t_1 > t_2 > \ldots$, which we denote $\mathsf{WF}(\succeq, >)$.

An effective corollary of Theorem 1 consists in discovering a well-founded ordering pair $(\succeq, >)$ for which $\rightarrow_R \subseteq \succeq$ and $u > v$ for all $\langle u, v \rangle \in \mathscr{D}$ to prove that $\rightarrow_{\mathsf{DPR}(\mathscr{D})}$ terminates.

Many efficient termination tools (see for instance [14, 17, 10]) use this criterion as a first step. Then, one is left with proving that there is no infinite dependency chain. In the following we describe the graph criterion which allows to split this proof into easier ones.

Dependency Pairs with graph. Not all DPs can follow another in a dependency chain, one may consider the graph of possible sequences of DPs. Note that since we restricted to finite TRSs, this graph is finite. Thus, each dependency chain corresponds to a *path* in this graph. Therefore if there is no infinite path *corresponding to a dependency chain* in the graph, then there is no infinite dependency chain.

Definition 8 (Dependency graph [2]). *Let R be a rewriting system. The* dependency graph *of R is the graph* $\mathcal{G} = (DP(R), A)$ *where* $\langle s, t \rangle \mapsto \langle s', t' \rangle \in A$ *if and only if there exists a substitution σ such that* $t\sigma \xrightarrow{\neq \Lambda \, *} s'\sigma$.

Remark 1. It is worth noticing that this graph $(\mathscr{D}, A)$ is not computable, so one uses a sound approximation i.e., a graph $(\mathscr{D}, A' \supseteq A)$ that contains it. Arts & Giesl proposed a simple yet efficient approximation, namely *connectability* (with **REN/CAP**) [2]. The approximation we choose to implement in our prototype corresponds to this simple one (see Section 4.2).

The (enhanced) graph refinement, as stated by Giesl, Arts and Ohlebusch is:

Theorem 2 (Dependency graph refinement [13]). *A TRS R terminates if and only if for each circuit C in the dependency graph there exists no infinite dependency chain of dependency pairs of C.*

Note that it is *not* sufficient to consider elementary cycles instead of circuits[3] (for a counterexample, see [4]).

Further note that proving in a proof assistant that his theorem can be applied to a particular termination problem amounts to proving that *all* cycles have been considered, which is difficult in practice. We provide hereafter an approach to avoid this problem.

2.3   Modelling Rewriting and Graphs in CoQ

The goal of our methodology and prototype is to be able to derive *with full automation*, from the definition of a rewrite system R, a proof certified (i.e. checked) by a skeptical proof assistant. Regarding termination proofs, our tool generates a lemma of the form `well_founded R` together with its proof.

We will reuse the model we introduced in [9]. We only recall here the notions and notations used in this paper.

We illustrate our approach using the CoQ proof assistant which is based on *type theory* and enjoys in particular the ability to define *inductive types* to express inductive

---

[3] This is why we use the word "circuit" instead of the original "cycle", cf. [6].

data types and inductive properties, and a very expressive tactic language. Tactics in CoQ unsafely produce proof terms which are safely validated at saving time by type checking the proof.

If R is the relation modelling a TRS $\mathcal{R}$, we should write R u t (which means $u < t$) when a term t rewrites to a term u. For the sake of readability we will use as much as possible the CoQ notation: t - [R] > u (and t - [R] *> u for $t \rightarrow^* u$) instead.

We use in this work a deep embedding for term algebras and shallow embedding for rewriting relations. In CoQ scripts below a term $f(x, y, z)$ will be denoted Term f [x;y;z], and $f(x, y, z) \rightarrow^*_R g(a, z)$ will be denoted Term f [x;y;z] - [R] *> Term g [a;z].

3 Formalizing Graph Refinements

Our project aims at making skeptical proof assistants and automated provers cooperate. Hence, our presentation of the graph refinement differs from the original one from Arts and Giesl [2] in order to fit our general scheme for proving properties on graphs. Such a general scheme could form a basis for a general trace language, similar to the processors setting [14]. For example, in our framework, Theorem 1 is expressed formally by the following inference rule:

$$\frac{\mathsf{SN}(\rightarrow_{\mathsf{DPR}(\mathsf{DP}(R))})}{\mathsf{SN}(\rightarrow_R)} \; \mathrm{DP}$$

The graph criterion consists in proving that there is no infinite dependency chain by proving that *circuits* in the graph cannot be crossed infinitely many times by a dependency chain. For the sake of simplicity, we consider *strongly connected parts* (finite *sets* of vertices) instead of circuits (finite *sequences* of arcs). In particular, a strongly connected part corresponds to a set of circuits. This choice is particularly convenient since our relation DPR is also parameterized by a set of pairs (i.e. vertices).

Definition 9. *Let P be a strongly connected part of a dependency graph, we denote by $\mathsf{SNG}(P)$ the property that there is no reduction in $\rightarrow_{\mathsf{DPR}(P)}$ such that each vertex of P is crossed infinitely many times. For $X = \bigcup_{0 \le i < k} P_i$, we denote $\mathsf{SNG}\{X\} \equiv \bigwedge_{0 \le i < k} \mathsf{SNG}(P_i)$.*

The main theorem of the graph criterion can be rephrased as follows:

Theorem 3 (graph criterion). *Let R be a rewriting system, let $\mathcal{G}$ be its dependency graph. Let $P_1, \ldots, P_k$ be the k scp of $\mathcal{G}$ (the $P_i \in \mathcal{SCP}(\mathcal{G})$ are the subgraphs of $\mathcal{G}$). $\mathsf{SN}(\rightarrow_{\mathsf{DPR}(\mathsf{DP}(R))})$ if and only if $\mathsf{SNG}\{\bigcup_{i=1}^k P_i\}$.*

Remark 2. If $\mathcal{G}$ is a sound approximation of the dependency graph of a TRS R, then only the *if* direction is true.

This theorem can in turn be expressed by the following inference rule:

$$\frac{\mathsf{SNG}\{\mathcal{SCP}(\mathcal{G})\}}{\mathsf{SN}(\rightarrow_{\mathsf{DPR}(\mathsf{DP}(R))})} \; \mathrm{GRAPH}$$

Where $\mathcal{G}$ is the (approximated) dependency graph of R. Note that the termination proof of each scp may be done using a different ordering. In practice this is expensive. Instead, we will gather scp into subsets of $\mathcal{SCP}(\mathcal{G})$, which will be recursively proved to be terminating separately. Actually, the graph criterion can be completed by the following rule for recursive splitting:

$$\frac{\mathsf{SNG}\left\{\,X_1\,\right\}\dots\mathsf{SNG}\left\{\,X_k\,\right\}}{\mathsf{SNG}\left\{\,X\,\right\}}\ \text{S{\scriptsize UB}G{\scriptsize RAPH}}$$

where $\bigcup_{1\leq i\leq k} X_i = X$.

Since the set of strongly connected *components* covers all scp of $\mathcal{G}$, one way to use the graph criterion is to prove that $\to_{\mathsf{DPR}(X_i)}$ terminates for all $X_i \in \mathcal{SCC}(\mathcal{G})$. A weak version of the graph criterion consists in providing *for each strongly connected component* an ordering pair that decreases strictly for all its vertices, and weakly for all rules of the initial system.

The whole termination proof for a system R by this weak graph criterion may be represented by a proof tree like:

$$\frac{\dfrac{\mathsf{WF}(\leq_1,<_1)\quad\to_R\subseteq\,\leq_1}{\to_{\mathsf{DPR}(\mathcal{G}_{C_1})}\subseteq\,<_1}\ \text{O{\scriptsize RD}}}{\mathsf{SNG}\left\{\,\mathcal{SCP}(C_1)\,\right\}}\dots\ \frac{\mathsf{WF}(\leq_k,<_k)\quad\to_R\subseteq\,\leq_k}{\dfrac{\to_{\mathsf{DPR}(\mathcal{G}_{C_k})}\subseteq\,<_k}{\mathsf{SNG}\left\{\,\mathcal{SCP}(C_k)\,\right\}}\ \text{O{\scriptsize RD}}}}{\cdots}$$

$$\frac{\dfrac{}{\mathsf{SN}(\to_{\mathsf{DPR}(\mathsf{DP}(R))})}\ \text{G{\scriptsize RAPH}}}{\mathsf{SN}(\to_R)}\ \text{DP}$$

Where $\mathcal{G}$ is the (approximated) dependency graph of R, and $C_1 \dots C_k$ are the strongly connected components.

The graph criterion in its strong version consists in partitioning the set of scp of $\mathcal{G}$ in parts smaller than scc. An efficient technique, due to Middledorp and Hirokawa [16], is to apply recursively the following steps for each scc C of $\mathcal{G}$:

1. choose a node $p = \langle t, u \rangle$ of C;
2. prove that each scp D containing p is such that $\mathsf{SNG}(D)$;
3. prove (recursively) that each scp of the remaining graph is SNG.

This technique can be formalized by the following application of S{\scriptsize UB}G{\scriptsize RAPH}:

$$\text{S{\scriptsize UB}G{\scriptsize RAPH}}\ \frac{\mathsf{SNG}\left\{\,\mathcal{SCP}(\mathcal{G}\backslash\langle t,u\rangle)\,\right\}\quad\mathsf{SNG}\left\{\,\{P \in \mathcal{SCP}(\mathcal{G})|\langle t,u\rangle \in P\}\,\right\}}{\mathsf{SNG}\left\{\,\mathcal{SCP}(\mathcal{G})\,\right\}}$$

where $\langle t, u \rangle \in \mathcal{G}$. Notice that the rule S{\scriptsize UB}G{\scriptsize RAPH} is applied correctly since $\mathcal{SCP}(\mathcal{G}\backslash\langle t,u\rangle) \cup \{P \in \mathcal{SCP}(\mathcal{G})|\langle t,u\rangle \in P\} = \mathcal{SCP}(\mathcal{G})$.

Usually step 3 is done by computing $\mathcal{G}_1 \dots \mathcal{G}_n$ (the scc of $\mathcal{G}\backslash\langle t, u\rangle$), and by applying recursively the S{\scriptsize UB}G{\scriptsize RAPH} rule to each $\mathcal{G}_i$ until one of the following happens:

- there is no more scc,
- one finds an ordering pair $(\leq, <)$ such that $\to_R\subseteq\,\leq$ and $\to_{\mathsf{DPR}(\mathcal{G})}\subseteq<$.

Usually step 2 is done by discovering a well-founded ordering pair $(\leq, <)$ such that $\to_R\subseteq\,\leq$, $\to_{\mathsf{DPR}(C\backslash\langle t,u\rangle)}\subseteq\,\leq$ and $t < u$. This is rule V{\scriptsize ERTEX}.

Finally a typical graph criterion application is illustrated by Figure 1.

$$\text{SUBGRAPH} \cfrac{\text{SUBGRAPH} \cfrac{\begin{array}{c} \vdots \\ \text{SNG}\left\{\,\mathcal{SCP}(\mathcal{G}_1)\,\right\} \end{array} \dots \begin{array}{c} \vdots \\ \text{SNG}\left\{\,\mathcal{SCP}(\mathcal{G}_k)\,\right\} \end{array}}{\text{SNG}\left\{\,\mathcal{SCP}(\mathcal{G}\backslash\langle t, u\rangle)\,\right\}} \qquad \cfrac{\text{WF}(\leq, <) \quad t < u \quad \rightarrow_R \subseteq\, \leq \quad \rightarrow_{\text{DPR}(\mathcal{G}\backslash\langle t, u\rangle)} \subseteq\, \leq}{\text{SNG}\left\{\,\{P \in \mathcal{SCP}(\mathcal{G})| \langle t, u \rangle \in P\}\,\right\}}\ \text{VERTEX}}{\text{SNG}\left\{\,\mathcal{SCP}(\mathcal{G})\,\right\}}$$

$$\text{where } \mathcal{G}_1 \dots \mathcal{G}_k \text{ are scc of } \mathcal{G}\backslash\langle t, u\rangle.$$

Fig. 1. Typical application of strong graph criterion

4 Mechanical Certification of the Graph Refinement

The key point of our approach is that the graph will be defined implicitly. We never actually model a graph, we just use a set of vertices and a relation between them to build it implicitly as we prove the relevant property on its parts, in a hierarchical fashion. Regarding termination proofs: vertices will be dependency pairs, a pair p_1 will be in relation with a pair p_2 if $p_1 p_2$ may occur in a dependency chain.

In the formal proof that is generated automatically by our technique, each rule applied corresponds to an independent lemma. We described the proof techniques relevant to some of these rules in a previous work [9]. Some of these lemmas are proved using generic theorems previously formalized in a deep embedding, some others are proved by generating directly a shallow embedded proof. Graph refinement rules are of the latter category. As explained in the introduction, the reason is that the premises of rules GRAPH and SUBGRAPH are computationally hard to deal with. For example to prove an application of rule SUBGRAPH one has not only to prove recursively SNG $\left\{\, X_1 \,\right\} \dots$ SNG $\left\{\, X_k \,\right\}$ *but also* to prove that $\bigcup_{1 \leq i \leq k} X_i = X$. Such completeness properties are known to be difficult to prove.

Instead of relying on a generic proof of GRAPH and SUBGRAPH, we generate a direct proof for each application of these rules. This proof is done by induction on the possible dependency chains in the initial graph X. In particular this induction follows a graph that is now *implicit*.

Suppose we have to prove an application of a graph refinement as shown in Figure 1. The goal of our methodology is to build from this tree (output by an automated tool), a formal proof of the property SNG $\left\{\, \mathcal{SCP}(\mathcal{G}) \,\right\}$. To that purpose, we[4] will generate two lemmas proved by different techniques:

- SNG $\left\{\, \mathcal{SCP}(\mathcal{G}\backslash\langle t, u\rangle) \,\right\} \Rightarrow$ SNG $\left\{\, \mathcal{SCP}(\mathcal{G}) \,\right\}$ proved by induction on the well-founded ordering $<$, and
- SNG $\left\{\, \mathcal{SCP}(\mathcal{G}\backslash\langle t, u\rangle) \,\right\}$ proved by a hierarchical decomposition of $\mathcal{G}\backslash\langle t, u\rangle$.

We describe this hierarchical decomposition in more details in the next section.

4.1 Hierarchical Decomposition of $\mathcal{SCC}(\mathcal{G})$

In order to prove SNG $\left\{\, \mathcal{SCP}(\mathcal{G}\backslash\langle t, u\rangle) \,\right\}$ in a shallow embedded way, we proceed as follows:

---

[4] More precisely "the tool in which this approach is implemented", since all this is done without any human interaction.

- Compute the DAG $(\mathcal{G}\backslash\langle t, u\rangle)^{SCC}$.
- Prove *successively*, for each sub-DAG S_i (rooted by $\mathcal{G}_i$) and *in a bottom-up fashion*:
  $(\bigwedge_{S\subset S_i} \mathsf{SNG}\{\mathcal{SCP}(S)\}) \Rightarrow \mathsf{SNG}\{\mathcal{SCP}(S_i)\}$ (See Figure 3 for an example).
  Therefore we can formalize this part of the proof as follows:

$$\text{SubGraph} \quad \frac{\bigwedge_{S_i} \left(\left(\bigwedge_{S\subset S_i} \mathsf{SNG}\{\mathcal{SCP}(S)\}\right) \Rightarrow \mathsf{SNG}\{\mathcal{SCP}(S_i)\}\right)}{\mathsf{SNG}\{\mathcal{SCP}(\mathcal{G}\backslash\langle t, u\rangle)\}}$$

Each proof of $(\bigwedge_{S\subset S_i} \mathsf{SNG}\{\mathcal{SCP}(S)\}) \Rightarrow \mathsf{SNG}\{\mathcal{SCP}(S_i)\}$ is done by proceeding the same way recursively with the proofs of $\mathsf{SNG}\{\mathcal{SCP}(\mathcal{G}_i)\}$ from Figure 1.

4.2 Formalization of Hierarchical Decomposition

Dependency chains. In our methodology dependency graphs and sub-graphs are not concrete. Instead, we will work directly on dependency chains which we model by inductive relations. Using inductive types, reasoning on all possible dependency chains can be done by induction on the definition of the relation.

For example suppose we have a scc SCC defined by the set of pairs:

$$\mathrm{SCC} = \{\langle \mathtt{plus(s\ x,y),plus(x,y)}\rangle, \langle \mathtt{plus(x,s\ y),plus(x,y)}\rangle\}$$

The corresponding dependency chain relation is generated as follows:

```
Inductive SCC : term → term → term:=
  SCC0: ∀V₀ V₁, x -[R]*> S(V₀) → y -[R]*> V₁
                      → plus(x,y) -[SCC]> plus(V₀,V₁)
| SCC1: ∀V₀ V₁, x -[R]*> V₀ → y -[R]*> s(V₁)
                      → plus(x,y) -[SCC]> plus(V₀,V₁)
```

Note that SCC y x is *exactly* equivalent to x $\rightarrow_{\text{DPR(SCC)}}$ y, in particular notice how head reduction by R is disallowed *by construction*. Further note that this relation is not constructively defined: for instance the set of terms x such that x -[R]*> S(V₀) is not defined explicitly and actually, it cannot be computed in general. There are several possible approximations of this set as noted in Remark 1. In our methodology the approximation lies, *during reasoning*, in the way we discard terms t that cannot reduce to u. Currently our generated proofs implement the simple connectability relation of Arts & Giesl [2].

Sub-DAGs of the dependency chains. A scp $\mathcal{G}_i$ of the graph $\mathcal{G}$ implicitly modelled by DPR is built by restraining the constructors of SCC to the set of pairs inside $\mathcal{G}_i$. For example let us consider the following graph corresponding to the relation above:

. The scp of SCC containing only pair $\langle\mathtt{plus(s\ x,y),plus(x,y)}\rangle$ is

modelled by the following relation, which corresponds to :

```
Inductive SCP₀ : term → term → term:=
  SCP₀0: ∀V₀ V₁, x -[R]*> S(V₀) → y -[R]*> V₁
                      → plus(x,y) -[SCP₀]> plus(V₀,V₁)
```

Finally, to prove $\bigwedge_{S_i} ((\bigwedge_{S \subset S_i} \mathsf{SNG} \{ \mathcal{SCP}(S) \}) \Rightarrow \mathsf{SNG} \{ \mathcal{SCP}(S_i) \})$ as explained above, we prove the following equivalent lemmas:

Lemma Acc_S0: $\forall$x y, SCC₀ x y → Acc SCC x.

. . .

Lemma Acc_Sn: $\forall$x y, SCCₙ x y → Acc SCC x.

The proof of `Acc_Si` may use any `Acc_Sj` for j<i. Note that the conclusion is about accessibility in the current graph instead of the well-foundedness of the sub-DAG S_i. Since all dependency chains starting in scc_i can only stay in S_i, those lemmas are equivalent to $\bigwedge_{S_i} ((\bigwedge_{S \subset S_i} \mathsf{SNG} \{ \mathcal{SCP}(S) \}) \Rightarrow \mathsf{SNG} \{ \mathcal{SCP}(S_i) \})$.

5   Examples

5.1   A Weak Graph Criterion Example

The example R_1 below is due to Arts and Giesl [1] and computes the sum of the elements of a list:

```
app(nil, k)  →  k
app(l, nil)  →  l
app(cons(x,l), k)  →  cons(x, app(l, k))
sum(cons(x,nil))  →  cons(x,nil)
sum(cons(x,cons(y,l)))  →  sum(cons(plus(x,y),l))
sum(app(l,cons(x,cons(y,k))))  →  sum(app(l,sum(cons(x,cons(y,k)))))
plus(0,y)  →  y
plus(s(x),y)  →s(plus(x,y)))
```

The dependency pairs of this system are the following:

```
1 : ⟨plus(s(x),y),plus(x,y)⟩
2 : ⟨sum(app(l,cons(x,cons(y,k)))),sum(cons(x, cons(y,k)))⟩
3 : ⟨sum(app(l,cons(x,cons(y,k)))),app(l,sum(cons(x,cons(y,k))))⟩
4 : ⟨sum(app(l,cons(x,cons(y,k)))),sum(app(l,sum(cons(x,cons(y,k)))))⟩
5 : ⟨sum(cons(x,cons(y,l))),plus(x,y)⟩
6 : ⟨sum(cons(x,cons(y,l))),sum(cons(plus(x,y),l))⟩
7 : ⟨app(cons(x,l),k),app(l,k)⟩
```

The (approximated) dependency graph, the induced DAG of scc and corresponding sub-DAGs may be found Figure 2 and Figure 3. Each scc is modelled by the corresponding sub-relation of DPR:

```
Inductive SCC₀ : term → term → Prop :=
| SCC₀₀ : ∀x₀ x₁ V₂ V₃,
      x₀ -[R]*> Term s [V₂] → x₁ -[R]*> V₃
   → Term plus [x₀;x₁] -[SCC₀]> Term plus [V₂;V₃].
```

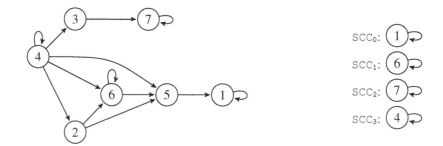

Fig. 2. Dependency graph of R_1 and its scc

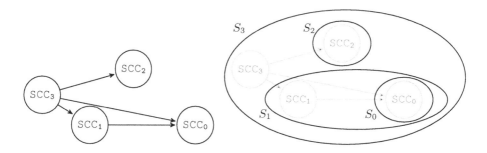

Fig. 3. DAG of scc of R_1 and corresponding sub-DAGs

```
Inductive SCC₁ : term → term → Prop :=
|  SCC₁₀ : ∀x₀ V₁ V₂ V₃,
      x₀ -[R]*> (Term cons [V₂; (Term cons (V₃;V₁))])
   → (Term sum [x₀])
        -[SCC₁]> Term sum [Term cons [Term plus [V₂; V₃];V₁]].

Inductive SCC₂ : term → term → Prop :=
|  SCC₂₀ : ∀x₀ x₁ V₀ V₁ V₂, x₀ -[R]*> (Term cons [V₂;V₁])
      → x₁ -[R]*> V₀
      → Term app [x₀;x₁] -[SCC₂]> Term app [V₁;V₀].

Inductive SCC₃ : term → term → Prop :=
|  SCC₃₀ : ∀x₀ V₀ V₁ V₂ V₃,
      x₀ -[R]*> Term app [V₁;Term cons [V₂;Term cons [V₃;V₀])]] →
   Term sum [x₀] -[SCC₃]>
   Term sum [Term app
               [V₁;Term sum [Term cons [V₂;term cons [V₃;V₀]]]]].
```

Each scc is proved to be terminating using a different ordering by classical induction (see [9]). Then the final proof of **SNG**($\mathcal{G}$) (well_founded DPR below) must be built. We show below how this is done by composing results on sub-DAGs in a bottom-up fashion as described in Section 4.2.

```
(*Now suppose scc are SNG and prove that then DPR is well-founded.*)
Hypothesis Well_Founded_SCC₀ : well_founded SCC₀.
Hypothesis Well_Founded_SCC₁ : well_founded SCC₁.
Hypothesis Well_Founded_SCC₂ : well_founded SCC₂.
Hypothesis Well_Founded_SCC₃ : well_founded SCC₃.

Lemma Acc_SCC₀ : ∀x y, SCC₀ x y → Acc DPR x.
Proof. (*well-founded induction on SCC₀.*) Qed.

Lemma Acc_SCC₁ : ∀x y, SCC₁ x y → Acc DPR x.
Proof. (*well-founded induction on SCC₁ + Acc_SCC₀.*) Qed.

Lemma Acc_SCC₂ : ∀x y, SCC₂ x y → Acc DPR x.
Proof. (*well-founded induction on SCC₂. *) Qed.

Lemma Acc_SCC₃ : ∀x y, SCC₃ x y → Acc DPR x.
Proof. (*well-founded induction on SCC₃ + Acc_SCC₁ + Acc_SCC₂.*) Qed.

Lemma Well_Founded_DPR : well_founded DPR.
Proof. (*case analysis on DPR + Acc_SCC₃.*) Qed.
```

5.2   A Strong Graph Criterion Example

The example below is also due to Arts and Giesl [3] and checks whether the first argument of evenodd is even or not.

```
not(true) →false
not(false) →true
evenodd(x,0) →not(evenodd(x,s(0)))
evenodd(0,s(0)) →false
evenodd(s(x),s(0)) →evenodd(x,0)
```

The dependency pairs of this new system is:

$$1 : \langle \text{evenodd}(x,0), \text{evenodd}(x,s(0)) \rangle$$
$$2 : \langle \text{evenodd}(s(x),s(0)), \text{evenodd}(x,0) \rangle$$
$$3 : \langle \text{evenodd}(x,0), \text{not}(\text{evenodd}(x,s(0))) \rangle$$

The corresponding (approximated) dependency graph is : and its only scc is SCC₀ :

The first step of the graph proof is very similar to the previous example one:

```
Inductive SCC₀ : term →term →Prop :=
| SCC₀₁ : ∀x₀ x₁ V₀,
    x₀ -[R]*> V₀ →x₁ -[R]*> Term 0 [] →
    Term evenodd [x₀;x₁] -[SCC₀]>
      Term evenodd [V₀;Term s [Term 0 []]]
```

```
| SCC₀₂ : ∀x₀ x₁ V₀,
    x₀ -[R]*> Term s [V₀] →x₁ -[R]*> Term s [Term 0 []] →
    Term evenodd [x₀;x₁] -[SCC₀]> Term evenodd [V₀;Term 0 []]
```

Variable Well_Founded_SCC₀ : well_founded SCC₀.

Lemma Acc_SCC₀ : ∀x y, SCC₀ x y →Acc DPR x.
Proof. (*well-founded induction on SCC₀.*) **Qed**.

Lemma Well_Founded_DPR : well_founded DPR.
Proof. (*case analysis on DPR + Well_Founded_SCC₀.*) **Qed**.

Despite the simplicity of this graph, our automated (termination) prover does use the enhanced version of the graph criterion in order to split the single component SCC₀. The pair 2 is strictly oriented by the discovered ordering pair while the pair 1 is only weakly oriented.

The proof of SNG $\{ \mathcal{SCP}(\text{SCC}_0) \}$ is obtained as follows :

```
Inductive SCC₀_large : term →term →Prop :=
| SCC₀_large₁ : ∀x₀ x₁ V₀,
    x₀ -[R]*> V₀ →x₁ -[R]*> Term 0 [] →
    Term evenodd [x₀;x₁] -[SCC₀]>
      Term evenodd [V₀;Term s [Term 0 []]]
```

Variable Well_Founded_SCC₀_large : well_founded SCC₀_large.

```
Inductive SCC₀_strict : term →term →Prop :=
| SCC₀_strict₁ : ∀x₀ x₁ V₀,
    x₀ -[R]*> Term s [V₀] →x₁ -[R]*> Term s [Term 0 []] →
    Term evenodd [x₀;x₁] -[SCC₀]> Term evenodd [V₀;Term 0 []]
```

Variable lt le : term →term →Prop.
Hypothesis lt_le_compat : ∀x y z, lt x y →le y z →lt x z.
Hypothesis wf_lt : well_founded lt.
Hypothesis SCC₀_strict_in_lt :
  Relation_Definitions.inclusion _ SCC₀_strict lt.
Hypothesis SCC_large_in_le :
  Relation_Definitions.inclusion _ SCC₀_large le.

Lemma Well_Founded_SCC₀ : well_founded SCC₀.
Proof. (*well-founded induction on lt and SCC₀_large+
         case analysis on SCC₀*) **Qed**.

6 Experiments

Our approach is implemented in a prototype which is an automated prover dedicated to termination, based on a restricted version of the termination engine of C*i*ME 2.04 [10]. We ran experiments using this technique on a 3GHz, 8GB, Debian-linux machine. Up to now it gives termination certificates for more than 550 problems of the Termination Problems DataBase (TPDB) 4.0^5 (i.e. $\sim$27% of the standard category, some problems of which being non-terminating). It is important to notice that the limiting factor in these results is *not* the certification process itself but the termination techniques used by the prototype! Actually, *all* proofs found by the prototype are certified by COQ.

Some interesting proofs should be highlighted here. We can certify problems with 159 nodes (TRS/TRCSR/PALINDROME_complete-noand_FR.trs, certified in 568.77 s), with 1015 edges (TRS/TRCSR/ExSec11_1_Luc02a_iGM.trs, certified in 113.49s), with 10 strongly connected components at top level (TRS/TRCSR/inn/Ex26_Luc03b_C.trs, certified in 78.80s) or even using 6 times graph splitting (TRS/TRCSR/Ex9_BLR02_iGM.trs.v, certified in 64.15s).

Of course, the certification time for those examples is not representative of the average certification time. It emphasizes what we are able to certify in the TPDB's worst cases. Further note that the certification time takes *all* the certification process into account (orderings, etc.) and not only the graph management. Yet, those times are still reasonable for such tricky examples.

The average certification time on all certified problems is 14s (83% of them are certified in less than 15s and 58% in less than 5s). This makes our approach exploitable in practice, for developments where (involved) proofs of well-foundedness are required.

7 Conclusion

We described an approach to deal efficiently with some graph analysis in skeptical proof assistants. This approach is based on proof scripts generation and uses well-founded induction and dependent inductive types. One of the key points is that induction on the paths in the graph can be simulated by judicious generation of intermediate lemmas that propagate the desired property along the arcs. It benefits from shallow embedding by handling the graph implicitly, without any concrete model. Thus, some premises that would be difficult to prove in a full deep embedding setting (like completeness, for instance) are avoided.

Regarding termination proofs, this technique is fully implemented in a prototype in the context of the A3PAT project; the prototype is available and can be tried online from the web-page of the project: http://a3pat.ensiie.fr. Note that our implementation takes benefit of the expressive tactic language of Coq. Experiments with our prototype illustrate the power of this approach, in particular it allows us to certify in COQ, *in a few seconds*, (scripts of) termination proofs that rely on the circuit analysis of graphs consisting of more than a thousand arcs. Another approach for certifying termination proofs is the Rainbow+Color approach [7], which is based on deep embedding only. Restricted to termination problems, the Color library models several orderings, among which powerful orderings induced by matrix interpretations. However it cannot handle graph criteria as involved as Theorem 2.

Although we illustrate our approach through its instantiation for termination proofs in COQ, it is general enough to adapt to other skeptical proof assistant (e.g. Isabelle/HOL [20]), provided that the targeted assistant is powerful enough to handle

---

[5] http://www.lri.fr/~marche/tpdb

inductive relations and associated reasoning tools, provides a mechanism to deal with well-founded induction and enjoys an expressive enough tactic language. Among the perspectives, a first one could be to implement this approach in other assistants than COQ. A second one could be to generalize this shallow model + external prover technique to deal with generic graph analysis and other properties that rely on it.

Acknowledgments

The authors would like to thank Christiane Goaziou for her help in improving the readability of this paper, and the anonymous referees for their comments.

References

1. Arts, T., Giesl, J.: Automatically Proving Termination Where Simplification Orderings Fail. In: Bidoit, M., Dauchet, M. (eds.) CAAP 1997, FASE 1997, and TAPSOFT 1997. LNCS, vol. 1214. Springer, Heidelberg (1997)
2. Arts, T., Giesl, J.: Termination of term rewriting using dependency pairs. Theoretical Computer Science 236, 133–178 (2000)
3. Arts, T., Giesl, J.: A collection of examples for termination of term rewriting using dependency pairs. Technical report, RWTH Aachen (September 2001)
4. Arts, T., Giesl, J.: Verification of Erlang Processes by Dependency Pairs. Application Algebra in Engineering, Communication and Computing 12(1,2), 39–72 (2001)
5. Baader, F., Nipkow, T.: Term Rewriting and All That. Cambridge University Press, Cambridge (1998)
6. Berge, C.: Graphs, 3rd edn. North-Holland mathematical library, vol. 6. North-Holland, Amsterdam (1991)
7. Blanqui, F., Coupet-Grimal, S., Delobel, W., Hinderer, S., Koprowski, A.: Color, a coq library on rewriting and termination. In: Geser, A., Sondergaard, H. (eds.) Extended Abstracts of the 8th International Workshop on Termination, WST 2006 (August 2006)
8. Contejean, É.: The Coccinelle library for rewriting,
   http://www.lri.fr/~contejea/Coccinelle/coccinelle.html
9. Contejean, É., Courtieu, P., Forest, J., Pons, O., Urbain, X.: Certification of automated termination proofs. In: Konev, B., Wolter, F. (eds.) FroCos 2007. LNCS (LNAI), vol. 4720, pp. 148–162. Springer, Heidelberg (2007)
10. Contejean, É., Marché, C., Monate, B., Urbain, X.: Proving termination of rewriting with CiME. In: Rubio, A. (ed.) Extended Abstracts of the 6th International Workshop on Termination, WST 2003, June 2003, pp. 71–73 (2003), http://cime.lri.fr
11. Dershowitz, N., Jouannaud, J.-P.: Rewrite systems. In: van Leeuwen, J. (ed.) Handbook of Theoretical Computer Science, vol. B, pp. 243–320. North-Holland, Amsterdam (1990)
12. Endrullis, J.: Jambox, http://joerg.endrullis.de/index.html
13. Giesl, J.: Thomas Arts, and Enno Ohlebusch. Modular Termination Proofs for Rewriting Using Dependency Pairs 34, 21–58 (2002)
14. Giesl, J., Schneider-Kamp, P., Thiemann, R.: Aprove 1.2: Automatic termination proofs in the dependency pair framework. In: Furbach, U., Shankar, N. (eds.) IJCAR 2006. LNCS (LNAI), vol. 4130. Springer, Heidelberg (2006)
15. Giesl, J., Thiemann, R., Schneider-Kamp, P., Falke, S.: Mechanizing and Improving Dependency Pairs. Journal of Automated Reasoning 37(3), 155–203 (2006)
16. Hirokawa, N., Middeldorp, A.: Automating the dependency pair method. In: Baader, F. (ed.) CADE 2003. LNCS (LNAI), vol. 2741, pp. 32–46. Springer, Heidelberg (2003)

17. Hirokawa, N., Middeldorp, A.: Tyrolean termination tool. In: Giesl, J. (ed.) RTA 2005. LNCS, vol. 3467, pp. 175–184. Springer, Heidelberg (2005)
18. Koprowski, A.: TPA., http://www.win.tue.nl/tpa
19. Kusakari, K., Nakamura, M., Toyama, Y.: Argument filtering transformation. In: Nadathur, G. (ed.) PPDP 1999. LNCS, vol. 1702, pp. 47–61. Springer, Heidelberg (1999)
20. Nipkow, T., Paulson, L.C., Wenzel, M.T.: Isabelle/HOL. LNCS, vol. 2283. Springer, Heidelberg (2002)
21. The Coq Development Team. The Coq Proof Assistant Documentation – Version V8.1 (February 2007), http://coq.inria.fr

Lightweight Separation

Holger Gast

Wilhelm-Schickard-Institut für Informatik
University of Tübingen
gast@informatik.uni-tuebingen.de

Abstract. Lightweight separation is a novel approach to automatic reasoning about memory updates in pointer programs. It replaces the spatial formulae of separation logic, which complicate automation, by independent assertions about the memory content and the memory layout. As a result, assertions about the content can be treated by existing reasoners. The effect of memory updates is evaluated using specialized tactics that prove disjointness of memory regions from a given memory layout.

1 Introduction

Separation logic [20, 17] has proven a powerful tool for specifying and verifying algorithms that work on mutable heap data structures. Its spatial formulae capture the structure and content of the heap precisely: $E \mapsto F$ holds on heaps with domain $\{E\}$ where F is stored at address E; spatial conjunction $P * Q$ holds on heaps that can be split such that P and Q hold for the disjoint parts. Specifications in separation logic are tight, i.e. a correct program never accesses memory outside of the heap specified in its precondition. A general frame rule then allows local specifications of procedures to be adapted at the call site.

Separation logic has a direct shallow embedding into higher-order logic [23, 1, 22, 2], but reasoning about programs requires a substantial amount of manual interaction, despite the sophisticated automation available in current proof assistants. Different reasons for this shortcoming have been identified in the literature. Appel [1] observes that separation logic is a linear logic, such that classical reasoners do not apply. Tuch [21] points out that the definition of spatial conjunction contains existential quantification, which is known to require human insight in proofs. Weber [23] argues that separation logic is not finitely axiomatizable [8], hence the goal can only be a comprehensive library of theorems.

We will investigate the pragmatic view that the spatial formulae of separation logic restrict access to those formulae that existing reasoners would handle well: the assertions about the content of memory are at least partially hidden below spatial conjunctions. Our goal is therefore to replace spatial formulae by assertions about the memory layout, such that the memory content can be specified independently in classical higher-order logic. This approach is lightweight in that it does not impose a structure on verification rules, assertions, and proofs in the way that separation logic does (e.g. [1, 2]).

O. Ait Mohamed, C. Muñoz, and S. Tahar (Eds.): TPHOLs 2008, LNCS 5170, pp. 199–214, 2008.

The approach poses the obvious challenge that each assertion about the memory content is potentially affected by each memory update in the program. We use conditional rewrite rules to evaluate the effects of memory updates:

$$\frac{\text{is-valid (block a len} \parallel \text{typed-block } \varGamma \text{ a' t)}}{\text{rd } \varGamma \text{ a' t (STORE a len M' M)} = \text{rd } \varGamma \text{ a' t M}}$$

In this rule, rd fetches a typed value from address a and STORE represents one update of len bytes at address a. M is the current, M' a previous memory state. The update can be discarded if the read and the modified memory regions do not overlap, which is expressed in the rule's premise. The disjointness in the premise will be derived from the given memory layout by specialized tactics.

This paper's contribution is a framework, developed in Isabelle/HOL, which uses memory layouts to simplify the effect of memory updates, to discharge side-conditions on the allocatedness of memory regions, and which maintains the layouts themselves through memory updates, allocation, and deallocation. We apply the approach to a low-level language with memory-allocated local variables, references to variables, and structured data types.

Organization of the Paper Section 2 surveys Isabelle's notation and simplifier. Section 3 describes the low-level language that serves as an application of our framework. Section 4 formalizes memory layouts and derives their properties. Section 5 describes the tactics for reasoning about the effects of memory updates. Section 6 applies the framework to the in-place list reversal algorithm. Section 7 surveys related work. Section 8 discusses future work and concludes.

2   Isabelle Syntax and Simplifier

The syntax of Isabelle/HOL follows standard mathematical conventions, with a few exceptions. In Isabelle's meta-logic, $[\![P_1; \dots; P_n]\!] \Longrightarrow Q$ denotes implication of Q by $P_1, \dots, P_n$. Meta-level forall quantification of P over $x_1, \dots, x_n$ is written $\bigwedge x_1, \dots, x_n.P.$ $s \equiv t$ is meta-level equality. The type of total functions from τ to τ' is written $\tau \Rightarrow \tau'$. $\tau \rightharpoonup \tau' = \tau \Rightarrow \tau'$ option is used for partial functions. Type variables are written $'a$, $'b$, ... Type $'a$ set denotes the sets over $'a$. $\{a ..< b\}$ is the half-closed interval $[a, b)$.

Isabelle's simplifier [19, §10] is a flexible rewriting engine whose operation is controlled by *simpsets*. We use three advanced elements of simpsets: the *premises*, *simplification procedures*, and the *subgoaler*. The premises of a simpset are theorems representing the local assumptions valid at the redex. When the simplifier is invoked on a goal $[\![P_1; \dots; P_n]\!] \Longrightarrow Q$, it simplifies each P_i using the remaining P_j as premises, and simplifies Q with premises $P_1, \dots, P_n$. Simplification procedures [19, §10.2.5] are ML functions that compute rewrite rules on demand. Simpsets associate simplification procedures with term patterns. When a redex matches a pattern, the simplifier calls the corresponding procedure with the current simpset and the redex. The procedure may then prove a theorem that the simplifier will apply as a rewrite rule. The subgoaler [19, §10.2.7] of a simpset is a specialized tactic for solving the premises of conditional rewrite rules. To apply a rule $[\![C_1; \ldots; C_n]\!] \Longrightarrow A \equiv B$, the simplifier matches A against the redex and and calls the subgoaler with the instantiated premises $C_1\sigma, \ldots, C_n\sigma$ and the current simpset.

3   A Low-Level Language

L_0 is a prototypical low-level language that shares many aspects with C, but ignores its semantic subtleties [16]. Its main features are memory-allocated local variables and an unrestricted address operator. Since our focus is on reasoning about memory updates, the presentation in this section is necessarily brief.

3.1   Types and Contexts

Types in L_0 are used to determine the size of data objects.

datatype ty = TVoid | TNat | TBool | TPtr ty | TStruct string

Contexts, denoted by Γ, store the definitions of structs, and the types and addresses of local variables. The representation of local variables as an association list simplifies reasoning about block-structured allocation and deallocation.

record ctx =
  ctx-structs :: "string ⇀ (string × ty) list"
  ctx-vars   :: "(string × (addr × ty)) list"

The size of types is given by inductively defined predicates tysz Γ t l and tyszs Γ F l, where t is a type, F is a list of fields, and l is the length as a natural number. Type TVoid has size 0, the primitive types have size 1. tyszs gives the size of a struct type as the sum of the field sizes. The auxiliary predicate field-data Γ t f d l t' yields the offset d, the length l, and the type t' for field f of struct type t in context Γ. A type t is *well-formed* in context Γ, written wf-ty Γ t, if it has a size. This definition allows the use of C-style incomplete types. The definition of tysz and tyszs as a least fixpoint ensures that well-formed types have no cyclic references. The predicate is-field Γ t f asserts that struct type t is well-formed in Γ and has a field f. The function sz-of-ty Γ t = (THE s. tysz Γ t s) facilitates reasoning about type sizes using the definite description operator from Isabelle/HOL. Likewise, addr-of Γ x and type-of Γ x, and sz-of Γ x yield the address, type, and size of local variable x in context Γ using the.

3.2   Memory Model

Memory is a partial mapping from addresses to numbers. Values are represented as lists of numbers [16, 22]. Memory operations are total functions memory ⇒ memory, but they mark memory invalid for accesses outside of the domain. A memory state is *well-formed* if it is valid and has a finite domain.

record memory =	**types**
m-dom :: addr set	addr = nat
m-cnt :: addr ⇒ nat	val = nat list
m-valid :: bool	wf-mem M ≡ m-valid M ∧ mem-finite M

The field m-valid is a history variable [21]: it does not influence the behaviour of memory operations, but records illegal accesses. Side-conditions in the Hoare-logic will guarantee that correct programs do not invalidate memory.

The operations fetch and store follow heap-list and heap-update from [22].

```
fetch :: addr ⇒ nat ⇒ memory ⇒ (nat list × memory)
fetch a 0 M = ([],M)
fetch a (Suc l) M = (let (vs',M') = fetch (a + 1) l M
                     in if a ∈ m-dom M
                        then (m-cnt M a # vs', M')
                        else (arbitrary # vs', M' (| m-valid := False |)))
store :: addr ⇒ nat list ⇒ memory ⇒ memory
store a [] M = M
store a (v # vs) M = (let M' = store (Suc a) vs M
                      in if a ∈ m-dom M'
                         then M' (| m-cnt := (m-cnt M') (a := v) |)
                         else M' (| m-valid := False |))
```

The function alloc abstracts over the allocation strategy using Hilbert's choice operator. Deallocation restricts the domain of the memory state.

```
alloc :: nat ⇒ memory ⇒ (addr × memory)
alloc l M ≡ (let c = (εa. { a ..< a + l } ∩ m-dom M = {})
             in (c, M (| m-dom := m-dom M ∪ { c..< c + l } |)))
dealloc :: addr ⇒ nat ⇒ memory ⇒ memory
dealloc a l M ≡ (if { a ..< a+l } ⊆ m-dom M
                    then M (| m-dom := m-dom M - { a ..< a+l } |)
                    else M (| m-valid := False |))
```

The terms fetch-val a l M, fetch-M a l M, alloc-a a M, and alloc-M a M abbreviate the first and second components, respectively, of the corresponding fetch and alloc.

3.3   State Updates for Forward-Reasoning

Separation logic is usually formulated with forward-style verification rules [17,20, 5,1]. Logics for forward reasoning have not been discussed extensively in the literature on mechanized Hoare logics, but Floyd's assignment axiom generalizes readily.

$$\{ P \} \; x := e \; \{ \exists x'. \; P[x'/x] \wedge x = e[x'/x] \}$$

The post-condition here is obtained by replacing references to x in both the pre-condition and the (side-effect-free) expression e by an existentially quantified variable x' denoting the old value of x. The generalization consists in inverse operators that cancel the effects of memory operators from Section 3.2. The definition of ALLOC is the definition of dealloc, and we omit it for brevity.

```
STORE :: addr ⇒ nat ⇒ memory ⇒ memory ⇒ memory
STORE a l M' M ≡ (if { a ..< a+l } ⊆ m-dom M
                  then M (| m-cnt := λb. if b ∈ { a ..< a + l }
                                         then m-cnt M' b else m-cnt M b |)
                  else M (| m-valid := False |))
ALLOC :: addr ⇒ nat ⇒ memory ⇒ memory
DEALLOC :: addr ⇒ nat ⇒ memory ⇒ memory
DEALLOC a l M ≡ (if m-dom M ∩ { a ..< a+l } = {}
                 then M (| m-dom := m-dom M ∪ { a ..< a+ l } |)
                 else M (| m-valid := False |))
```

3.4   Language, Semantics, and External Syntax

Expressions in L_0 imitate low-level intermediate languages. Type annotations at all operators and constants determine the size of the handled values. Variable types are kept in the context. Primitive operators are given explicitly by values of types prim1 and prim2, which also contain the parameter and return types.

EVar string	EDeref ty exp	EPrim1 prim1 exp
ECnat nat	EAddr exp	EPrim2 prim2 exp exp
ENil ty	EAcc exp ty string	ENew ty
	EAssign ty exp exp	EFree ty exp

The statement language is standard and is omitted for brevity. It includes if, while, and block statements. Expressions can be executed as statements. Block statements contain local variable declarations as a list of names and types.

L_0 uses a standard big-step semantics for expressions and statements. Evaluation of expressions as l-values and r-values (in the sense of C, see [16]) is defined inductively by relations $\Gamma \vdash_l M, e \to a, M'$ and $\Gamma \vdash_r M, e \to v, M'$, where a is the address of the l-value, v is the r-value of type val, M is the old memory state and M' is the new memory state. The statement semantics is given by the relation $\Gamma \vdash_{sm} M, s \to M'$.

In the following, we highlight two properties that are crucial for the Hoare logic. Like any fetch operation, pointer dereferencing for an r-value may invalidate memory. to-ptr lifts a value of type val to type addr by taking its first element.

$$\begin{aligned}
&\llbracket \ \Gamma \vdash_r M0, e \to vs', M1; \\
&\quad (vs,M2) = \text{fetch (to-ptr vs') (sz-of-ty } \Gamma \text{ t) M1} \\
&\rrbracket \Longrightarrow \Gamma \vdash_r M0, \text{EDeref t e} \to vs,M2
\end{aligned}$$

Local variables are memory-allocated. The execution of block statements uses alloc-var and dealloc-var, defined in terms of alloc and dealloc. Declaration d is a pair of name and type; the type determines the size of the allocated memory.

$$\begin{aligned}
&\llbracket \ (x,M1) = \text{alloc-var } \Gamma0 \text{ d M0}; \\
&\quad \Gamma1 = \Gamma0 \ (\!\!(\text{ ctx-vars} := x \ \# \ \text{ctx-vars } \Gamma0 \ )\!\!); \\
&\quad \Gamma1 \vdash_{sm} M1, \text{SBlock decls sm} \to M2; \\
&\quad M3 = \text{dealloc-var } \Gamma1 \text{ x M2} \\
&\rrbracket \Longrightarrow \Gamma0 \vdash_{sm} M0, \text{SBlock (d } \# \text{ decls) sm} \to M3
\end{aligned}$$

L_0 expressions are very verbose. We therefore define an external syntax [19, §6.1] for statements, expressions, and types, and register a parse-translation function [19, §8.6] that creates the internal representation. It includes a type-checker that resolves overloading of primitive operators and computes type annotations by Hindley/Milner type inference. Assertions in Hoare triples are also pre-processed: names of declared local variables are replaced by a read access to the variables. A corresponding print-translation reverses these effects for display.

3.5   Hoare Logic

The following memory access functions are used in assertions. rd and rdv yield a value from address a and variable x, respectively. tyval asserts that a value has the correct length for a given type.

rd Γ a t $\equiv$ λM. fetch-val a (sz-of-ty Γ t) M
rdv Γ x $\equiv$ λM. rd Γ (addr-of Γ x) (type-of Γ x) M
tyval Γ vs t $\equiv$ (length vs = sz-of-ty Γ t)

The inverse operator STORE is lifted to STORE-TYPED Γ a t and STORE-VAR Γ x. Analogous lifted versions exist for ALLOC and DEALLOC. The inverse operator ADD-VAR x Γ cancels the addition of x to Γ; DEL-VAR (x,t) a Γ re-inserts the variable x with type t and address a.

The assertions in Hoare triples for statements are of type ctx$\Rightarrow$memory$\Rightarrow$bool. Post-conditions on expressions have access to the computed result [13]. Well-formedness of memory is an implicit invariant of all correct programs.

$\models$l { P } e { Q } $\equiv$ ($\forall \Gamma$ M M' a. wf-mem M $\longrightarrow$ P Γ M
$\longrightarrow$ Γ $\vdash$l M, e $\rightarrow$ a, M' $\longrightarrow$ wf-mem M' $\wedge$ Q Γ M' a)
$\models$r { P } e { Q } $\equiv$ ($\forall$ Γ M M' vs. wf-mem M $\longrightarrow$ P Γ M
$\longrightarrow$ Γ $\vdash$r M, e $\rightarrow$ vs, M' $\longrightarrow$ wf-mem M' $\wedge$ Q Γ M' vs)
$\models$sm { P } s { Q } $\equiv$ ($\forall \Gamma$ M M'. wf-mem M $\longrightarrow$ P Γ M
$\longrightarrow$ Γ $\vdash$sm M,s $\rightarrow$ M' $\longrightarrow$ wf-mem M' $\wedge$ Q Γ M')

Sound verification rules are obtained as lemmata about Hoare triples. We illustrate those aspects that concern memory reasoning. Pointer dereferencing for an r-value shows the expected side-condition on the allocatedness of the read region. The relation $\triangleright$ is introduced in Section 4.1.

⟦ $\models$r { P0 } e { P1 };
$\forall \Gamma$ M vs. P1 Γ M vs $\longrightarrow$ M $\triangleright$ typed-block Γ (to-ptr vs) t
⟧ $\Longrightarrow$ $\models$r { P0 } EDeref t e { $\lambda \Gamma$ M vs. $\exists$vs'. vs = rd Γ (to-ptr vs') t M $\wedge$ P1 Γ M vs'}

Assignment introduces the inverse operator STORE into the post-condition. The side-condition concerns the allocatedness of the modified memory region.

⟦ $\models$l { P0 } e1 { P1 };
$\forall$a. ($\models$r { $\lambda \Gamma$ M. P1 Γ M a } e2 { P2 a } $\wedge$
$\forall \Gamma$ M vs. P2 a Γ M vs $\longrightarrow$ M $\triangleright$ typed-block Γ a t $\wedge$ tyval Γ vs t)
⟧ $\Longrightarrow$ $\models$l { P0 } EAssign t e1 e2
{ $\lambda \Gamma$ M a. $\exists$M' vs. P2 a Γ (STORE a (sz-of-ty Γ t) M' M) vs $\wedge$ vs = rd Γ a t M }

Block-statements allocate and deallocate local variables. The side-condition on the allocatedness of x could be replaced by the invariant that local variables remain allocated throughout their lifetime. For L_0, we have opted for simplicity.

⟦ $\models$sm { $\lambda \Gamma$ M. type-of Γ x = t $\wedge$ P0 (ADD-VAR x Γ) (ALLOC-VAR Γ x M) }
SBlock decls s { P2 };
$\forall \Gamma$ M. P2 Γ M $\longrightarrow$ M $\triangleright$ var-block Γ x
⟧ $\Longrightarrow$ $\models$sm { P0 } SBlock ((x,t) # decls) s
{ $\lambda \Gamma$ M. $\exists$a. P2 (DEL-VAR (x,t) a Γ) (DEALLOC-VAR (DEL-VAR (x,t) a Γ) x M) }

4   Formalizing Memory Layouts

We split the presentation of lightweight separation into two parts. This section contains the definition and properties of covers and their use in formalizing memory layouts. The next Section 5 describes the tactics that use these properties to automate memory reasoning.

4.1 Covers

A cover is a predicate on address sets; it is *well-formed* if it yields true for a single set. We will call address sets *memory regions*, and speak of *the* memory region covered by a well-formed cover. A cover is *valid* if it covers some region.

> **types** cover = addr set $\Rightarrow$ bool
> wf-cover A $\equiv$ ($\forall$S S'. A S $\longrightarrow$ A S' $\longrightarrow$ S = S')
> is-valid A $\equiv$ ($\exists$S. A S)

Well-formed covers are interchangeable with the regions that they cover. It is this property which will enable many manipulations on memory layouts that would be invalid on spatial formulae of separation logic.

The relations $M \doteq A$, read "M is covered by A" and $M \triangleright A$, read "A is allocated in M" allow covers to be treated as descriptions of the memory layout. Both relations hold only on valid memory, because no assertion must be made about invalid, hence corrupt memory.

> M $\doteq$ A $\equiv$ m-valid M $\wedge$ A (m-dom M)
> M $\triangleright$ A $\equiv$ m-valid M $\wedge$ ($\exists$S. S $\subseteq$ m-dom M $\wedge$ A S)

The *subcover* relation $A \preceq B$ asserts that A covers a part of the region covered by B and that validity of B implies validity of A. It is both reflexive and transitive for arbitrary covers and antisymmetric for well-formed covers.

> A $\preceq$ B $\equiv$ $\forall$S. B S $\longrightarrow$ ($\exists$S'. S' $\subseteq$ S $\wedge$ A S')

Lemma (1) below reduces proving allocatedness of memory regions to proving a subcover relation if a memory layout is given. Lemma (2) uses antisymmetry to derive a complete layout B from a given layout A. The significance of the fact is-valid A in the third premise will be clarified in Section 4.2.

$$[\![\, M \doteq A; \, B \preceq A \,]\!] \implies M \triangleright B \tag{1}$$

$$[\![\, M \doteq A; \, B \preceq A; \, \text{is-valid A} \implies A \preceq B; \, \text{wf-cover A} \,]\!] \implies M \doteq B \tag{2}$$

The *matchcover* relation $A \prec B$ is a weaker variant of subcover. It asserts that validity of B implies validity of A, but does not relate their covered regions. The tactics will employ it to reason about disjointness independently of allocatedness.

> A $\prec$ B $\equiv$ $\forall$S. B S $\longrightarrow$ ($\exists$S'. A S')

The following Lemma (3) reduces proving validity to proving a matchcover relation if a memory layout is given. Since valid layout expressions (Section 4.2) can capture disjointness, this lemma is central to lightweight separation.

$$[\![\, M \doteq A; \, B \prec A \,]\!] \implies \text{is-valid B} \tag{3}$$

4.2 Layout Expressions

The memory layout is captured by nested covers, which we call *layout expressions*. The base of the layout expressions are blocks, which cover an interval of addresses. The empty cover covers the empty region.

> block a l $\equiv$ λS. S = {a ..< a + l}   Empty $\equiv$ λA . A = {}

A typed-block Γ a t covers the block at a with the size of t; a var-block Γ x covers the memory region of variable x. Note that all blocks are well-formed and valid.

The following combinators of type cover⇒cover⇒cover build nested layout expressions. A cover that is not constructed by one of them is a *layout block*.

"A ‖ B ≡ λS. ∃S1 S2. A S1 ∧ B S2 ∧ S = S1 ∪ S2 ∧ S1 ∩ S2 = {}"
"A \ B ≡ λS. ∃S1 S2. A S1 ∧ B S2 ∧ S2 ⊆ S1 ∧ S = S1 - S2"
"A | B ≡ λS. ∃S2. A S ∧ B S2 ∧ S ∩ S2 = {}"

The cover $A \parallel B$, read *disjoint*, covers the union of the disjoint memory regions covered by A and B. Note that it parallels spatial conjunction of separation logic, but does not consider the memory content. As an operator, ‖ is both associative and commutative. $A \parallel B$ is well-formed if both A and B are well-formed.

The operators *clip* $A \setminus B$ and *weak disjointness* $A \mid B$ capture memory deallocation. The cover $A \setminus B$ asserts that the memory region covered by B has been deallocated, but that region is still covered by A. The cover $A \mid B$, on the other hand, asserts that the memory region covered by A does not contain the region covered by B.[1] Both $A \mid B$ and $A \setminus B$ are well-formed if A and B are well-formed.

The tactics of Section 5 use clipping as an intermediate step towards the layout after deallocation. The following lemmata provide the basis for removing clip. (4) shows that removing A from A yields an empty block, which is, of course, disjoint from A. The Lemmata (5) allow this replacement to occur recursively, on both side of ‖ and on the left, the allocated side, of |. Note that the deallocated part C is kept with the already deallocated D in this last case.

$$\llbracket \text{wf-cover } C; \text{ is-valid } C; A = C \rrbracket \implies A \setminus C = \text{Empty} \mid C \tag{4}$$

$$\begin{array}{l} \llbracket \text{wf-cover } C; C \preceq A; (A \setminus C) = A' \rrbracket \implies ((A \parallel B) \setminus C) = (A' \parallel B) \mid C \\ \llbracket \text{wf-cover } C; C \preceq B; B \setminus C = B' \rrbracket \implies ((A \parallel B) \setminus C) = (A \parallel B') \mid C \\ \llbracket \text{wf-cover } C; C \preceq A; (A \setminus C) = A' \rrbracket \implies ((A \mid D) \setminus C) = (A' \mid (D \parallel C)) \mid C \end{array} \tag{5}$$

Several applications of weak disjointness in a row are redundant.

$$\begin{array}{l} ((A \mid C) \parallel B) \mid C = (A \parallel B) \mid C \\ (A \parallel (B \mid C)) \mid C = (A \parallel B) \mid C \\ ((A \mid C) \mid (B \parallel C)) \mid C = (A \mid (B \parallel C)) \mid C \end{array} \tag{6}$$

Proving allocatedness and disjointness is reduced to proving subcovers by Lemmata (1) and (3). The right-hand side then is a known memory layout, and the left-hand side must be proven to be a fragment of it. Lemmata (7) is used to focus on parts of the right-hand-side. Lemma (8) allows subcovers on disjoint regions to be proven independently. Lemma (9) shows that the covered region is immaterial for matchcover.

$$\frac{A \preceq A \parallel B}{B \preceq A \parallel B} \qquad A \preceq A \mid B \qquad \frac{A \prec A \parallel B}{B \prec A \parallel B} \qquad \frac{A \prec A \mid B}{B \prec A \mid B} \tag{7}$$

$$\llbracket A \preceq A'; B \preceq B' \rrbracket \implies A \parallel B \preceq A' \parallel B' \tag{8}$$

$$A \parallel C \prec A \mid C \tag{9}$$

---

[1] Weak disjointness avoids a well-known incompleteness [5] of separation logic. The triple $\{E \mapsto F * E' \mapsto F'\}$ dispose(E) $\{\text{emp} * E' \mapsto F'\}$ is provable, but looses information, since postcondition does not imply $E \neq E'$. Operator $A \mid B$ is defined to maintain just this lost information.

Lemmata (10) and (11) parallel (7) and (8) for the left-hand-side. Their repeated application removes weak disjointness from the third premise of (2). Note that cover B in (2) will not contain the combinator | during forward reasoning.

$$\llbracket \text{ is-valid } (A \mid B); \text{ is-valid } A \Longrightarrow A \preceq A'; \text{ wf-cover } A \rrbracket \Longrightarrow A \mid B \preceq A' \qquad (10)$$

$$\llbracket \text{ is-valid } (A \parallel B); \text{ is-valid } A \Longrightarrow A \preceq A'; \text{ is-valid } B \Longrightarrow B \preceq B' \rrbracket \Longrightarrow A \parallel B \preceq A' \parallel B'$$
$$\tag{11}$$

5   Simplification of Memory Updates

This section describes the specialized tactics that simplify memory updates in verification conditions. Sections 5.1 and 5.2 reduce this problem to proving sub-covers and eliminating clip. Sections 5.3– 5.5 describe the implemented tactics.

5.1   A Framework of Modifiers and Accessors

The Hoare logic introduces several *memory operators*, which take memory states to memory states, and several *memory accessors*, which read a particular value from memory. We define two types:

    mem-op = memory ⇒ memory
    'a mem-acc = memory ⇒ 'a

The following two predicates characterize the behaviour of memory operators and memory accessors using covers: modifies asserts that mop does not change the memory state outside the region covered by A, accesses asserts that the result value does not depend on the memory state outside of region A.

    modifies mop M A ≡ eqv-outside A M (mop M)
    accesses ac M A ≡ ∀ M'. eqv-inside A M M' ⟶ ac M = ac M'

The equivalence relation eqv-inside $A\ M\ M'$ checks that m-dom $M\ a = $ m-dom $M'\ a$ and m-cnt $M\ a = $ m-cnt $M'\ a$ for all addresses a covered by A. eqv-outside does the same for all addresses not covered by A. With these definitions, we have:

$$\llbracket \begin{array}{l} \text{modifies mop M A; accesses mac M B;} \\ \text{wf-cover A; wf-cover B;} \\ \text{is-valid } (A \parallel B) \end{array} \rrbracket \Longrightarrow \text{mac (mop M)} = \text{mac M} \qquad (12)$$

Lemma (12) serves as a generator for rewrite rules that simplify memory update operators in assertions. For each accessor and operator from Sections 3.3 and 3.5, we prove a lemma of the following form.

    lemma STORE-TYPED-modifies[sepmod]:
      "modifies (STORE-TYPED Γ a t M') M (typed-block Γ a t)"
    lemma rd-accesses[sepacc]:
      "accesses (rd Γ a t) M (typed-block Γ a t)"

The attributes sepmod and sepacc declare the lemmata as modifier and accessor theorems, respectively. The framework resolves each pair of modifier/accessor theorems against the first two premises of Lemma (12), solves the wf-cover premises by theorems with attribute `sepwfcover`, and uses the result as a conditional rewrite rule like (13). The implemented tactics (Section 5.5) register a subgoaler that proves the premise from a given memory layout.

$$\text{is-valid (typed-block } \Gamma \text{ a t } \| \text{ typed-block } \Gamma' \text{ a' ta)}$$
$$\Longrightarrow \text{ rd } \Gamma' \text{ a' ta (STORE-TYPED } \Gamma \text{ a t M' M)} = \text{rd } \Gamma' \text{ a' ta M} \tag{13}$$

5.2   Simplifying the Memory Layouts

The memory state M obviously appears in assertions $M \doteq A$, such that these terms need to be simplified as well. For the left-hand-side, we can derive rewrite rules for the inverse operators from Sections 3.3 and 3.5 based on these:

block a l $\preceq$ A $\Longrightarrow$ (STORE a l M' M $\doteq$ A) = (M $\doteq$ A)
(ALLOC a l M $\doteq$ A) = (M $\doteq$ A $\|$ block a l)
(DEALLOC a l M $\doteq$ A) = (M $\doteq$ A $\setminus$ block a l)

Note how allocation and deallocation directly influence the layout expressions describing the memory layout, while STORE has the obvious proof obligation as a premise. The clip operator will be removed in Section 5.4.

The assertion $M \doteq A$ may contain, for example, list-cover (Section 6), and therefore M itself, within A. We can prove a `sepacc` lemma for list-cover, but no simplification can take place, since the memory layout $M \doteq A$ is not available as a premise during simplification of A. The solution is to register a simplification procedure (Section 2) that produces rewrite rules $M \doteq A \equiv M \doteq A'$, where the memory operators in A have been removed in A'. The procedure selects each layout block B of A in turn, and computes $\hat{A}$ with $A = \hat{A} B$. It then proves:

$$M \doteq A = (\exists C.M \doteq \hat{A} C \wedge C = B) \overset{(\dagger)}{=} (\exists C.M \doteq \hat{A} C \wedge C = B') = M \doteq \hat{A} B'$$

The equality ($\dagger$) is resolved using variants of the Isabelle/HOL congruence rules for $\exists$ and $\wedge$. The result is a goal of the form

$$M \doteq \hat{A} C \Longrightarrow (C = B) = (C = B')$$

Now the rewrites from Section 5.1 can be applied to B. Any disjointness derivable from A that does not involve B is also derivable from $\hat{A} C$. Layout blocks like list-cover, however, access their own covered area. Lemma (14) provides the missing rewrite rules, using again pairs of modifier/accessor theorems.

$$[\![\text{ modifies mop M A; } \bigwedge \text{M'. accesses mac M' (mac M');}$$
$$\text{wf-cover A; } \bigwedge \text{M'. wf-cover (mac M');}$$
$$\text{is-valid (A } \| \text{ C)}$$
$$]\!] \Longrightarrow (C = \text{mac (mop M)}) = (C = \text{mac M}) \tag{14}$$

5.3   Proving Subcover Relations

The basis of automation of memory reasoning is the tactic `prove_subcover`, which solves goals of the form $A \preceq B$ and $A \prec B$. The procedure is motivated by example goals that arise from applications of Lemmata (3), (1), and (2).

$$A \parallel B \preceq B \parallel C \parallel A \parallel F \tag{15}$$

$$A \parallel B \parallel E \preceq ((A \parallel B) \mid C) \parallel E \mid F \tag{16}$$

$$A \parallel B \prec B \parallel C \mid (A \parallel F) \tag{17}$$

$$A \parallel B \parallel E \parallel F \prec ((A \parallel B) \mid C) \parallel E \mid F \tag{18}$$

$$A \parallel B \parallel C \prec ((A \parallel B) \mid C) \parallel E \mid F \tag{19}$$

In the following, we leave applications of transitivity and reflexivity implicit. Goal (15) has only $\parallel$ operators on both sides of the subcover relation. It suffices to rearrange the layout blocks using commutativity and associativity into $A \parallel B \parallel \mathsf{Empty} \preceq A \parallel B \parallel C \parallel F$ and to apply (8) repeatedly to achieve trivial goals. Goal (16) is resolved like Goal (15), after the weak disjointness operators on the right have been removed by Lemmata (7) and (8). Goal (17) exploits the top-level weak disjointness; after an application of (9), the goal has the form of (15). Goal (18) is a combination of Goals (17) and Goal (16), in that the nested weak disjointness can be discarded while the top-level one must be exploited. Goal (19) is solved by bringing the inner weak disjointness to the top using Lemmata (7), and then proceeding in (17).

Note also that the following Goal (20) can *not* be derived, because block E may have been allocated in the region covered by the deallocated C.

$$C \parallel E \prec ((A \parallel B) \mid C) \parallel E \mid F \tag{20}$$

Applications of Lemma (2) during forward resolution can lead to subcover goals with weak disjointness on the left, as in Goal (21) below. However, these occurrences are removed by Lemmata (10) and (11).

$$\mathsf{is\text{-}valid}\ (B \parallel C \mid (A \parallel F)) \implies B \parallel C \mid (A \parallel F) \preceq B \parallel C \tag{21}$$

These examples motivate the following procedure for proving $A \preceq B$ or $A \prec B$.

1. *Remove weak disjointness on the left-hand-side.* (Goal (21))
2. *Match the layout blocks of A with the layout blocks of B.*
   Determine the layout blocks $A_1, \ldots, A_n$ and $B_1, \ldots, B_m$ of A and B. For each $i = 1 \ldots n$, find a block B_{j_i} such that theorem $A_i \preceq B_{j_i}$ can be proven. Store also the paths p_i and q_{j_i} of A_i and B_{j_i} from the root of the layout expression. Fail if $j_i = j_{i'}$ for some i, i'.
3. *Drop common the prefix of $q_{j_1}, \ldots, q_{j_n}$ from B.* (Goal (19))
4. *Convert $\prec$ to $\preceq$.* (Goal (17))
5. *Remove weak disjointness on the right-hand-side.* (Goals (16), (20))
6. *Align matching layout blocks on left- and right-hand-side.*
   Use commutativity and associativity and the path information from Step 2.
7. *Solve the goal recursively.*
   Use (Lemma 8) and the theorems from Step 2.

Step 2 contains the opportunity for automatic unfolding. Suppose that A_i and $A_{i'}$ are different fields in the same struct covered by B_j. By definition of the struct, we can prove $A_i \preceq B_j$ and $A_{i'} \preceq B_j$. The the resulting equality $j_i = j_{i'}$ then serves as a trigger for unfolding of B_j into its constituent fields.

5.4 Simplifying Clip

The clip operator $A \setminus B$ leaves implicit which part covered by A has been deallocated. The cover must be simplified into some $A' \,|\, B$ in which the part of A matching B does no longer occur in A'. Here are typical examples:

$$(A\|B\|C\|D)\setminus A = (B\|C\|D)\,|\,A \qquad (22)$$

$$(A\|B|C)\setminus A = (B|(C\|A)) \qquad (23)$$

$$(A\|B|(C\|D))\setminus A = (B|(C\|D\|A)) \qquad (24)$$

In Example (22), one of the allocated blocks is removed. Example (23) demonstrates that the term $(B\,|\,C\,|\,A)$ is not a solution, because it loses the information that A and C are disjoint. Example (24) applies this insight again.

A simplification procedure `clipproc` computes the above rewrite rules as follows. Given a cover $A \setminus B$, it recursively searches for the layout block B within A. When called on $C \setminus A$ and A is found in C, the search returns two theorems: $C \setminus B = C' \,|\, B$ and $B \preceq C$. In the base case, it uses Lemma (4) and reflexivity of subcover. In the recursion step, it resolves the two theorems obtained from the recursive call against the premises of one of the Lemmata (5).

5.5 Implemented Tactics

We have implemented the lightweight separation framework in the form of four tactics. Tactic `sep` invokes the Isabelle simplifier with the conditional rewrite rules from Section 5.1, the simplification procedure from Section 5.4, and a special subgoaler. The subgoaler resolves the is-valid-premise of a rewrite rule (13) by Lemma (3), resolves that lemma's first premise with a premise from the passed simpset, and proves the remaining matchcover relation by the procedure from Section 5.3. Tactic `ctx` registers only the simplification procedure from Section 5.2. Tactic `lift` rewrites the goal with rules declared as `lift`, among them the lifts from STORE to STORE-TYPED and STORE-VAR from Section 3.5.

Tactic `hoare` implements a verification condition generator for forward reasoning. It repeatedly applies one verification rule from Section 3.5 and then applies tactics `lift`, `ctx`, and `sep` to the resulting post-condition. It stops if some inverse operator cannot be removed, if a non-trivial side-condition is to be proven, and after the `if` and `while` rules have introduced the outcome of the test into the precondition. At these points, the user can apply arbitrary tactics before resuming the verification by another invocation of `hoare`.

6 Example: List Reversal

The algorithm for the reversal of a singly-linked list has been used frequently for comparing different approaches to the verification of pointer programs [6, 14, 1, 21]. Figure 1 states the specification and algorithm in L_0 (Section 3.4). The VERIFY keyword triggers the parse translation from the external syntax, the

lemma list-reversal:
"∀ XS. VERIFY VAR p : * struct node; VAR q : * struct node;
⊨sm { node-known Γ ∧ M $\doteq$ p ∥ q ∥ list-cover Γ p M ∧ list-vals Γ p XS M }
BEGIN
  q = nil;
  {INV node-known Γ ∧ M $\doteq$ p ∥ q ∥ list-cover Γ p M ∥ list-cover Γ q M ∧
     (∃ PS QS. list-vals Γ p PS M ∧ list-vals Γ q QS M ∧ XS = (rev QS) @ PS) }
  WHILE p != nil DO
  BEGIN VAR t : * struct node;
    t = (*p). next;
    (*p). next = q;
    q = p;
    p = t
  END
END
{ M $\doteq$ p ∥ q ∥ list-cover Γ q M ∧ list-vals Γ q (rev XS) M }"

Fig. 1. List Reversal

subsequent VAR declarations are used to type-check the program. The outermost forall quantifier introduces the auxiliary variable XS [14].

The assertions in the example demonstrate the independent treatment of memory layout and memory content in lightweight separation. The content of a list is captured in the standard way (e.g. [14, 21]) by an inductively defined predicate list-vals Γ p XS M, where the values of the HOL list XS are stored in the data fields of the singly-linked list starting at p. Note that the predicate can be used freely in HOL terms, e.g. below an existential quantifier in the invariant.

The assertions about the memory layout require a cover for the nodes of a linked list. The struct node with fields data (L_0-type nat) and next (L_0-type *struct node) represents list nodes. The predicate node-known states that struct node is defined in context Γ. The functions node-data-rd and node-next-rd read the fields' content. The cover for the list is given by the following inductive definition.

⟦ p = nil; S = {} ⟧ $\implies$ list-cover Γ p M S
⟦ p ≠ nil; (typed-block Γ (to-ptr p) (TStruct "node")
         ∥ list-cover Γ (node-next-rd Γ (to-ptr p) M) M) S
⟧ $\implies$ list-cover Γ p M S

To use the framework from Section 5.1, we show that list-cover is a well-formed cover and as a memory operator accesses the addresses covered by itself; list-vals accesses the corresponding list-cover.

node-known Γ $\implies$ accesses (list-cover Γ p) M (list-cover Γ p M)
node-known Γ $\implies$ accesses (list-vals Γ p vs) M (list-cover Γ p M)

The verification of the algorithm then consists of 17 steps, 7 of which are invocations of the **hoare** tactic. The main proof obligations are straightforward. After the existentially quantified PS has been instantiated to XS, Isabelle's **force** method proves that the loop invariant holds initially. Maintenance of the loop invariant is proven by **sep** for the memory layout and by **force** for the memory content. Note that the temporary variable t has been allocated and deallocated, and that the list-cover in the memory layout has been simplified.

Within the loop body, the side-conditions on allocatedness (Section 3.5) and the applications of `ctx` and `sep` (Section 5) require the memory layout in different levels of detail: For some, the individual `node` fields need to be exposed, others use the entire list-cover. The user must therefore fold or unfold these structures manually using equalities like the following.

```
node-known Γ ⟹
  typed-block Γ a (TStruct "node") =
    field-block Γ a (TStruct "node") "data" ∥ field-block Γ a (TStruct "node") "next"
```

When the automatic unfolding mechanism sketched in Section 5.3 is implemented, the proof is reduced to 12 lines, 5 of which are applications of `hoare`, and the others solve the natural proof obligations described above.

7 Related Work

Appel [1] provides tactics to manipulate spatial conjunctions in forward verification. His approach is to lift pure assertions, which do not access the heap, to assertions about the empty heap. The tactics facilitate access to such pure spatial conjuncts: they extract equalities for rewriting, apply lemmata to specific conjuncts, and solve trivial implications between spatial formulae by rearranging conjuncts. The general problem remains that the application of Hoare rules places syntactic restrictions on the pre-condition, and these restrictions must be resolved by hand. The example verifications in [1, Section 5] therefore still require a number of technical insights.

Tuch et al. [22] use separation logic for reasoning about C programs by lifting an untyped memory to a typed view, where distinct types occupy disjoint memory regions [7]. They also lift proof obligations about raw memory updates to goals in separation logic [22, §5.2], but do not provide specific support for reasoning in separation logic. Tuch [21] extends the approach to structured types and provides rewrite rules that exploit disjointness of parts within a structured value. He reports using 67 lines for the verification of the list-reversal algorithm.

A different line of research aims at completely automatic verification [5, 4, 15] and shape analysis [9, 11, 3] using separation logic. These approaches focus on the shape of data structures and restrict formulae to the form $\Pi \wedge \Sigma$, where Π is a pure conjunction of pointer equalities and inequalities, and Σ is a spatial conjunction. Automation is achieved by a decidable entailment relation between the restricted formulae (e.g. [5, §4]). Unfortunately, the restrictions also preclude general assertions, such as about the values stored in a linked list.

While separation logic aims at describing the layout of heaps explicitly, many languages maintain the invariant that the data objects of different types and the different fields of the same struct do not overlap on the heap [7]. This insight has proven very successful for mechanized reasoning about pointer programs [10, 6, 14]. However, the assertions there need to contain explicit inequalities between pointers that must be reasoned about manually. Recently, this model has been complemented by a static analysis that discharges many of the remaining disjointness conditions [12]. Since the invariant cannot be circumvented even locally [22], these approaches do not apply to low-level programs.

8   Conclusion

Lightweight separation replaces the spatial formulae of separation logic by layout expressions that capture the disjointness of memory regions in classical higher-order logic. Assertions about the memory content likewise remain classical predicates, such that existing reasoners can be applied. A framework of specialized tactics removes memory updates from verification conditions and discharges side-conditions on allocatedness.

This paper has introduced the method of lightweight separation, focussing on the automation of memory reasoning. Two extensions will be described in companion papers. First, the formalization of (nested) structured datatypes with automatic unfolding (Sections 5.3 and 6). Second, the extension of L_0 by procedures, which require the following variant of the frame rule [20]: preconditions of procedures contain the memory layout $M \doteq R \parallel A$, where A is the region accessed by the procedure and R is an auxiliary variable [18] covering the remaining memory. Pre- and post-condition contain eqv-inside R M_0 M with auxiliary variable M_0. The framework generates additional simplification rules by instantiating the following lemma with modifier theorems:

$$\begin{aligned} & [\![\ \bigwedge M. \ \text{modifies mop M A}; \\ & \quad \text{is-valid } (A \parallel B); \\ & \quad \text{wf-cover A; \ wf-cover B} \\ &]\!] \implies \text{eqv-inside B M (mop M')} = \text{eqv-inside B M M'} \end{aligned}$$

Since the procedure does not modify R, the post-condition eqv-inside R M_0 M can be proven. By instantiating M_0 with the pre-state, and proving the arising tautology in the precondition, the caller gets the assertion that the post-state is unchanged from the pre-state within R, which can in turn be used to simplify the memory operators introduced by the procedure call.

References

1. Appel, A.W.: Tactics for separation logic (January 2006),
   http://www.cs.princeton.edu/~appel/papers/septacs.pdf
2. Appel, A.W., Blazy, S.: Separation logic for small-step C minor. In: Schneider, K., Brandt, J. (eds.) TPHOLs 2007. LNCS, vol. 4732, pp. 5–21. Springer, Heidelberg (2007)
3. Berdine, J., Calcagno, C., Cook, B., Distefano, D., O'Hearn, P., Wies, T., Yang, H.: Shape analysis of composite data structures. In: Damm, W., Hermanns, H. (eds.) CAV 2007. LNCS, vol. 4590. Springer, Heidelberg (2007)
4. Berdine, J., Calcagno, C., O'Hearn, P.W.: Smallfoot: Modular automatic assertion checking with separation logic. In: de Boer, F.S., Bonsangue, M.M., Graf, S., de Roever, W.-P. (eds.) FMCO 2005. LNCS, vol. 4111. Springer, Heidelberg (2006)
5. Berdine, J., Calcagno, C., O'Hearn, P.W.: Symbolic execution with separation logic. In: Yi, K. (ed.) APLAS 2005. LNCS, vol. 3780, pp. 52–68. Springer, Heidelberg (2005)
6. Bornat, R.: Proving pointer programs in Hoare logic. In: Mathematics of Program Construction (2000)

7. Burstall, R.: Some techniques for proving correctness of programs which alter data stuctures. In: Meltzer, B., Michie, D. (eds.) Machine Intelligence, vol. 7. Edinburgh University Press (1972)

8. Calcagno, C., Yang, H., O'Hearn, P.W.: Computability and complexity results for a spatial assertion language for data structures. In: FST TCS 2001: Proceedings of the 21st Conference on Foundations of Software Technology and Theoretical Computer Science, London, UK, 2001, pp. 108–119. Springer, London (2001)

9. Distefano, D., O'Hearn, P.W., Yang, H.: A local shape analysis based on separation logic. In: Hermanns, H., Palsberg, J. (eds.) TACAS 2006 and ETAPS 2006. LNCS, vol. 3920, pp. 287–302. Springer, Heidelberg (2006)

10. Filliâtre, J.-C., Marché, C.: Multi-prover verification of C programs. In: Davies, J., Schulte, W., Barnett, M. (eds.) ICFEM 2004. LNCS, vol. 3308. Springer, Heidelberg (2004)

11. Gotsman, A., Berdine, J., Cook, B.: Interprocedural shape analysis with separated heap abstractions. In: Yi, K. (ed.) SAS 2006. LNCS, vol. 4134, pp. 240–260. Springer, Heidelberg (2006)

12. Hubert, T., Marché, C.: Separation analysis for deductive verification. In: Heap Analysis and Verification (HAV 2007), Braga, Portugal (March 2007)

13. Kowaltowski, T.: Axiomatic approach to side effects and general jumps. Acta Informatica 7, 357–360 (1977)

14. Mehta, F., Nipkow, T.: Proving pointer programs in higher-order logic. Inf. Comput. 199(1-2), 200–227 (2005)

15. Nguyen, H.H., David, C., Qin, S., Chin, W.-N.: Automated verification of shape and size properties via separation logic. In: Cook, B., Podelski, A. (eds.) VMCAI 2007. LNCS, vol. 4349, pp. 251–266. Springer, Heidelberg (2007)

16. Norrish, M.: C formalised in HOL. PhD thesis, University of Cambridge, Technical Report UCAM-CL-TR-453 (1998)

17. O'Hearn, P.W., Reynolds, J.C., Yang, H.: Local reasoning about programs that alter data structures. In: Fribourg, L. (ed.) CSL 2001 and EACSL 2001. LNCS, vol. 2142, pp. 1–19. Springer, Heidelberg (2001)

18. von Oheimb, D., Nipkow, T.: Hoare Logic for NanoJava: Auxiliary Variables, Side Effects, and Virtual Methods Revisited. In: Eriksson, L.-H., Lindsay, P.A. (eds.) FME 2002. LNCS, vol. 2391. Springer, Heidelberg (2002)

19. Paulson, L.C.: Isabelle – A Generic Theorem Prover. LNCS, vol. 828. Springer, Heidelberg (1994)

20. Reynolds, J.C.: Separation logic: A logic for shared mutable data structures. In: Proceedings of the 17th Annual IEEE Symposium on Logic in Computer Science (LICS 2002) (2002)

21. Tuch, H.: Structured types and separation logic. In: 3rd International Workshop on Systems Software Verification (SSV 2008) (February 2008)

22. Tuch, H., Klein, G., Norrish, M.: Types, bytes, and separation logic. In: Hofmann, M., Felleisen, M. (eds.) Proc. 34th ACM SIGPLAN-SIGACT Symposium on Principles of Programming Languages (POPL 2007), Nice, France (January 2007)

23. Weber, T.: Towards mechanized program verification with separation logic. In: Marcinkowski, J., Tarlecki, A. (eds.) CSL 2004. LNCS, vol. 3210. Springer, Heidelberg (2004)

Real Number Calculations and Theorem Proving
Validation and Use of an Exact Arithmetic

David R Lester

School of Computer Science, University of Manchester
Manchester M13 9PL, United Kingdom
dlester@cs.man.ac.uk

Abstract. When handling proofs of properties in the real world we
often need to assert that one numeric quantity is greater than another.
When these numeric quantities are real-valued, it is often tempting to
get out the calculator to calculate the values of the expressions and then
enter the results directly into the theorem prover as "facts" or axioms,
since formally proving the desired properties can often be very tiresome.
Obviously, such a procedure poses a few risks.

An alternative approach, presented in this paper, is to prove the correctness of an arbitrarily accurate calculator for the reals. If this calculator is expressed in terms of the underlying integer arithmetic operations
of the theorem-prover's implementation language, then there is a reasonable expectation that a practical evaluator of real-valued expressions
may have been constructed.

Obviously, there are some constraints imposed by computability theory. It is well known, for example, that it is not possible to determine the
sign of a computable real in finite time. We show that for all practical
purposes, we need not worry about such fussy details. After all, mathematicians have – throughout the centuries – been prepared to make such
calculations without being overly punctilious about the computability of
the operations they were performing!

We report on the experience of validating and using a real number
calculator in PVS.

Keywords: Computable Reals, Exact Arithmetic, Higher-order Logic,
PVS, Theorem Proving.

1 Introduction

My initial motivation to get involved with theorem proving was the recogni-
tion that without guarantees on its correctness, any work on exact arithmetic
was valueless. What is the point in generating arbitrarily accurate answers to
questions involving real arithmetic, if these answers cannot then be relied upon?
This paper presents a small arbitrary precision calculator written and validated
in PVS. To provide interest, the operations implemented include trigonometric
functions as well as exponential and logarithm. The implementation strategy is
that of "Fast Cauchy Sequences"; with a better quality range reduction than

O. Ait Mohamed, C. Muñoz, and S. Tahar (Eds.): TPHOLs 2008, LNCS 5170, pp. 215–229, 2008.

presented in this paper the technique is reasonably competitive in the world of Exact Arithmetic. The library has exactly 500 theorems and will be distributed as part of the NASA library for PVS; all proofs and type-checks are complete and there are no AXIOMs. This is the culmination of work commenced in 1998, and which has proceeded rather fitfully ever since. Initially, a certain amount of work was undertaken to place the NASA library into a non-axiomatic footing. The new material in this paper is that for $\sin(x)$, $\cos(x)$, $\exp(x)$ and $\log(x)$. Previous work on basic operations is in [LG03] and that on power series and inverse trigonometric functions is in [Les03].

2 Real Arithmetic and Computability

One of the obvious problems encountered when we seek to implement real arithmetic on a computer is this: there are more real numbers than there are computer programs. There are only countably many computer programs (and hence representable real numbers), but Cantor showed that there are uncountably many real numbers.

To address this issue much work has been done to impose a computability structure onto, firstly, the real numbers (the so-called "Computable Reals" [Tur36, Tur37]), and to then investigate the consequences to real analysis (or "Computable Analysis" [Grz55a, Grz55b, Grz57, Grz59, BB85, PER89]).

In a non-constructive theorem-proving setting (such as PVS [ORS92], or Isabelle/HOL [Pau94]) we can quite happily prove theorems about real numbers and almost never have to face the issue of computability or constructivity. The exceptions to this happy state of affairs involve actually calculating the values of arithmetic expressions, and comparing their results. At this point we would like to use the power of computers to perform these calculations. An obvious alternative approach is to work in a constructive setting using a theorem-prover such as Coq, as has been done in [Ber07, Jul08].

For those who'd like to see a formal definition of computable reals, it is as follows [PER89].

Definition 1. The real number x is *Computable* if, and only if, there exists three computable functions over the naturals $n, d, s : \mathbb{N} \to \mathbb{N}$, such that for all $k \geq n : \mathbb{N}$

$$\left| x - (-1)^{s(k)} \frac{n(k)}{d(k)} \right| < 2^{-n}$$

In other words a real number x is computable precisely when there is a computable sequence of rational approximations that converge to x, and for which the nth approximation is within 2^{-n} of the value of x.

For the work presented in this paper, the above definition is not strictly required, as we instead relate our representation directly to the reals rather than to the computable reals, relying on the underlying integer arithmetic functions to be computable. However, the limitations on computability implied by our definition inform the practicality of the methods we implement.

Because we use the rounding operation so frequently, we use the following notation.

Definition 2. Rounding is defined as $\lceil x \rceil = \lfloor x + \frac{1}{2} \rfloor$.

In practical terms, then, here is what might realistically be able to implement.

Theorem 1. *Every natural number $i \in \mathbb{N}$ is a computable real.*

Proof. Take $n(k) = i$, $d(k) = 1$, and $s(k) = 0$ in Definition 1.

Theorem 2. *If x and y are computable reals then the following operations are computable and give computable real answers: $x + y$, $x - y$, xy, $\sin(x)$, $\cos(x)$, $\arctan(x)$ (and hence $\pi = 4\arctan(1)$ is computable), $\exp(x)$, $\max(x, y)$, and $\min(x, y)$. In addition, if $y \neq 0$ we can calculate x/y, if $x \geq 0$ we can calculate $\sqrt{(x)}$ and if $x > 0$ we can calculate $\log(x)$.*

Proof. We give a few of these proofs, to give a flavour. We begin with addition; if x and y are computable reals then they each have an associated computable sequence of rational approximations $x_k = (-1)^{s_x(k)} \frac{n_x(k)}{d_x(k)}$ and $y_k = (-1)^{s_y(k)} \frac{n_y(k)}{d_y(k)}$. Let $z_k = \lfloor \frac{x_{k+1} + y_{k+1}}{2} \rceil$, then $|x + y - z_k| < 2^{-k}$ as required. In addition z_k gives rise to associated computable functions $n(k)$, $d(k)$, and $s(k)$.

For $\max(x, y)$ we simply compare approximations. Let $z_k = \max(x_k, y_k)$, then $|\max(x, y) - z_k| < 2^{-k}$ as required.

For $\sin(x)$ we choose a Taylor series approximation $P(x)$ that is within $2^{-(k+1)}$ of $\sin(x)$. Since $|\frac{d}{dx}\sin(x)| \leq 1$, and provided $|\frac{d}{dx}P(x)| \leq 2$ on the open interval $(x_{k+2} - 2^{-(k+2)}, x_{k+2} + 2^{-(k+2)})$ – which can be arranged by taking a sufficiently good approximating polynomial – then

$$|\sin(x) - P(x_{k+2})| < 2^{-k}$$

as required.

At this point we might be feeling pretty smug about our prospects. The following theorem removes this unfounded optimism and also dictates the nature of computable analysis.

Theorem 3. *If x and y are computable reals then the following operations are not computable: $x = y$, $x \leq y$ and $\lfloor x \rfloor$.*

A formal proof requires us to take account of the computability structure, however for our purposes, an informal argument will suffice. Suppose that the approximations for x and y are the same for the first n cases. Then we are unable to answer either $x = y$ or $x \neq y$ in the affirmative. Instead we must determine whether the $n + 1^{\text{st}}$ terms for x and y differ. The significance of the final part is that printing an answer accurate to 10 decimal places may be thought of as calculating $\lfloor 10^{10} x \rfloor 10^{-10}$.

In practice these restrictions impose minimal constraints on calculations. We can replace the calculation of $x \leq y$ with $x \leq_{\epsilon} y$, which is true when $x \leq y - \epsilon$

and false when $y + \epsilon \leq x$ and is either true or false when $|x - y| < \epsilon$. Provided we use this operation in a suitable context this causes no problem. For example, provided F and G give the same answer in the interval $x \pm \epsilon$, the program

$$\texttt{if } x \leq_\epsilon y \texttt{ then } F(x) \texttt{ else } G(x) \texttt{ endif}$$

is computable; it should be noted that many algorithms from Numerical Analysis have this property, since they are intended to work with inaccurate floating-point representations. With regard to printing, it would be a harsh task-master indeed who was prepared to quibble that $2.999999\ldots9999999$ was not an acceptable answer to the calculation of $(\sqrt{3})^2$. As we shall see, even this infelicity can usually be avoided. But note carefully: we cannot completely evade this constraint, as it is a fundamental theoretical limitation on computable arithmetic.

The more common problem is how to evaluate something like x/y? Clearly, if we can prove that y is non-zero then we can calculate x/y. And the best way to calculate that $y \neq 0$? Evaluate it! Thus if we are prepared to exchange a total function x/y for a partial one, our problems are over. In other words we now have a system for numeric calculations which can guarantee that any answers it provides are correct, but which might not terminate for certain arguments. Contrast this with the situation involving floating-point numbers: our calculations execute quickly, none of the basic operations fail to terminate because explicit division by 0 is trapped, but where even simple properties of the real numbers such as associativity are no longer true.

3   Validation of an Exact Arithmetic

Various attempts have been made to do this before. Ménissier-Moran provides a hand proof of an exact arithmetic [Mén94], Harrison [Har97a, Har97b, Har99, Har00] provides a formal verification of mainly floating point arithmetic. Muñoz et al. provide a formal presentation of interval arithmetic [DMM05, ML05], and Lester [LG03, Les03] provides a PVS validation of a restricted set of operations for an exact arithmetic.

There are two justifications for using PVS [ORS92] in this work; firstly, in conjunction with the NASA libraries, there is extensive coverage of the principal functions needed for this paper and their properties. In addition, it is possible to exploit the integer arithmetic operations of the underlying LISP system to execute some of the specification. It is interesting to speculate on whether one can rely on this process being undertaken correctly; as far as the system presented in this paper is concerned we only need the basic integer operations of addition, subtraction, multiplication and division/remainder along with comparisons and string conversions to work correctly.

There is currently no computability library defined in PVS. Obviously, this makes any attempt to restrict attention to just the computable reals problematic. For practical real number calculations this problem is something of a non-issue. Observe that calculating the value of a non-computable real presents something of a problem: if we have an algorithmic way to calculate increasingly accurate rational approximations to the number, then it will in fact be computable!

So, instead, we choose to deal with reals whose computability is guaranteed because they are the results of applying computable operations to the integers, e.g. expressions such as $\exp(\sin(1)/191)\log(672)$. Again – as there is currently no machinery to define computability in PVS – we do not attempt to prove that the operations are in fact computable. Instead, we note that any attempt to define non-computable operations will result in the non-termination of the system.

The key property is:

x: VAR $\mathbb{R}$
p: VAR $\mathbb{N}$
c: VAR $\mathbb{N} \rightarrow \mathbb{Z}$

cauchy_prop(x, c): bool $= \forall\ p$: $c(p) - 1\ <\ x2^p\ \wedge\ x2^p\ <\ c(p) + 1$

cauchy_real?(c): bool $= \exists\ x$: cauchy_prop(x, c)

cauchy_real: NONEMPTY_TYPE $=$ (cauchy_real?) CONTAINING $(\lambda\ p : 0)$

In other words: the computable real x is represented by a function c from $\mathbb{N} \rightarrow \mathbb{Z}$ with the property that for all n: $|x - \frac{c(n)}{2^n}| < 2^{-n}$. Note that every real x has an associated function c, satisfying the Cauchy property above: take $c(n) = \lfloor x2^n \rfloor$; observe very carefully – from Theorem 3 – that this function is not in general computable.

In the argot of the computer arithmetic community the functions c with the above cauchy property – or more correctly $\lambda n : \frac{c(n)}{2^n}$ – are known as *Fast Cauchy Sequences*. The important point about the Fast Cauchy Sequences is that their storage requirements grow linearly with the accuracy required, whereas working directly with the implied representation of the previous section (functions of type $\mathbb{N} \rightarrow \mathbb{Q}$) places no constraints on the storage requirements of each rational in the representation.

The PVS fragment above also defines a type cauchy_real, which restricts the functions from $\mathbb{N} \rightarrow \mathbb{Z}$, to those for which there is a real number satisfying the cauchy property. We have similar definitions for the non-negative reals and positive reals amongst many others.

Note that every integer i is uniquely represented by $\lambda n : i2^n$, but that this uniqueness property is not in general true even for the rationals: consider $\frac{1}{3}$. Both $\lambda n : \lfloor \frac{2^n}{3} \rfloor$ and $\lambda n : \lceil \frac{2^n}{3} \rceil$ are acceptable representations because $\frac{2^n}{3}$ is not an integer and hence either floor or ceiling may be selected.

We can define an addition operation for our representation.

cx,cy: VAR cauchy_real
x, y: VAR $\mathbb{R}$
p: VAR $\mathbb{N}$

cauchy_add(cx, cy): cauchy_real

$$= \lambda\ p:\ \lfloor(cx(p+2)\ +\ cy(p+2))/4\rceil$$

add_lemma: LEMMA cauchy_prop(x,cx) $\wedge$ cauchy_prop(y,cy)
$\Rightarrow$ cauchy_prop($x+y$, cauchy_add(cx,cy))

This is typical of the approach. We define a function that manipulates the representations of the input computable reals, in this case: cauchy_add. If we wish to (or are able to) we can show that this function is computable. As we shall see, these functions are normally simple rational calculations for which we will have no qualms about their computability. We then show that if the cauchy property holds for the input computable reals, then the output computable real also has the required cauchy property.

Before attempting this, one needs a viable theorem proving system which is capable of dealing with some of the more everyday transcendental functions such as sin, cos, arctan, exp and log. After all, without some such functions the provision of an exact arithmetic becomes trivial: the algebraic numbers are a countable computable field.

4   An Example in Raw PVS

Before discussing the proofs, let's look at the system in action. Suppose we wish to use the system to calculate an approximation to π, how do we proceed? First, we declare as a LEMMA the bounds of our calculation, in this case forty decimal places of π.

```
new_pi_bnds:    LEMMA % 40 dp
       3.1415926535897932384626433832795028841971 < π ∧
  π <  3.1415926535897932384626433832795028841972
```

This theorem takes about 30 seconds to prove on a reasonable 1GHz laptop. That's because the underlying LISP system is not using a modern integer representation such as GMP, and our power series calculations are not as efficient as they might be. Running the same calculation in Haskell under the Hugs interpreter the time is only 1 second.

The proof of this result is in two stages. First, we claim that |cauchy_pi $(140)/2^{140} - \pi| < 1/2^{140}$. By using pi_lemma, which merely states that the cauchy property holds between cauchy_pi and π, we know that this is true for any natural number, not just 140. For reference

cauchy_pi(140) = 4378741080330103799233250808471022728399425.

To ease the exposition, we have used the PVS renaming command for some of the expressions. The rational cauchy_pi(140)/2^{140} has been named APPROX, the rational $1/2^{140}$ has been named EPS, and we have also named the lower and upper bounds we are seeking to establish LO and HI respectively.

And now, because we have been lucky, it turns out that both LO + EPS < APPROX and APPROX < HI − EPS are true. This is where the heavy-duty evaluation is taking place. Note these are simple integer arithmetic operations that are being undertaken within the theorem-prover. It is this process of evaluation that has been previously referred to as using the "integer arithmetic of the underlying LISP system".

Had the approximation APPROX been close to one or other bound then we would have needed a better approximation. To do this, we merely increase the value of the precision from 140 to a higher value, thereby reducing the approximation error EPS.

A mildly compressed version of the output proof trace of PVS is as follows:
new_pi_bnds:

> {1} 3.14159265358979323846264338327950288841971 $< \pi \wedge$
> $\pi <$ 3.14159265358979323846264338327950288841972

Case splitting on abs(cauchy_pi(140)/$2^{140} - \pi$) $< 1/2^{140}$,
we get 2 subgoals:
new_pi_bnds.1:

> {-1} abs(cauchy_pi(140)/2 $\wedge$ 140 $- \pi$) $<$ 1/2 $\wedge$ 140
> {1} 3.14159265358979323846264338327950288841971 $< \pi \wedge$
> $\pi <$ 3.14159265358979323846264338327950288841972

Using APPROX to name and replace cauchy_pi(140)/2^{140}, EPS to name and replace $1/2^{140}$, LO to name and replace 3.14159265358979323846264338327950288841971, and HI to name and replace 3.14159265358979323846264338327950288841972,
new_pi_bnds.1:

> {-1} abs(APPROX $- \pi$) $<$ EPS
> {1} LO $< \pi \wedge \pi <$ HI

Case splitting on LO + EPS $<$ APPROX AND APPROX $<$ HI $-$ EPS, and flattening,
we get 2 subgoals:
new_pi_bnds.1.1:

> {-1} LO + EPS $<$ APPROX
> {-2} APPROX $<$ HI $-$ EPS
> {-3} abs(APPROX $- \pi$) $<$ EPS
> {1} LO $< \pi \wedge \pi <$ HI

Expanding the definition of abs, simplifying, rewriting, and recording with decision procedures,
This completes the proof of new_pi_bnds.1.1.
Hiding formulas: -1, 2,

`new_pi_bnds.1.2:`

$$\mathbin{\vert\!\!\!\overline{}}$$
$$\{1\}\ \mathrm{LO+EPS}\ \ <\ \ \mathrm{APPROX}\ \wedge\ \mathrm{APPROX} < \mathrm{HI-EPS}$$

Evaluating formula 1 in the current sequent,
This completes the proof of `new_pi_bnds.1.2`.
Hiding formula: 2,
`new_pi_bnds.2:`

$$\mathbin{\vert\!\!\!\overline{}}$$
$$\{1\}\ \mathrm{abs}(\mathrm{cauchy\_pi}(140)/2\ {}^{\wedge}\ 140 - \pi) < 1/2\ {}^{\wedge}\ 140$$

Applying pi_lemma, expanding the definition of cauchy_prop, instantiating the top quantifier in - with the terms: 140,
`new_pi_bnds.2:`

$$\{\text{-}1\}\quad \mathrm{cauchy\_pi}(140) - 1\ <\ \pi \times 2\hat{\ }140\ \wedge$$
$$\pi \times 2\hat{\ }140\ <\ 1 + \mathrm{cauchy\_pi}(140)$$
$$\overline{}$$
$$\{1\}\ \mathrm{abs}(\mathrm{cauchy\_pi}(140)/2\ {}^{\wedge}\ 140 - \pi)\ <\ 1/2\ {}^{\wedge}\ 140$$

Rewriting with div_mult_pos_lt1 and div_mult_pos_lt2, expanding abs and Simplifying, rewriting, and recording with decision procedures,
This completes the proof of `new_pi_bnds.2`.
Q.E.D.

5   Transcendental Functions: Sine and Cosine

The new material in this paper is that concerning the transcendental functions $\sin(x)$ and $\cos(x)$ and of $\ln(x)$ and $\exp(x)$, as [LG03] showed how to evaluate power series in an exact arithmetic setting, and [Les03] applied this technique to evaluate $\arctan(x)$.

Assuming that one has selected a sensible representation of the numbers, the key to efficient evaluation of exact arithmetic lies in choosing a good range reduction which restricts the domain over which the associated power series is evaluated. The reason is simple: over a small interval even the most complicated continuous functions can be accurately evaluated with a low-order polynomial approximation (ideally just a linear interpolation). As the interval size increases, the order of the approximating polynomial will also increase, leading to poor performance. In this paper we choose a fairly large domain, as this simplifies the proof and presentation. For a realistic implementation, competitive in the annual exact arithmetic competition, a much more sophisticated scheme needs to be used [BEIR00].

For $\sin(x)$ and $\cos(x)$ we use the traditional Taylor series approximation with x restricted to the domain $-\frac{3\pi}{16} < x < \frac{3\pi}{16}$. Notice that $0 \le x^2 < \frac{1}{2}$ and that in the Taylor series approximation for $\sin(x)$ and $\cos(x)$ each coefficient is less than or equal to 1. Therefore, for p bits of accuracy we need at most $p+2$ terms of the power series and we perform the calculations involved in the summation also at $p+2$ bits of accuracy.

cauchy_sin_drx(csnx): cauchy_real =
    λ p: $\lfloor$cauchy_powerseries(csnx, cauchy_sin_series, $p+2$)$(p+2)/4\rceil$

cauchy_cos_drx(csnx): cauchy_real =
    λ p: $\lfloor$cauchy_powerseries(csnx, cauchy_cos_series, $p+2$)$(p+2)/4\rceil$

sin_drx_lemma: LEMMA cauchy_prop(snx, csnx) $\wedge$ snx $\neq$ 0 $\Rightarrow$
    cauchy_prop(sin(sqrt(snx))/sqrt(snx), cauchy_sin_drx(csnx))

cauchy_sin_dr(csx): cauchy_real =
    cauchy_mul(csx, cauchy_sin_drx(cauchy_mul(csx, csx)))

cauchy_cos_dr(csx): cauchy_real =
    cauchy_cos_drx(cauchy_mul(csx, csx))

sin_dr_lemma: LEMMA
    cauchy_prop(sx, csx) $\Rightarrow$ cauchy_prop(sin(sx), cauchy_sin_dr(csx))

cos_dr_lemma: LEMMA
    cauchy_prop(sx, csx) $\Rightarrow$ cauchy_prop(cos(sx), cauchy_cos_dr(csx))

The critical part now is to find a computable way to divide up the domain so that we only ever apply sin and cos to arguments in the interval $(-\frac{3\pi}{16}, \frac{3\pi}{16})$. For both $\sin(x)$ and $\cos(x)$ we calculate the integer closest to $k = \lfloor \frac{4x}{\pi} \rceil$ evaluated to two bits accuracy. It is this approximation that causes the argument interval to be wider than the expected $(-\frac{\pi}{8}, \frac{\pi}{8})$. Defining $y = x - \frac{k\pi}{4} \in (-\frac{3\pi}{16}, \frac{3\pi}{16})$, and knowing the remainder mod 8 of k we can reconstruct $\sin(x)$ and $\cos(x)$ in terms of $\sin(y)$, $\cos(y)$ and $\sqrt{\frac{1}{2}} = \sin(\frac{\pi}{4}) = \cos(\frac{\pi}{4})$. The desired cauchy properties then follow.

cauchy_sin(cx): cauchy_real =
  LET p_2 = cauchy_div2n(cauchy_pi, 2),
      s_2 = cauchy_sqrt(cauchy_div2n(cauchy_int(1), 1)),
      k = round(cauchy_div(cx, p_2)(2)/4),
      n = rem(8)(k),
      cy = cauchy_sub(cx, cauchy_mul(p_2, cauchy_int(k))),
      s = cauchy_sin_dr(cy),
      c = cauchy_cos_dr(cy)
    IN IF n = 0 THEN s
      ELSIF n = 1 THEN cauchy_mul(s_2, cauchy_add(c, s))
      ELSIF n = 2 THEN c
      ELSIF n = 3 THEN cauchy_mul(s_2, cauchy_sub(c, s))
      ELSIF n = 4 THEN cauchy_neg(s)

      ELSIF $n = 5$ THEN cauchy_neg(cauchy_mul(s_2, cauchy_add(c, s)))
      ELSIF $n = 6$ THEN cauchy_neg(c)
      ELSE cauchy_neg(cauchy_mul(s_2, cauchy_sub(c, s))) ENDIF

cauchy_cos(cx): cauchy_real =
  LET p_2 = cauchy_div2n(cauchy_pi, 2),
      s_2 = cauchy_sqrt(cauchy_div2n(cauchy_int(1), 1)),
      k = round(cauchy_div(cx, p_2)(2)/4),
      n = rem(8)(k),
      cy = cauchy_sub(cx, cauchy_mul(p_2, cauchy_int(k))),
      s = cauchy_sin_dr(cy),
      c = cauchy_cos_dr(cy)
    IN IF $n = 0$ THEN c
      ELSIF $n = 1$ THEN cauchy_mul(s_2, cauchy_sub(c, s))
      ELSIF $n = 2$ THEN cauchy_neg(s)
      ELSIF $n = 3$ THEN cauchy_neg(cauchy_mul(s_2, cauchy_add(c, s)))
      ELSIF $n = 4$ THEN cauchy_neg(c)
      ELSIF $n = 5$ THEN cauchy_neg(cauchy_mul(s_2, cauchy_sub(c, s)))
      ELSIF $n = 6$ THEN s
      ELSE cauchy_mul(s_2, cauchy_add(c, s)) ENDIF

sin_lemma: LEMMA
  cauchy_prop(x, cx) $\Rightarrow$ cauchy_prop($\sin(x)$, cauchy_sin(cx))

cos_lemma: LEMMA
  cauchy_prop(x, cx) $\Rightarrow$ cauchy_prop($\cos(x)$, cauchy_cos(cx))

Since they're now easy to implement and prove, $\sec(x)$, $\csc(x)$, $\tan(x)$, $\cot(x)$, $\arcsin(x)$ and $\arccos(x)$ are also provided.

6   Transcendental Function: Natural Logarithm

We begin by restricting the domain for $\ln(x)$ to $x \in [\frac{1}{2}, \frac{3}{2}]$ (or ssx $\in [-\frac{1}{2}, \frac{1}{2}]$), and calculating the natural logarithm by Taylor series expansion.

cauchy_ln_drx(cssx)(p): $\mathbb{Z}$ =
    $\lceil$cauchy_powerseries(cssx, cauchy_ln_series, $p + 2$)($p + 2$)/4$\rceil$

ln_drx_lemma: LEMMA cauchy_prop(ssx, cssx) $\Rightarrow$
    cauchy_prop($\ln(1 + \text{ssx})$, cauchy_ln_drx(cssx))

From this we can calculate values for $\ln(\frac{1}{2})$, $\ln(2)$ and $\ln(\sqrt{2})$. Using these values we calculate a value for $\ln(\text{cmx})$ with cmx $\in [\frac{1}{4}, \frac{9}{4}]$. We note first that if $3 \leq \text{cmx}(2) \leq 5$ then the argument supplied lies in the interval: $(\frac{1}{2}, \frac{3}{2})$, and we can use our previous approximation to $\ln(\text{cmx})$. Otherwise, the value of cmx must lie in either of the intervals $[\frac{1}{4}, \frac{3}{4})$ or $(\frac{5}{4}, \frac{9}{4}]$. In either case $\sqrt{\text{cmx}} \in [\frac{1}{2}, \frac{3}{2}]$, and we use the identity $\ln(\text{cmx}) = 2 * \ln(\sqrt{\text{cmx}})$ to establish the required cauchy property.

cauchy_ln_dr(cmx): cauchy_real =
    IF $3 \leq \text{cmx}(2) \wedge \text{cmx}(2) \leq 5$
    THEN cauchy_ln_drx(cauchy_sub(cmx, cauchy_int(1)))
    ELSE cauchy_mul2n(cauchy_ln_drx(cauchy_sub(cauchy_sqrt(cmx),
                                                cauchy_int(1))), 1)
    ENDIF

ln_dr_lemma: LEMMA
    cauchy_prop(mx, cmx) $\Rightarrow$ cauchy_prop(ln(mx), cauchy_ln_dr(cmx))

Next, we perform the classic range reduction for $\ln(x)$ [CW80, Mul97], that is: determine y and n such that $x = y2^n$ with $y \leq \frac{9}{4}$, in which case $\ln(x) = \ln(y) + n\ln(2)$. Provided $x \geq \frac{1}{4}$, we have two cases. If $\text{cx}(2) \leq 8$ then $x \leq \frac{9}{4}$ as required. Otherwise we use the identity $\ln(\frac{x}{2^n}) + n\ln(2) = \ln(x)$ choosing n so that $\frac{x}{2^n} \leq \frac{9}{4}$. Notice carefully the use of the recursively defined function floor_log2 : $[\mathbb{N} \rightarrow \mathbb{N}]$. Using $\lfloor \ln(t) \rfloor$ here would break the executability of the PVS specification.

cauchy_lnx(cx): $[\mathbb{N} \rightarrow \mathbb{Z}]$ =
    LET $t = \text{cx}(2)$ IN
        IF $t \leq 8$
            THEN cauchy_ln_dr(cx)
            ELSE LET $n = \text{floor\_log2}(t) - 1$ IN
                    cauchy_add(cauchy_ln_dr(cauchy_div2n(cx, n)),
                               cauchy_mul(cauchy_int(n), cauchy_ln2))
        ENDIF

ln_lemma_x: LEMMA
    cauchy_prop(x, cx) $\Rightarrow$
    cauchy_prop(ln(x), cauchy_lnx(cx))

Finally, to obtain a function for the whole range (including the currently missing interval $(0, \frac{1}{4})$), we use the identity $-\ln(\frac{1}{x}) = \ln(x)$ should it be needed.

cauchy_ln(pcx): cauchy_real =
    IF $\text{pcx}(2) \leq 2$
        THEN cauchy_neg(cauchy_lnx(cauchy_inv(pcx)))
        ELSE cauchy_lnx(pcx)
    ENDIF

ln_lemma: LEMMA
    cauchy_prop(px, pcx) $\Rightarrow$ cauchy_prop(ln(px), cauchy_ln(pcx))

7   Transcendental Function: Exponential

This is now relatively easy. For sx $\in (-1, 1)$ the standard Taylor series performs adequately.

cauchy_exp_dr(csx)(p): $\mathbb{Z}$ =
    $\lfloor$cauchy_powerseries(csx, cauchy_exp_series, $p + 3)(p + 2)/4\rceil$

exp_dr_lemma: LEMMA
    cauchy_prop(sx, csx) $\Rightarrow$ cauchy_prop(exp(sx), cauchy_exp_dr(csx))

The classic range reduction for $\exp(x)$ now involves exploiting the identity $\exp(x) = 2^n \exp(x - n \ln(2))$

cauchy_exp(cx): $\begin{bmatrix} \text{nat} \rightarrow \text{int} \end{bmatrix}$ =
    LET n = cauchy_div(cx, cauchy_ln2)(0),
        cy = cauchy_sub(cx, cauchy_mul(cauchy_int(n), cauchy_ln2))
    IN IF $n < 0$ THEN cauchy_div2n(cauchy_exp_dr(cy), $-n$)
        ELSIF $n > 0$ THEN cauchy_mul2n(cauchy_exp_dr(cy), n)
                    ELSE cauchy_exp_dr(cy) ENDIF

exp_lemma: LEMMA
    cauchy_prop(x, cx) $\Rightarrow$ cauchy_prop(exp(x), cauchy_exp(cx))

Since they're now easy to implement and prove, hyperbolic and inverse hyperbolic functions are also provided.

8   Printing the Answer

In the version presented here, in PVS, we have no easy way to determine the value of a calculation. Instead we have provided a mechanism by which the value may be compared in a fixed inequality. If instead we had a mechanism to print out the results, how would we ensure accuracy? The simple technique is to evaluate an expression to $n + 1$ decimal's accuracy.

Suppose that we wish to print out just three decimal places, and that these are estimated to be 0.129. Provided that the calculation to 4 decimal places is not precisely 0.1295 we will be able to round up or down correctly. Obviously – and this is a fundamental limitation – if the result really is 0.1295000000... then no amount of extra evaluation will be able to determine whether to round up or down. However we could count ourselves unlucky to have hit this precise problem: selecting one further decimal place of accuracy would have obviated the problem entirely.

9   An Oversight

One problem discovered as a result of directly executing the PVS specification was that the integer square-root of n had been defined as $\lfloor \sqrt{n} \rfloor$. Obviously this caused the evaluation to fail as there is no LISP evaluation available to evaluate this definition. An earlier version of the proof had been more explicit in that it had constructively defined this operation. When it was reinstated, the system evaluated integer square-root correctly, but due to the algorithm being perhaps the worst possible (starting at $i = 0$, iterate until $i^2 \geq n$), the evaluation only works for very small values of n.

A better version of integer square-root using bisection or Newton iteration is in the process of being constructed.

10   Conclusion

One question that a number of reviewers posed is: to what extent can this technique be used to automatically determine inequalities? I'm afraid that the only honest answer is another question: to what extent is an approximate answer satisfactory? If an exact answer is required then Theorem 3 already provides the result: in general – using computable operators on computable reals – we cannot determine which of $x < y$, $x = y$ or $x > y$ is true. However, we *can* define automated semi-decision procedures for $x \neq y$ and $x < y$ (naïvely, increase the precision until the inequality is proved); this was what we did in the interval-based version described in [ML05]. Note carefully that the converses $x = y$ and $x \leq y$ cannot be proved using this technique. Another alternative, if an approximate answer (accurate to any particular ϵ) is acceptable, is to evaluate $x - y$ to accuracy ϵ: this technique can be easily automated, and will always terminate. Notice that we are discussing the general situation: *i.e.* general computable operators applied to arbitrary computable reals. For the particular case of elementary functions a better result is possible, see, for example, [AP06, AP07].

One of the problems which came to light in performing this work is that partial functions are second class citizens in PVS. By this we do not mean that they are not provided (they are), nor that they are tricky to reason about (they aren't). Rather, we might like to represent a non-terminating evaluation by a recursive partial function, and this is tricky. Our particular problem occurs with division, where we need to show that the divisor is non-zero. In the PVS system described, we will need to provide a *proof* that the divisor is non-zero, leading to a system that is a curious mix of theorem-proving and calculation.

Perhaps because of anticipating the use of a theorem-prover to double-check the results, the specification was found to be defective in various minor ways. Nineteen bugs were detected by the use of PVS validation. Typically one or two too few terms of Taylor series expansion were used or evaluation was one or two bits too inaccurate for the desired accuracy of the output. One of these was detected by testing. One other bug involved an over-enthusiastic cut-and-paste for the cases in $\cos(x)$!

No claim is made that the bounds provided in this paper are optimal. Indeed it is reasonably obvious that a better quality implementation can be obtained by using better range reduction on the transcendental functions.

It is interesting that the decision to define integer square-root in such an inefficient way came to be a problem. In my defence, it should be pointed out that it was never originally intended that the PVS specification should be executable, and with that in mind I had reworked the specification with a non-constructive definition, only to return to the original formulation.

An interesting possibility is now open to us: it is now possible to commence validation of floating-point algorithms by relating them to the "correct" values they should have as determined by this system. If this is combined with a validating compiler, it ought to be possible to at least partially automate this process. Although there has been much previous work on aspects of this, such as Harrison's work on floating point trigonometric functions [Har00] or Moore, Lynch and Kaufmanns' work on floating point division algorithms [MLK98], I envisage a system which automatically determines and correctly handles the condition number of an algorithm and determines when the floating point results are credible. As an example, consider an algorithm to invert a non-singular matrix using Gaussian Elimination with pivoting. This algorithm is essentially the same in floating point and exact real arithmetic, but the results can be wildly different depending on how close the matrix (or its inverse) is to singularity.

References

[AP06]   Akbarpour, B., Paulson, L.C.: Towards automatic proofs of inequalities involving elementary functions. In: PDPAR 2006: Pragmatics of Decision Procedures in Automated Reasoning, pp. 27–37 (2006)

[AP07]   Akbarpour, B., Paulson, L.C.: Extending a resolution prover for inequalities on elementary functions. In: Dershowitz, N., Voronkov, A. (eds.) LPAR 2007. LNCS (LNAI), vol. 4790, pp. 47–61. Springer, Heidelberg (2007)

[BB85]   Bishop, E., Bridges, D.S.: Constructive Analysis. Grundlehren der Mathematischen Wissenschaften, vol. 279. Springer, Berlin (1985)

[BEIR00] Bajard, J.-C., Ercegovac, M., Imbert, L., Rico, F.: Fast evaluation of elementary functions with combined shift-and-add and polynomial methods. In: 4th Conference on Real Numbers and Computers, pp. 75–87 (2000)

[Ber07]   Bertot, Y.: Affine functions and series with co-inductive real numbers. Mathematical Structures in Computer Science 17(1) (2007)

[CW80]   Cody Jr., W.J., Waite, W.: Software Manual for the Elementary Functions. Prentice-Hall, Englewood Cliffs (1980)

[DMM05]   Daumas, M., Melquiond, G., Muñoz, C.: Guaranteed proofs using interval arithmetic. In: Montuschi, P., Schwarz, E. (eds.) Proceedings of the 17th Symposium on Computer Arithmetic, Cape Cod, Massachusetts (2005)

[Grz55a] Grzegorczyk, A.: Computable functionals. Fundamenta Mathematicae 42, 168–202 (1955)

[Grz55b] Grzegorczyk, A.: On the definition of computable functionals. Fundamenta Mathematicae 42, 232–239 (1955)

[Grz57]   Grzegorczyk, A.: On the definitions of computable real continuous functions. Fundamenta Mathematicae 44, 61–71 (1957)

[Grz59]   Grzegorczyk, A.: Some approaches to constructive analysis. In: Heyting, A. (ed.) Constructivity in mathematics. Studies in Logic and the Foundations of Mathematics, pp. 43–61. North-Holland, Colloquium, Amsterdam (1957)

[Har97a]   Harrison, J.: Floating point verification in HOL light: the exponential function. Technical Report 428, University of Cambridge Computer Laboratory (1997)

[Har97b]   Harrison, J.: Verifying the accuracy of polynomial approximations in HOL. In: Proceedings of the 10th International Conference on Theorem Proving in Higher Order Logics, Murray Hill, New Jersey, pp. 137–152 (1997)

[Har99]   Harrison, J.: A machine-checked theory of floating point arithmetic. In: Bertot, Y., Dowek, G., Hirschowitz, A., Paulin, C., Théry, L. (eds.) 12th International Conference on Theorem Proving in Higher Order Logics, Nice, France, pp. 113–130 (1999)

[Har00]   Harrison, J.: Formal verification of floating point trigonometric functions. In: Hunt, W.A., Johnson, S.D. (eds.) Proceedings of the Third International Conference on Formal Methods in Computer-Aided Design, Austin, Texas, pp. 217–233 (2000)

[Jul08]   Julien, N.: Certified exact real arithmetic using co-induction in arbitrary integer base. In: Functional and Logic Programming Symposium, Saratoga, NY. LNCS, vol. 4989, pp. 48–63. Springer, Heidelberg (2008)

[Les03]   Lester, D.: Using PVS to validate the inverse trigonometric functions of an exact arithmetic. In: Proceedings of the Seminar on Numerical Software with Result Verification, Dagstuhl, Germany, pp. 259–273 (2003)

[LG03]   Lester, D., Gowland, P.: Using PVS to validate the algorithms of an exact arithmetic. Theoretical Computer Science 291, 203–218 (2003)

[Mén94]   Ménissier, V.: Arithmétique Exacte. PhD thesis, Université Pierre et Marie Curie, Paris (December 1994)

[ML05]   Muñoz, C., Lester, D.: Real number calculations and theorem proving. In: 18th International Conference on Theorem Proving in Higher Order Logics, Oxford, England, pp. 239–254 (2005)

[MLK98]   Moore, J.S., Lynch, T., Kaufmann, M.: A mechanically checked proof of the correctness of the kernel of the amd5k86 floating-point division algorithm. IEEE Transactions on Computers 47(9), 913–926 (1998)

[Mul97]   Muller, J.-M.: Elementary Functions. Birkhauser, Basel (1997)

[ORS92]   Owre, S., Rushby, J.M., Shankar, N.: PVS: A prototype verification system. In: Kapur, D. (ed.) CADE 1992. LNCS, vol. 607, pp. 748–752. Springer, Heidelberg (1992)

[Pau94]   Paulson, L.C.: Isabelle. LNCS, vol. 828. Springer, Saratoga (1994)

[PER89]   Pour-El, M.B., Ian Richards, J.: Computability in Analysis and Physics. Springer, Berlin (1989)

[Tur36]   Turing, A.M.: On computable numbers, with an application to the "Entscheidungsproblem". Proceedings of the London Mathematical Society 42(2), 230–265 (1936)

[Tur37]   Alan, M., Turing, A.M.: On computable numbers, with an application to the "Entscheidungsproblem". A correction. Proceedings of the London Mathematical Society 43(2), 544–546 (1937)

A Formalized Theory for Verifying Stability and Convergence of Automata in PVS*

Sayan Mitra and K. Mani Chandy

California Institute of Technology
Pasadena, CA 91125
{mitras,mani}@caltech.edu

Abstract. Correctness of many hybrid and distributed systems require stability and convergence guarantees. Unlike the standard induction principle for verifying invariance, a theory for verifying stability or convergence of automata is currently not available. In this paper, we formalize one such theory proposed by Tsitsiklis [27]. We build on the existing PVS metatheory for untimed, timed, and hybrid input/output automata, and incorporate the concepts about fairness, stability, Lyapunov-like functions, and convergence. The resulting theory provides two sets of sufficient conditions, which when instantiated and verified for particular automata, guarantee convergence and stability, respectively.

1 Introduction

Verification of many classes of systems require proofs for stability and convergence. For example, the requirement that a hybrid control system regains equilibrium in the face of disturbances is a stability property; the requirement that a set of mobile agents get arbitrarily close to the centroid of their initial positions through interaction is a convergence property. To best of our knowledge, existing frameworks that formalize automata in higher-order logics do not define these notions nor do they provide sufficient conditions for verifying them. In this paper we present a PVS [23] metatheory for stating and verifying stability and convergence properties. This theory extends the PVS interface for the Tempo toolkit [1,4]; thus, along with invariance properties and implementation relations, now we can also prove stability and convergence of automata, within the same framework.

In a 1987 paper [27] Tsitsiklis analyzed stability and convergence of a general class of models which he called Asynchronous Iterative Processes (henceforth, AIPs). An AIP consists of a set X, and a finite collection of functions or "operators" $T_k : X \to X$, $k \in \{1, \ldots, K\}$. Given an initial point $x_0 \in X$, an execution is obtained by choosing an arbitrary sequence of T_k's, and iteratively applying them to x_0. An AIP is *stable* around a given point $x^* \in X$, with respect to a given topological structure $\mathscr{T}$ on X, if for every neighborhood set $U \in \mathscr{T}$

---

* The work is funded in part by the Caltech Information Science and Technology Center and AFOSR MURI FA9550-06-1-0303.

O. Ait Mohamed, C. Muñoz, and S. Tahar (Eds.): TPHOLs 2008, LNCS 5170, pp. 230–245, 2008.

containing x^*, there exists another neighborhood set $V \in \mathscr{T}$, such that every execution starting from V remains within U. An infinite execution is said to be *fair* if every operator T_k is applied infinitely many times. An AIP *converges* to x^* with respect to a topological structure $\mathscr{T}$ around x^*, if for every $U \in \mathscr{T}$ containing x^*, there exists $n \in \mathbb{N}$, such that for every fair execution all the states obtained after applying the first n operations, are in U. In general, neither of these properties imply the other (see, Figure 1 for examples). In [27] the author provides sufficient conditions for proving stability and convergence of AIPs in terms of Lyapunov-like functions [16]. Moreover, under some weak assumptions about the topological structures, it turns out that these conditions are also necessary.

AIPs generalized to infinite (and possibly uncountable) set of operations subsume the classes of discrete, timed, and hybrid automata, and therefore, the sufficient conditions for proving convergence and stability of AIPs apply to these classes as well. In this paper, we formalize Tsitsiklis' theory of stability and convergence in PVS. We build on the existing PVS metatheory for untimed, timed, and hybrid I/O automata [3,15] which is integrated with the Tempo toolkit [1]. The preexisting theory defines reachable states and implementation relations for automata and provides theorems for inductively verifying invariant properties and simulation relations. We extend this metatheory as follows:

(a) A real-valued distance function d on pairs of states is introduced as a parameter to the theory; for a given state s^*, the sublevel sets of the function $d(s^*, .)$ define a topological structure around s^*.
(b) Infinite executions of automata and fairness conditions are defined.
(c) Stability and convergence of automata are defined with respect to a given state (or a set of states), a distance function, and a fairness condition.
(d) A set of theorems are stated and proved which provide sufficient conditions for verifying convergence and stability.

The new metatheory can be downloaded from `http://ist.caltech.edu/~` `mitras/research/pvs/convergence/`. In order to apply the theory to particular automata, the user has to supply a Lyapunov-like function and check that this function satisfies the criteria prescribed in the theorems. This check can be performed by analyzing the state transitions of the automaton and one need not reason about infinite executions. We illustrate the application of the theory on a simple distributed coordination protocol.

A key issue in formalizing this metatheory was to reconcile fairness of AIPs with the notion of fair executions of automata. Recall that an AIP-execution is fair, if *every* operator is applied infinitely often. This definition is too strong for automata, where each operator corresponds to a *specific* state transition. A fair execution for an automaton typically does not have every state transition occurring infinitely many times, instead it has some set of transitions represented infinitely often. More precisely, fairness of an automaton is defined with respect to a collection $\mathcal{F} = \{F_1, F_2, \ldots, F_n\}$, where each F_i is a set of transitions. An execution α is said to be $\mathcal{F}$-fair if every $F_i \in \mathcal{F}$ is represented (or scheduled) in α infinitely often. In our theory, we formalize this weaker notion of fairness, and

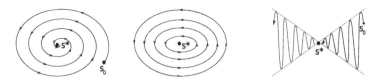

Fig. 1. Stable & convergent (*left*), stable & nonconvergent (*middle*), and convergent & unstable (*right*) executions. In the last case, the execution from s_0 converges, but executions starting from in the left neighborhood of s^* diverge.

hence, the sufficient conditions we obtain are stronger than those in [27]. If each F_i is defined to be a singleton transition then our conditions reduce to Tsitsiklis'. There are several other relatively minor differences between the original theory and our PVS formalization; these are discussed in Section 6.

2   Related Work

Convergence of general sequences has been formalized in the PVS and Isabelle libraries for differential calculus [10], real-analysis [11], and topology [13]. There are several formalizations of automata in higher-order logics of theorem provers including Isabelle/HOL [21,22,20,25], PVS [3,7,19], and Coq [8,24]. These theories formalize reachable states, invariant properties, acceptance conditions, and abstraction relations, but neither stability nor convergence. Our theory builds on the PVS formalization of input/output automata [17] presented in [3] and its subsequent extensions to timed and hybrid I/O automata [12] that were presented in [19,18].

Literature on stability and convergence (also called asymptotic stability) of purely discrete or continuous systems is extensive. Control theory textbooks, such as [16], typically provide conditions for checking stability of processes evolving in Euclidean space. Stability conditions for hybrid and switched systems is an active area of research; we refer the reader to [14] for an overview.

Although convergence is distinct from termination, constructing proofs for both these properties rely on existence of Lyapunov-like functions from the state space of the automaton to some well-ordered set. There is a large body of literature on proving termination of programs and recursive functions using theorem provers (see, e.g., [5,26]). This research direction focuses on automatically finding Lyapunov-like functions that prove termination. This connection is further discussed in Section 6.

3   Automata and Executions

3.1   Preliminaries

In this paper, we present our PVS metatheory using standard set theoretic notations, in as much of a syntax-free manner as possible. The set theoretic definitions, in most cases, correspond in an obvious way to the definitions in PVS's simply typed higher-order logic. Wherever necessary we note special constructs that are necessary for this translation.

We denote the set of boolean constants by $\mathbb{B} = \{true, false\}$, the set of natural numbers by $\mathbb{N} = \{0, 1, \ldots, \}$, and the set of reals by $\mathbb{R}$. For a set A, A^ω is defined as the set of infinite sequences of elements in A indexed by $\mathbb{N}$. For $a^\omega \in A^\omega, i \in \mathbb{N}$, we denote the ith element by a_i^ω. In the rest of this section we summarize the relevant definitions and theorems from the existing metatheory of [3].

3.2   Formalization of Automata

An automaton $\mathscr{A}$ is a nondeterministic, labeled transition system. Formally, it is a quintuple consisting of:

(a) a nonempty set S,
(b) a nonempty set A,
(c) a nonempty subset S_0 of S,
(d) a function $E : [S, A \to \mathbb{B}]$, and
(e) a function $T : [S, A \to S]$.

Elements of S, S_0, and A are called *states, starting states* and *actions*, respectively. E and T are called the *enabling predicate* and the *transition function* of $\mathscr{A}$. The actions are labels for state transitions. For $s \in S$ and $a \in A$, $E(s, a)$ holds if and only if the transition labeled by a can be applied to s. In this case, a is said to be *enabled* at s. At any state s, multiple actions may enabled. However, once an action a is fixed the post-state of the transition s' is uniquely determined. Specifically, $s' = T(a, s)$, if a is enabled at s, or else $s' = s$.

The set of actions can be uncountably infinite and indeed actions can label functions for discrete transitions as well as continuous evolution. We refer the reader to [15] for models of timed and hybrid systems in this formalism.

The PVS metatheory formalizing automata is parameterized by $\mathcal{S}, \mathcal{A}, \mathcal{S}_0, \mathcal{E}$ and $\mathcal{T}$, where $\mathcal{S}$ and $\mathcal{A}$ are uninterpreted type parameters, and $\mathcal{S}_0, \mathcal{E}$, and $\mathcal{T}$ are parameters with the appropriate type constraints[1]. To apply the metatheory to specific systems, the parameters are instantiated with concrete types and function definitions. The following example shows such an instantiation.

Example 1. We model a distributed algorithm in which a set of N agents start at arbitrary positions on a line and through interactions converge to a point. Specifically, any two agents interact at any time; when they do, their positions are atomically updated so that they move towards each other by some fraction of the distance between them.

Table 1 provides concrete definitions for the types and the functions for modeling this protocol as an automaton. N is a constant natural number. L is a constant real in the range $(0, 1)$. I is the type $\{0, \ldots, N - 1\}$. The states type is an array of $\mathbb{R}_{\geq 0}$ indexed by I. For any state s and $i \in I$, the ith component of

---

[1] Each parameter corresponds to its unscripted version in our presentation of the automaton theory. E.g., the set of states S is modeled as the type $\mathcal{S}$ in PVS.

the array is denoted by $s[i]$. The state s_0 is an arbitrary but constant element of S. The predicate S_0 defines a state s to be a starting state if and only if it equals s_0. The action type is defined using a (datatype) constructor called *interact*. This construct means that for every $i, j \in I$, $r \in [L, 1-L]$ $interact(i, j, r)$ is an action; and nothing else is an action. The set of actions is uncountably infinite because of the real parameter r.

The enabling predicate $E(s, a)$ returns true for any action a and state s, which means that agents i and j can interact always. Finally, for action a of the form $interact(i, j, r)$, the state transition function $T(a, s)$ returns a state s' that is identical to s except that the ith and the jth values of s' are $s[i] + r.(s[i] - s[j])$ and $s[j] - r.(s[i] - s[j])$, respectively. Informally, for given i, j, each choice of r defines a different proportion by which agents i and j move towards each other. For example, $interact(i, j, \frac{1}{2})$ causes $s[i]$ and $s[j]$ to move to their mid-point.

Table 1. An instance of automaton metatheory

S : Type $:= array[I \mapsto \mathbb{R}_{\geq 0}]$

s_0 : Const S

$S_0(s : S) : bool := (s = s_0)$

A : Datatype $interact(i, j : I, r : [L, 1-L])$

$E(s : S, a : A) : bool :=$ true

$T(s : S, a : A) : S := [\text{Case } a \equiv interact(i, j, r) :$
$\quad s \text{ With } [(i) := s[i] + r(s[j] - s[i]), (j) := s[j] - r(s[j] - s[i])]]$

3.3 Executions, Reachability, and Invariance

The semantics of an automaton $\mathscr{A}$ is defined in terms of its executions. An *execution fragment* of $\mathscr{A}$ is a (possibly infinite) alternating sequence of states and actions $s_0, a_0, s_1, a_1, s_2, \ldots$, such that for each i, $E(s_i, a_i)$ holds and $s_{i+1} = T(a_i, s_i)$. An execution fragment is an *execution* if s_0 is a starting state. The *length* of a finite execution is the number of actions in it. For $s \in S$ and a natural number n, $Reach\_rec(s, n)$ returns true if and only if there exists an execution of length n that ends in the state s. The reachability predicate on states is defined recursively as follows:

$$Reach\_rec(s, n) := \begin{cases} s \in S_0 & n = 0 \\ \exists\ s_1 \in S, a \in A\ (E(a, s_1) \wedge & \text{otherwise.} \\ \quad s = T(a, s_1) \wedge\ Reach\_rec(s_1, n-1)) \end{cases}$$

$$Reach(s) := \exists\ n \in \mathbb{N},\ Reach\_rec(s, n).$$

An *invariant* of $\mathscr{A}$ is a predicate on its states that holds in all reachable states. Invariants are useful for capturing safety requirements, such as, multiple processes *never* access a critical resource simultaneously. The following theorem formalizes Floyd's induction principle [9] in this framework.

Theorem 1. *Suppose* $G : [S \to \mathbb{B}]$ *is a predicate on* S, *and*

A1. $\forall\, s \in S_0, G(s)$, *and*
A2. $\forall\, s \in S, a \in A, Reach(s) \;\wedge\; E(a, s) \;\wedge\; G(s) \Rightarrow G(T(a, s))$.

Then $\forall\, s \in S, Reach(s) \Rightarrow G(s)$, *that is,* G *is an invariant predicate.*

This theorem has been employed for verifying safety properties of untimed [3], timed [15], and hybrid automata [28]. Features of this verification method that make it attractive are: (a) It suffices to check that the predicate G is preserved over individual actions, and hence, the check breaks down into a case analysis of actions. (b) This structure facilitates partial automation of proofs using customized proof strategies [2].

4   Formalizing Stability and Convergence of Automata

In this section we present the extensions to the PVS metatheory for stability and convergence verification. In order to define stability and convergence properties, first, we have to explicitly define arbitrary prefixes of infinite executions of the automaton. Given a state $s \in S$, an infinite sequence of actions $a^\omega \in A^\omega$, and $n \in \mathbb{N}$, the recursively defined $Trans(s, a^\omega, n)$ function returns the state that is obtained by applying the first n actions in a^ω to s.

$$Trans(s, a^\omega, n) = \begin{cases} s & \text{if } n = 0 \\ T(a^\omega_{n-1}, Trans(s, a^\omega, n-1)) & \text{if } E(a^\omega_{n-1}, Trans(s, a^\omega, n-1)) \\ Trans(s, a^\omega, n-1) & \text{otherwise} \end{cases}$$

Note that s, a^ω, and n, uniquely determine an execution fragment $s_0, a^\omega_0, s_1, \ldots,$ a^ω_{n-1}, s_n of length n, where $s_i = Trans(s, a^\omega, i)$, for each $i < n$.

Stability and convergence of automaton $\mathscr{A}$ to a state $s^* \in S$ are defined with respect to a topological structure around s^*. This topological structure is formalized using a real-valued function. The metatheory can be easily generalized to define stability (and convergence) with respect to arbitrarily defined topological structures around a point; this is discussed further in Section 6.

Definition 1. *A* distance function d *for a state* $s^* \in S$ *is a real-valued function* $d : [S, S \to \mathbb{R}_{\geq 0}]$, *such that for all* $s \neq s^*$, $d(s^*, s) > d(s^*, s^*)$. *A distance function* d *for a set of states* $S^* \subseteq S$ *is a real-valued function* $d : [2^S, S \to \mathbb{R}_{\geq 0}]$, *such that for all* $s \notin S^*, s' \in S^*$, $d(S^*, s) > d(S^*, s')$.
For $\epsilon > 0$ *and* $s \in S$, ϵ-ball *around* s, *is the set*

$$B_\epsilon(s) := \{s_1 \in S \mid d(s_1, s) \leq \epsilon\}.$$

In our theory, d is not required to satisfy identity, symmetry, and triangle inequality, properties that are usually attributed to metrics. The ϵ-balls around a given state s define a topological structure around s. In the new PVS metatheory we add S^* and d as theory parameters, in addition to the six parameters enumerated in Section 3.2.

4.1 Stability

Informally, automaton $\mathscr{A}$ is stable if every execution fragment that starts close to the equilibrium state s^* remains close to s^*, where closeness is defined in terms of the ϵ-balls of some distance function for s^*.

Definition 2. *Let $\mathscr{A}$ be an automaton $\langle S, A, S_0, E, T \rangle$, s^* be a state in S, and d be a distance function for s^*. $\mathscr{A}$ is (s^*, d)-stable if*

$$\forall \, \epsilon > 0, \exists \, \delta > 0, \, \forall s \in S, a^\omega \in A^\omega, n \in \mathbb{N}, s \in B_\delta(s^*) \Rightarrow Trans(s, a^\omega, n) \in B_\epsilon(s^*).$$

Note that stability is independent of the starting states of the automaton. For a nonempty set $S^* \subseteq S$, let d be a distance function for S^*. The definitions for the ϵ-balls around S^* (denoted by $B_\epsilon(S^*)$) and (S^*, d)-stability of $\mathscr{A}$ are analogous to Definition 2.

The coarseness of the topological structure around s^* (or S^*), and hence, the meaning of stability depends on the function d. For example, suppose $d(s^*, s) := 0$ if $s^* = s$, and $d(s^*, s) := 1$, otherwise. Then, $\mathscr{A}$ is trivially (s^*, d)-stable. On the other hand, if the set of states S is an Euclidean space and d is the Euclidean metric on S, then d defines an uncountable set of distinct ϵ-balls around s^*. And in this case (s^*, d)-stability of $\mathscr{A}$ depends on E and T.

4.2 Sufficient Conditions for Stability

In this section we present the sufficient conditions for proving stability of automaton $\mathscr{A}$. The proofs of all the theorems presented in Section 4 have been completed in PVS and are available as part of the metatheory. Here we present summaries of these PVS proofs.

Let T be a set and $<$ be a total order on T, and f be a function that maps S to T. The range of f is denoted by Rng_f, and the *p-sublevel set* is defined as $L_{f,p} := \{s : S \mid f(s) \leq p\}$. We omit the subscript f when the function is clear from the context. The following theorem gives a sufficient condition for proving stability of an automaton in terms of a Lypunov-like function.

Theorem 2. *Let S^* be a nonempty subset of S and d be a distance function for S^*. Suppose there exists a totally ordered set $(T, <)$ and a function $f : [S \to T]$ that satisfies the following conditions:*

B1. $\forall \, \epsilon \geq 0, \, \exists \, p \in T$, *such that* $L_p \subseteq B_\epsilon(S^*)$.
B2. $\forall \, p \in T, \, \exists \, \epsilon \geq 0$, *such that* $B_\epsilon(S^*) \subseteq L_p$.
B3. $\forall \, s \in S, a \in A, \, E(a, s) \Rightarrow f(T(a, s)) \leq f(s)$.

Then $\mathcal{A}$ is (S^*, d)-stable.

B1 requires that every ϵ-ball around S^* contains a p-sublevel set L_p. B2 is symmetric; it requires that every sublevel set contains an ϵ-ball. B3 states that the value of the function f does not increase if an action a is applied to state where it is enabled.

Proof: Let us fix an $\epsilon > 0$. We have to show that there exists a $\delta > 0$, such that any execution fragment that starts in $B_\delta(S^*)$ remains within $B_\epsilon(S^*)$. There exists $p \in T$, such that $L_p \subseteq B_\epsilon(S^*)$ (by B1), and there exists a $\eta \geq 0$, such that $B_\eta(S^*) \subseteq L_p \subseteq B_\epsilon(S^*)$ (by B2). Set $\delta = \eta$, and fix an $s \in B_\delta(S^*)$, $a^\omega \in A^\omega$. We show by induction that every state in the execution fragment starting from s and corresponding to a^ω remains within $B_\epsilon(S^*)$.

Base case: We know that $s \in B_\delta(S^*) \subseteq B_\epsilon(S^*)$.
Inductive step: Let s' be the state $Trans(s, a^\omega, j)$, $0 \leq j \leq n - 2$. By the induction hypothesis, $s' \in B_\delta(S^*) \subseteq L_p$. If a_j^ω is not enabled at s', then $s_{j+1} = s' \in B_\delta(S^*) \subseteq B_\epsilon(S^*)$. Otherwise, $s_{j+1} = T(a_m, s')$. As $f(s_{j+1}) \leq f(s')$ (by B3), and it follows that $s_n \in L_p \subseteq B_\epsilon(S^*)$.

4.3 Fairness

An automaton $\mathcal{A}$ is said to *converge* to s^* with respect to distance function d, if for every infinite execution $s_0, a_0, s_1, \ldots, a_{n-1}, s_n \ldots$, $d(s^*, s_n) \to 0$ as $n \to \infty$. This captures the informal notion that every execution of the automaton gets closer and closer to s^*.

In typical applications, the above definition of convergence is too strong because it quantifies over all infinite executions—including those in which some set of actions never occur. For instance, consider an infinite execution α for the automaton of Example 1 that starts from a state s_0 with distinct values for all $s_0[i]$'s, and in which agent 0 never interacts with any other agent. It is easy to see that such an execution does not converge. On the other hand, a different infinite execution $\alpha' = s_0, a_0, s_1, \ldots$, converges to s^*, where $s^*[i] = \frac{1}{N} \sum_{i=1}^{N} s_0[i]$, provided for every $i, j \in I$, infinitely many $interact(i, j, *)$ actions occur in α'. In fact, an infinite execution in which for every $i \in I$, $interact(i, (i+1) \mod N, *)$ occurs infinitely often, also converges to s^*. This suggests that the convergence of $\mathcal{A}$ can be studied under different sets of assumptions about the occurrence of the actions. This motivates the following definition of fairness.

Definition 3. *A* fairness condition $\mathcal{F}$ *for the set of actions A is a finite collection $\{F_i\}_{i=1}^n$, $n \in \mathbb{N}$, where each F_i is a nonempty subset of A. An infinite sequence of actions $a^\omega \in A^\omega$ is $\mathcal{F}$-fair if*

$$\forall F \in \mathcal{F}, n \in \mathbb{N}, \exists m \in \mathbb{N}, m > n, \text{ such that } a_m^\omega \in F.$$

An infinite execution $\alpha = s_0, a_0, s_1, a_1, \ldots$ is said to $\mathcal{F}$-fair if the corresponding sequence of actions $a_0, a_1, \ldots$ is $\mathcal{F}$-fair.

In other words, an execution is not $\mathcal{F}$-fair if there exists $F \in \mathcal{F}$ such that no action from F ever appears in some suffix of α.

Definition 4. *Given fairness conditions $\mathcal{F}_1$ and $\mathcal{F}_2$ for the set of actions A, $\mathcal{F}_1$ is said to be* weaker *than $\mathcal{F}_2$, denoted by $\mathcal{F}_1 \le \mathcal{F}_2$, if $\forall\, F_1 \in \mathcal{F}_1,\, \exists\, F_2 \in \mathcal{F}_2$, such that $F_2 \subseteq F_1$.*

The next lemma states that an $\mathcal{F}_2$-fair execution is also $\mathcal{F}_1$-fair, if $\mathcal{F}_1$ is a weaker fairness condition than $\mathcal{F}_2$.

Lemma 1. *Let $\mathcal{F}_1, \mathcal{F}_2$ be fairness conditions for the set of actions A. If $\mathcal{F}_1 \le \mathcal{F}_2$, then every $\mathcal{F}_2$-fair execution is $\mathcal{F}_1$-fair.*

4.4   Convergence

Having introduced fairness of executions, we now modify the previously suggested definition of convergence as follows. Informally, automaton $\mathscr{A}$ converges to s^* with respect to distance function d and a fairness condition $\mathcal{F}$, if every $\mathcal{F}$-fair execution converges to s^*.

Definition 5. *Let $\mathscr{A}$ be an automaton $\langle S, A, S_0, E, T \rangle$, s^* be an element of S, d be a function for s^*, and $\mathcal{F}$ be a fairness condition for A. $\mathscr{A}$ is $(s^*, d, \mathcal{F})$-convergent, if $\forall\, s_0 \in S_0, \epsilon > 0, a^\omega \in A^\omega$*

$$\text{if } a^\omega \text{ is } \mathcal{F}\text{-fair then } \exists n \in \mathbb{N}, \forall m \in \mathbb{N},\ m > n \Rightarrow Trans(s, a^\omega, m) \in B_\epsilon(s^*).$$

For a nonempty subset of states $S^* \subseteq S$, the definition of $(S^*, d, \mathcal{F})$-convergence is analogous to Definition 5. For $s \in S, a^\omega \in A^\omega$, we define $R_f(s, a^\omega)$ as the set of values in T that can be reached from s by applying some prefix of a^ω.

$$R_f(s, a^\omega) := \{ p \in T \mid \exists\, n \in \mathbb{N}, Trans(s, a^\omega, n) \in L_{f,p} \}.$$

The next theorem gives sufficient conditions for proving convergence of automaton $\mathscr{A}$ in terms of a Lyapunov-like function.

Theorem 3. *Let S^* be a nonempty subset of S, d be a distance function for S^*, and $\mathcal{F}$ be a fairness condition on A. Suppose there exists a totally ordered set $(T, <)$ and a function $f : S \to T$ that satisfies the following conditions:*

C1. $\forall\, p, q \in T, p < q \Rightarrow L_p \subsetneq L_q$.
C2. $\forall\, \epsilon > 0, \exists\, p \in T$, such that $L_p \subseteq B_\epsilon(S^*)$.
C3. $\forall\, s \in S, a \in A, (Reach(s) \wedge E(a, s)) \Rightarrow f(T(a, s)) \le f(s)$.
C4. $\forall\, p \in T, L_p \ne S^*$ implies $\exists\, F \in \mathcal{F}$, such that $\forall\, a \in F, s \in L_p, Reach(s) \Rightarrow (E(a, s) \wedge f(T(a, s)) < f(s))$.
C5. $\forall\, s \in S_0$ and $\mathcal{F}$-fair sequence $a^\omega \in A^\omega$, $R' \subseteq R_f(s, a^\omega)$, if R' is lower bounded then it has a smallest element.

Then $\mathscr{A}$ is $(S^*, d, \mathcal{F})$-convergent.

Some remarks about the hypothesis of the theorem are in order. C1 implies that every sublevel set of the function f is distinct. C2 requires that for any $\epsilon > 0$, there exists a p-sublevel set of f that is contained within the ϵ-ball around S^*. This is identical to condition B1. C3 requires that the function f is nonincreasing over all transitions from reachable states. This is a weaker version of B3. C4 requires that for any sublevel set L_p that is not equal to the convergence set S^*, there exists a fair set of actions $F \in \mathcal{F}$, such that any action $a \in F$ strictly decreases the value of f—possibly by some arbitrarily small amount. C5 requires that for all s, a^ω, every lower-bounded subset of $R_f(s, a^\omega)$ has a smallest element. This is a weaker assumption than requiring $R_f(s, a^\omega)$ to be well-ordered. Instead of C5 it is sometimes easier to prove that the set T is well-ordered.

Before proving Theorem 3, we state a set of intermediate lemmas that are used in the proof. In the following, S^* is a subset of S, $\mathcal{F}$ is a fairness condition on A, $(T, <)$ is a total order, and f is a function $S \to T$ satisfying conditions C1-5. We make the following assumption, without any loss of generality.

$$\exists\, p^* \in T, \text{ such that } \forall\, s \in S, \text{ if } s \in S^* \text{ then } f(s) = p^*, \text{ otherwise } f(s) > p^*.$$

This is without loss of generality because for any given f' we can define $f(x) := p^* = \inf_{s \in S^*} f(s)$ if $x \in S^*$ and $f'(x) := f(x)$ otherwise. Then f satisfies the assumption and we work with it instead of f'.

Lemma 2. *For all $a^\omega \in A^\omega, n \in \mathbb{N}, p \in T, s \in L_p$: $Trans(s, a^\omega, n) \in L_p$.*

Lemma 3. *For all $s \in S$, $s \in S^*$ iff $L_{f(s)} = S^*$.*

Lemma 4. *For all $s_0 \in S_0$ and $\mathcal{F}$-fair action sequence a^ω, $Rng_f = R_f(s_0, a^\omega)$.*

Proof: Let us fix an a^ω, s_0, and f. We abbreviate $R_f(s_0, a^\omega)$ as R. From its definition it is clear that $R \subseteq Rng_f$, so, we have to show that for every $p \in Rng$, there is an n such that $f(Trans(s_0, a^\omega, n)) \leq p$. Let us fix a value of $p \in Rng$, and suppose for the sake of contradiction that $p \notin R$. We know that $f(s_0) < p$, because otherwise $Trans(s_0, a^\omega, 0) = s_0 \in L_p$, that is, p would be in R. We consider two subcases:

Case 1: R has a lower bound. R has a smallest element, say p_{min} (by C5). If $p_{min} \leq p$ then $p \in R$, so, we consider the case where $p < p_{min}$. There exists p^* such that $f(s) = p^*$ for every $s \in S^*$, and $p^* < f(s)$ outside S^* (by Lemma 3).

Case 1.1: $p_{min} \leq p^*$. Then $p < p^*$ and this contradicts our assumption that p^* is the smallest value attained by f.

Case 1.2: $p^* < p < p_{min}$: Since $L_p \neq S^*$, there exists an $F \in \mathcal{F}$, such that for every $a \in F$ and for every reachable state s in L_p, a is enabled at s, and $f(T(a, s)) < p$ (by C4). Also, there exists n_0 such that $Trans(s_0, a^\omega, n_0) \in L_{p_{min}}$ (by definition of p_{min}). Since a^ω is an $\mathcal{F}$-fair sequence, there exists an m, $m > n_0$, such that $a_m^\omega \in F$. We define

$s' := Trans(s_0, a^\omega, m - 1)$. It follows that $s' \in L_{p_{min}}$ (by Lemma 2), that is, $f(T(a_m^\omega, s')) < p_{min}$. As $f(T(a_m^\omega, s')) \in R$, this contradicts our assumption that p_{min} is the smallest element in R.

Case 2: R does not have a lower bound. Then there exists $q \in R$, such that $q < p$, and by C1, $L_q \subsetneq L_p$. That is, there exists n, such that $Trans(s_0, a^\omega, n) \in L_p$, and therefore, contrary to our assumption $p \in R$.

Proof of Theorem 3: Let us fix $\epsilon \geq 0$, f satisfying the conditions in the hypothesis, $s_0 \in S_0$, and an $\mathcal{F}$-fair sequence $a^\omega = a_0, a_1, \ldots$. There exists $p \in Rng_f$, such that $L_{f,p} \subseteq B_\epsilon(S^*)$ (by C2). There exists $n \in \mathbb{N}$, such that $Trans(s_0, a^\omega, n) \in L_{f,p} \subseteq B_\epsilon(S^*)$ (by Lemma 4). It follows that for all $m > n$, $Trans(s_0, a^\omega, m) \in L_{f,p} \subseteq B_\epsilon(S^*)$ (by Lemma 2).

Corollary 1. *If $\mathscr{A}$ is $(S^*, d, \mathcal{F}_1)$-convergent and $\mathcal{F}_1 \leq \mathcal{F}_2$ then $\mathscr{A}$ is $(S^*, d, \mathcal{F}_2)$-convergent.*

4.5 Special Case

In certain applications the function d which defines the topological structure around s^* can itself be used as the Lyapunov-like function for proving convergence. The obvious advantage of doing so is that C2 follows automatically. We provide a restricted version of Theorem 3 which can be applied in such cases.

Corollary 2. *Let S^* be a nonempty subset of S and $\mathcal{F}$ be a fairness condition on A. We define $f : S \to \mathbb{R}_{\geq 0}$ as $f(s) := d(S^*, s)$. Suppose there exists a strictly decreasing sequence $p^\omega \in \mathbb{R}_{\geq 0}{}^\omega$ of valuations of f that converges to 0, such that:*

D1. $\forall\, i, j \in \mathbb{N}, i > j \Rightarrow L_{p_i} \subsetneq L_{p_j}$.
D2. $\forall\, s \in S, a \in A, i \in \mathbb{N}\ (Reach(s)\ \wedge\ E(a,s)\ \wedge\ s \in L_{p_i}) \Rightarrow T(a,s) \in L_{p_i}$.
D3. $\forall\, i \in \mathbb{N},\ p_i \neq 0$ *implies* $\exists\, F \in \mathcal{F}$, *such that* $\forall\, a \in F, s \in L_{p_i},\ Reach(s) \Rightarrow$
$\quad (E(a,s)\ \wedge\ T(a,s) \in L_{p_{i+1}})$.

Then $\mathscr{A}$ $(S^*, d, \mathcal{F})$-convergent.

Proof: : We check that the defined function f satisfies the conditions in the hypothesis of Theorem 3. C1 follows from D1. C2 follows from the convergence of the sequence p^ω. C3 follows from D2. C4 follows from D3 and the strictly decreasing property of the sequence p^ω. It remains to show C5.

Let us fix $s \in S_0$ and a $\mathcal{F}$-fair sequence $a^\omega \in A^\omega$, and let $p_{min} > 0$ be a lower bound for a subset $R' \subseteq R_f(s, a^\omega)$. Suppose, for the sake of contradiction, that R' does not have a smallest element. Then there exists a $i \in \mathbb{N}$ such that $p_i \in R'$, and for all $j > i$, $p_j < p_{min}$. There exists $s \in L_{p_i}$, such that $s' = Trans(s_0, a^\omega, m)$, for some $m \in \mathbb{N}$ (by definition of R'). There exists $F \in \mathcal{F}$, such that for all $a \in F, s \in L_{p_i}$, if s is reachable then $E(a,s)$ and $T(a,s) \in L_{p_{i+1}}$ (by D3). We fix such an F. As a^ω is an $\mathcal{F}$-fair sequence, there exists $k > m$, such that $a_k \in F$. Let $s' = Trans(s_0, a^\omega, k)$. As s' is reachable and so is $T(a_k, s')$. Since $f(T(a_k, s')) \leq p_{i+1} < p_i$, it contradicts our assumption.

5 An Application

In this section, we illustrate the application of our convergence theory to verify the convergence of the protocol introduced in Example 1. Recall, the set of states S is defined as arrays of $\mathbb{R}_{\geq 0}$ indexed by I, where $I = \{0, 1, \ldots, N-1\}$. For $i \in I$, $s[i]$ corresponds to the value of the ith participating agent at state s. The starting state s_0 is an arbitrary but constant element of S. We define a real-valued constant M that corresponds to the average of the initial values of the agents: $M := \frac{1}{N} \sum_{j=0}^{N-1} s_0[j]$. Let $s^* \in S$ be defined as the constant array:

$$s^* : \text{Const } S := [M, \ldots, M]$$

We would like to prove convergence of this protocol to s^*, and in order to do so we have to first define some notion of neighborhood around s^* and a fairness condition for the actions of this automaton. We define the neighborhood around s^* with the standard Euclidean distance between any state $s \in S$ and s^*.

$$d(s^*, s) := \sum_j (s[j] - s^*[j])^2$$

Next, we specify the fairness condition $\mathcal{F}$. Informally, we require that no two sets of participating agents are partitioned perpetually. That is, an action sequence a^ω is $\mathcal{F}$-fair, if for every $n \in \mathbb{N}$, and for every pair of disjoint subsets $J, K \subset I$, there exists $m > n$, such that $a_m = interact(j, k, r)$ for some $j \in J, k \in K$ and $r \in [L, 1 - L]$. Formally, let J, K be any two subsets of I. We define

$$F_{J,K} := \{interact(j, k, r) \mid j \in J, \ k \in K, \ r \in [L, 1 - L]\},$$
$$\mathcal{F} := \{F_{J,K} \mid J \cap K = \emptyset\}.$$

Our goal is to prove that automaton $\mathscr{A}$ is $(s^*, d, \mathcal{F})$-convergent. To this end we invoke Corollary 2. In this case $S^* = \{s^*\}$. We define

$$f(s) := s(s^*, s) = \sum_j (s[j] - M)^2,$$

and the sequence p^ω, as $p_i = M\beta^i$, where $\beta := (1 - \frac{2L(1-L)}{N^3})$. It is easy to check that p^ω converges to 0 and that f satisfies condition D1. In the remainder of this section we check that f satisfies D2 and D3.

We define $sorted(s)$ as a derived variable that returns the sorted version of the array s. That is, $sorted(s) : [I \to \mathbb{R}_{\geq 0}]$ has the following ordering property. For all $i, j \in I$, $sorted(s)(i) < sorted(s)(j)$ if and only if $s[i] > s[j]$ or $s[i] = s[j]$ and $i > j$. That is, $sorted(s)[k]$ is the kth largest element of s.

Prior to checking D2 and D3 the following invariant properties are verified using Theorem 1. The first invariant follows from the property of the protocol that $\sum_{i=0}^{N-1} s[i] = MN$ remains constant in all reachable states.

Invariant 4 *For every $s \in S$, Reach(s) implies $\sum_{i=0}^{N-1}(sorted(s)[i] - M) = 0$.*

Invariant 5 *For every $s \in S$, Reach(s) and $s \neq s^*$ implies*

$$\sum_{i=0}^{N-2} sorted(s)[i] - sorted(s)[i+1] \geq \sqrt{\frac{f(s)}{N}}.$$

Proof: Since $s \neq s^*$, $f(s) > 0$. There exists $j \in I$ such that $(s[j] - M)^2 \geq f(s)/N$ (by definition of $f(s)$). Let us fix such a j. Since $(s[j] - M)^2 \geq f(s)/N$ we conclude that $(sorted(s)[0] - M)^2 \geq f(s)/N$. We assume that $(sorted(s)[0] - M)^2 > 0$; proof for the negative case is symmetric. From Invariant 4, $\sum_{i=0}^{N-1}(sorted(s)[i] - M) = 0$ and the positivity of $sorted(s)[0] - M$ it follows that there exists $k \in I$, such that $sorted(s)[k] - M < 0$. Thus, $sorted(s)[N-1] - M < 0$. $\sum_{i=0}^{N-2}(sorted(s)[i] - sorted(s)[i+1])$

$$= sorted(s)[0] - sorted(s)[N-1]$$

$$= sorted(s)[0] - M - (sorted(s)[N-1] - M) \geq \sqrt{\frac{f(s)}{N}}.$$

Proposition 1. *f satisfies D2.*

Proof: We have to show that from every reachable state $s \in S$ and for any action $a \in A$, $f(T(a,s)) \leq f(s)$. Every action a of A, is of the form $interact(j,k,r)$, where $r \in [L, 1-L]$, and $L \in (0,1)$. We fix $j, k \in I$ and a reachable state $s \in S$, and define $\delta := s[k] - s[j]$. From the definition of the transition function T for action $interact(j,k,r)$ we know:

$$T(interact(j,k)s)[i] = \begin{cases} s[i] + \delta r & \text{if } i = j \\ s[i] - \delta r & \text{if } i = k \\ s[i] & \text{otherwise.} \end{cases} \tag{1}$$

Thus, $f(s) - f(T(interact(j,k,r),s)) = 2\delta^r(1-r)$, and since $r \in [L, 1-L]$, $f(s) - f(T(interact(j,k,r),s)) \geq 2.\delta^2.L.(1-L) \geq 0$

Proposition 2. *f satisfies D3.*

Proof: Consider any reachable state s such that $s \neq s^*$. It suffices to show that there exists $F \in \mathcal{F}$, such that for any action $a \in F$ $f(T(a,s)) \leq \beta f(s)$, where β has been defined to be $1 - \frac{2L(1-L)}{N^3}$. From Invariant 5 we know that $\sum_{i=0}^{N-2} sorted(s)[i] - sorted(s)[i+1] > \sqrt{\frac{f(s)}{N}}$. Since each term in the summation is nonnegative, we conclude that there exists $k \in I$, $sorted(s)[k] - sorted(s)[k+1] \geq \frac{1}{N}\sqrt{\frac{f(s)}{N}}$. We fix such a k and define two subsets of I:

$$A = \{j \in I \mid s[j] \geq s[k]\}$$
$$B = \{j \in I \mid s[j] \leq s[k-1]\}$$

Since A and B are disjoint subsets of I, $F_{A,B} \in \mathcal{F}$. Now we show that for any action $a \in F_{A,B}$, $f(T(a,s)) \leq \beta f(s)$. Let $a = interact(j,k,r)$, where $j \in A$ and $k \in B$. From Proposition 1 it follows that

$$f(s) - f(T(interact(j,k,r),s)) \geq 2L(1-L)(s[j] - s[k])^2$$

$$\geq \frac{2f(s)L(1-L)}{N^3} \quad \text{(by definition of } A, B)$$

$$f(T(interact(j,k,r),s)) \leq \left[1 - \frac{2L(1-L)}{N^3}\right] f(s).$$

6   Discussions

Comparison with Tsitsiklis' theory. Apart from the more general notion of fairness that we have formalized, our theory for stability and convergence differs from that presented in [27] in the following ways.

Specifying topologies. In [27] closeness to the point of convergence s^* is defined in terms of a topological structure called a *neighborhood system* around s^*. A neighborhood system around s^* is a collection $\mathcal{U}$ of subsets of S that satisfies the following conditions: (i) $s^* \in U$, $\forall U \in \mathcal{S}$. (ii) For all $s \in S, s \neq s^*$, there exists $U \in \mathcal{U}$ such that $s \notin U$. (iii) $\mathcal{U}$ is closed under finite intersections and arbitrary unions. For most natural definitions for the distance function d, the ϵ-balls of d satisfy conditions $(i), (ii)$ and (iii). We decided to use this functional specification of the neighborhood sets because (a) it is concise, and (b) in many applications there exists an inherent metric with respect to which we prove convergence (or stability). Introducing neighborhood systems in the style of [27] would require us make relatively minor modifications to B1, B2, and C2.

Reachability. In [27] reachability conditions for states of AIPs are not introduced. Consequently, C4 and C5 in Theorem 3 are weaker than the corresponding conditions in [27]. This is because we require the f to be nonincreasing (decreasing, resp.) only from the reachable states. Thus, invariant properties proved using Theorem 1, can be used to verify these conditions.

Convergence and termination. The general method for proving termination is based on finding a function which decreases along every transition of an automaton. If the co-domain of the function is wellfounded, the automaton must terminate, because there are no infinite descending chains.

The standard definition of termination—that of an automaton or a program executing a finite sequence of transitions and then stopping with an answer—is not directly applicable to reactive system models where the automaton runs forever producing an infinite sequence of outputs. Thus, we redefine termination as follows: given a subset of states S^T, $\mathscr{A}$ *terminates* at S^T if for every execution $s_0, a_1, s_1, \ldots$, there exits $n \in \mathbb{N}$, such that for all $m > n$, $s_m \in S^T$. If we set $S^* = S^T$, then this definition of termination is equivalent to the definition of convergence if one allows ϵ to be 0. With this interpretation, termination is a stronger property than convergence. Indeed, many distributed systems, such as the consensus protocol of Example 1, only convergence (for $\epsilon > 0$) can be guaranteed and not termination.

7   Conclusions

We have formalized fairness, stability, and convergence within an existing PVS framework for verifying untimed, timed, and hybrid automata. The theory provides a very general set of sufficient conditions for proving stability and convergence. These conditions can be checked using the PVS prover or using other tools. For example, the nonincreasing condition for convergence C3 can be checked with a model-checker. The theory extends the PVS interface for the Tempo toolkit, and hence, enables us to now verify invariance, implementation, convergence, and stability, all within the same software framework.

Currently we are applying the proposed metatheory to verify timed and hybrid system models; in particular, convergence of asynchronous pattern formation algorithms for mobile agent systems [6]. We plan on developing PVS strategies that exploit the common structural properties in these models and automate convergence and stability proofs.

References

1. Tempo toolset, version 0.2.2 beta (January 2008),
   http://www.veromodo.com/tempo/
2. Archer, M.: PVS Strategies for special purpose theorem proving. Annals of Mathematics and Artificial Intelligence 29(1/4) (February 2001)
3. Archer, M., Heitmeyer, C., Sims, S.: TAME: A PVS interface to simplify proofs for automata models. In: Proceedings of UITP 1998 (July 1998)
4. Archer, M., Lim, H., Lynch, N., Mitra, S., Umeno, S.: Specifying and proving properties of timed I/O automata using Tempo. Design Automation for Embedded Systems (to appear, 2008)
5. Bulwahn, L., Krauss, A., Nipkow, T.: Finding lexicographic orders for termination proofs in Isabelle/HOL. In: Schneider, K., Brandt, J. (eds.) TPHOLs 2007. LNCS, vol. 4732, pp. 38–53. Springer, Heidelberg (2007)
6. Chandy, K.M., Mitra, S., Pilotto, C.: Formations of mobile agents with message loss and delay (preprint) (2007),
   http://www.ist.caltech.edu/~mitras/research/2008/asynchcoord.pdf
7. Devillers, M.: Translating IOA automata to PVS. Technical Report CSI-R9903, Computing Science Institute, University of Nijmegen (February 1999),
   http://www.cs.ru.nl/research/reports/info/CSI-R9903.html
8. Filliâtre, J.: Finite automata theory in Coq: A constructive proof of kleene's theorem. Technical report, LIP -ENS, Research Report 97-04, Lyon (February 1997)
9. Floyd, R.: Assigning meanings to programs. In: Symposium on Applied Mathematics. Mathematical Aspects of Computer Science, pp. 19–32. American Mathematical Society (1967)
10. Gottliebsen, H.: Transcendental functions and continuity checking in PVS. In: Aagaard, M.D., Harrison, J. (eds.) TPHOLs 2000. LNCS, vol. 1869, pp. 197–214. Springer, Heidelberg (2000)

11. Harrison, J.: Theorem Proving with the Real Numbers. Springer, Heidelberg (1998)
12. Kaynar, D.K., Lynch, N., Segala, R., Vaandrager, F.: The Theory of Timed I/O Automata. Synthesis Lectures on Computer Science. Morgan Claypool, Technical Report MIT-LCS-TR-917 (November 2005)
13. Lester, D.: NASA langley PVS library for topological spaces, http://shemesh.larc.nasa.gov/fm/ftp/larc/PVS-library/topology-details.html
14. Liberzon, D.: Switching in Systems and Control. Systems and Control: Foundations and Applications. Birkhauser, Boston (2003)
15. Lim, H., Kaynar, D., Lynch, N., Mitra, S.: Translating timed I/O automata specifications for theorem proving in PVS. In: Pettersson, P., Yi, W. (eds.) FORMATS 2005. LNCS, vol. 3829. Springer, Heidelberg (2005)
16. Luenberger, D.G.: Introduction to Dynamic Systems: Theory, Models, and Applications. John Wiley and Sons, Inc, New York (1979)
17. Lynch, N., Tuttle, M.: An introduction to Input/Output automata. CWI-Quarterly 2(3), 219–246 (1989)
18. Mitra, S.: A Verification Framework for Hybrid Systems. PhD thesis, Massachusetts Institute of Technology, Cambridge, MA 02139 (September 2007)
19. Mitra, S., Archer, M.: PVS strategies for proving abstraction properties of automata. Electronic Notes in Theoretical Computer Science 125(2), 45–65 (2005)
20. Müller, O.: I/O automata and beyond: Temporal logic and abstraction in Isabelle. In: Proceedings of the 11th International Conference on Theorem Proving in Higher Order Logics, London, UK, pp. 331–348. Springer, London (1998)
21. Nipkow, T., Slind, K.: I/O automata in Isabelle/HOL. In: Smith, J., Dybjer, P., Nordström, B. (eds.) TYPES 1994. LNCS, vol. 996, pp. 101–119. Springer, Heidelberg (1995)
22. Müller, O.: A Verification Environment for I/O Automata Based on Formalized Meta-Theory. PhD thesis, Technische Universität München (September 1998)
23. Owre, S., Rajan, S., Rushby, J., Shankar, N., Srivas, M.: PVS: Combining specification, proof checking, and model checking. In: Alur, R., Henzinger, T.A. (eds.) CAV 1996. LNCS, vol. 1102, pp. 411–414. Springer, Heidelberg (1996)
24. Paulin-Mohring, C.: Modelisation of timed automata in Coq. In: Kobayashi, N., Pierce, B.C. (eds.) TACS 2001. LNCS, vol. 2215, pp. 298–315. Springer, Heidelberg (2001)
25. Paulson, L.C.: Mechanizing UNITY in Isabelle. ACM Transactions on Computational Logic 1(1), 3–32 (2000)
26. Rohwedder, E., Pfenning, F.: Mode and termination checking for higher-order logic programs. In: Riis Nielson, H. (ed.) ESOP 1996. LNCS, vol. 1058, pp. 296–310. Springer, Heidelberg (1996)
27. Tsitsiklis, J.N.: On the stability of asynchronous iterative processes. Theory of Computing Systems 20(1), 137–153 (1987)
28. Umeno, S., Lynch, N.A.: Safety verification of an aircraft landing protocol: A refinement approach. In: Bemporad, A., Bicchi, A., Buttazzo, G. (eds.) HSCC 2007. LNCS, vol. 4416, pp. 557–572. Springer, Heidelberg (2007)

Certified Exact Transcendental Real Number Computation in Coq*

Russell O'Connor

Institute for Computing and Information Science
Faculty of Science
Radboud University Nijmegen
r.oconnor@cs.ru.nl

Abstract. Reasoning about real number expressions in a proof assistant is challenging. Several problems in theorem proving can be solved by using *exact* real number computation. I have implemented a library for reasoning and computing with complete metric spaces in the Coq proof assistant and used this library to build a constructive real number implementation including elementary real number functions and proofs of correctness. Using this library, I have created a tactic that automatically proves strict inequalities over closed elementary real number expressions by computation.

1 Introduction

Mathematics increasingly relies on computation for proofs. Because software is often error prone, proofs depending on computation are sometimes considered suspect. Recently, people have used proof assistants to verify these kinds of mathematical theorems [7]. Real number computation plays an essential role in some of these problems. These proofs typically require finding a rational approximation of some real number expression to within a specified error or proving a (strict) inequality between two real number expressions. Two examples of such proofs are the disproof of Merten's conjecture [15] and the proof of Kepler's conjecture [8]. Certified real number computation also has other applications including verifying properties of hybrid automata.

Proof assistants based on dependent type theory, such as Coq [17], allow one to develop a constructive theory of real numbers in which approximations of real numbers can be evaluated by the system. Functions on real numbers compute what accuracy is needed from their input to satisfy the requested accuracy for their output. Rather than accumulating rounding errors, the resulting approximations are guaranteed to be within the accuracy requested. One can develop a constructive theory of real numbers that yields efficient functions by taking care to ensure the computational aspects of the proofs are efficient. This paper illustrates how to develop such an efficient constructive theory. We begin reviewing some results that are detailed in a previous publication [14]:

---

* This document has been produced using TeX_MACS (see http://www.texmacs.org)

O. Ait Mohamed, C. Muñoz, and S. Tahar (Eds.): TPHOLs 2008, LNCS 5170, pp. 246–261, 2008.

- A theory of metric spaces is developed (Section 3) that is independent of the real numbers. An operation for completing metric spaces is defined (Section 3.2), and this operation is seen to be a monad.
- This theory of complete metric spaces is used to define the real numbers (Section 4). A key idea is to first define elementary functions over the rational numbers, and then, once the functions are shown to be uniformly continuous, lift these functions to the real numbers by using the monad operations.

A large library of mathematical results called CoRN has previously been developed at Radboud University Nijmegen [3]. Its collection of proofs includes both the fundamental theorem of algebra and the fundamental theorem of calculus. I extended this library by formalizing this theory of complete metric spaces. The new results detailing how this theory was formalized in Coq are covered (Section 5):

- The formalization was designed with efficient execution in mind (Section 5.1).
- Care was needed to efficiently approximate infinite series (Section 5.2).
- The technique of proof by reflection is used to verify a definition of π (Section 5.3).
- Elementary functions are proved correct by showing that they are equivalent to their corresponding functions defined in the CoRN library (Section 5.4).
- This theory is put to use by developing a tactic that uses computation to automatically verify strict inequalities over closed real number expressions (Section 5.5).

This formalization will be part of the next version of the CoRN library, which will be released at the same time Coq 8.2 is released.

1.1 Notation

The propositions true and false are denoted by $\top$ and $\bot$ respectively. The type of propositions is written as $\star$. In Coq this type is `Prop`.

The type $\mathbb{Q}^+$ denotes the strictly positive rational numbers, and I will use similar notation for other number types. The type $\mathbb{Q}^+_\infty$ denotes $\mathbb{Q}^+ + \{\infty\}$.

Functions taking multiple arguments will be curried as in $f : A \Rightarrow B \Rightarrow C$; however, for readability, I will often use mathematical notation when applying parameters, $f(x, y)$, even though it should technically be written as $f(x)(y)$.

I denote the function f iterated n times as $f^{(n)}$.

Because constructive mathematics has a classical interpretation, all the theorems in this paper can also be understood as theorems of classical analysis. Although some of the definitions I use are somewhat different from the usual classical definitions, they are still equivalent (under classical logic) to their classical counterparts.

2 Background

The real numbers are typically defined as a Cauchy sequence of rational numbers. A sequence $x : \mathbb{N} \Rightarrow \mathbb{Q}$ is Cauchy when

$$\forall \varepsilon : \mathbb{Q}.0 < \varepsilon \Rightarrow \exists N : \mathbb{N}.\forall m : \mathbb{N}.N \le m \Rightarrow |x_m - x_N| \le \varepsilon.$$

The function mapping ε to N is the modulus of convergence. It tells you how far into the sequence you must reach in order to get good rational approximations to the real number that x represents.

By using the constructive existential, one ensures that the value of N is computable from ε. This results in the constructive real numbers. One can compute approximations of constructive real numbers to within any given precision.

Real numbers are usually created from Cauchy sequences (which often arise from Taylor series). Perhaps this is why the Cauchy sequence definition is common. On the other hand, approximation is the fundamental operation for *consuming* real numbers. This suggests an alternative definition of real numbers based on how they are consumed. One can define a real number as a regular function of rational numbers. A regular function of rational numbers is a function $x : \mathbb{Q}^+ \Rightarrow \mathbb{Q}$ such that

$$\forall \varepsilon_1 \varepsilon_2.|x(\varepsilon_1) - x(\varepsilon_2)| \le \varepsilon_1 + \varepsilon_2.$$

Regular functions are a generalization of regular sequences, which Bishop and Bridges use to define the real numbers [1]. With regular functions, x directly represents the function that approximates a real number to within ε. The regularity condition ensures that the approximations are coherent.

Regular functions and Cauchy sequences can be used to construct more than just the real numbers. They can be used to construct the completion of any metric space.

3   Metric Spaces

Usually a metric space X is defined by a metric function $d : X \times X \Rightarrow \mathbb{R}$; however, this assumes that the real numbers have already been defined. Instead, one can define a metric space based on a ball relation $\beta_\varepsilon(a, b)$, that characterizes when $d(a, b) \le \varepsilon$. Partial application, $\beta_\varepsilon(a)$, yields a predicate that represents the set of points inside the closed ball of radius ε around a. The following axioms characterize a ball relationship $\beta : \mathbb{Q}^+ \Rightarrow X \Rightarrow X \Rightarrow \star$.

1. $\beta_\varepsilon(a, a)$
2. $\beta_\varepsilon(a, b) \Rightarrow \beta_\varepsilon(b, a)$
3. $\beta_{\varepsilon_1}(a, b) \Rightarrow \beta_{\varepsilon_2}(b, c) \Rightarrow \beta_{\varepsilon_1 + \varepsilon_2}(a, c)$
4. $(\forall \delta : \mathbb{Q}^+.\beta_{\varepsilon + \delta}(a, b)) \Rightarrow \beta_\varepsilon(a, b)$

Axioms 1 and 2 state that the ball relationship is reflexive and symmetric. Axiom 3 is a form of the triangle inequality. Axiom 4 states that the balls are closed. Closed balls are used because their proof objects usually have no computational content and can be ignored during evaluation. For some metric spaces, such as the real numbers, open balls are defined with existential quantifiers and their use would lead to unnecessary computation [4].

Two points are considered identical if they are arbitrarily close to each other.

$$(\forall \varepsilon.\beta_\varepsilon(a,b)) \Leftrightarrow a \asymp b$$

This can be considered either the definition of equivalence in X, or if X comes with an equivalence relationship, then it can be considered a fifth axiom.

In Coq, a metric space X is a dependent record containing

1. a type (called the carrier)
2. a ball relation on that type
3. a proof that this ball relation satisfies the the above axioms.

The second projection function B returns the ball relation component of the metric space. I will write the metric space parameter in a superscript, as in B^X. I will not distinguish between a metric space and its carrier, so X will denote either a metric space or its carrier depending on the context.

Sometimes an extended ball relation $\check{B}^X : \mathbb{Q}^+_\infty \Rightarrow X \Rightarrow X \Rightarrow \star$ will be used where $\check{B}^X_\infty(a,b)$ always holds and reduces to $B^X_\varepsilon(a,b)$ when $\varepsilon < \infty$.

3.1 Uniformly Continuous Functions

A uniformly continuous function allows one to approximate the output from an approximation of the input. The usual definition for a function $f : X \Rightarrow Y$ to be uniformly continuous is

$$\forall \varepsilon.\exists \delta.\forall ab.B^X_\delta(a,b) \Rightarrow B^Y_\varepsilon(f(a),f(b)).$$

The function mapping ε to δ is what Bishop and Bridges [1] call the **modulus of continuity** and is denoted by μ_f. (This is the inverse of what mathematicians usually call the modulus of continuity.)

It is advantageous to use a more general notion of modulus of continuity that can return ∞. This is used for bounded functions when the requested accuracy is wider than the bound on the function. For example, $\mu_{\sin}(\varepsilon) = \infty$ for $1 \le \varepsilon$ because $\sin(x)(\varepsilon) = 0$ for all x. We also pull out the modulus of continuity in order to reason about it directly. Thus, we define a function $f : X \Rightarrow Y$ to be **uniformly continuous with modulus** $\mu_f : \mathbb{Q}^+ \Rightarrow \mathbb{Q}^+_\infty$ when

$$\forall ab\varepsilon.\check{B}^X_{\mu_f(\varepsilon)}(a,b) \Rightarrow B^Y_\varepsilon(f(a),f(b)).$$

In Coq, a uniformly continuous functions is a dependent record containing

1. a function f between two metric spaces
2. a modulus of continuity for f
3. a proof that f is uniformly continuous with the given modulus.

This means that μ is really the second projection function. Again, I will not distinguish between the uniformly continuous function f and its actual function.

I will denote the type of uniformly continuous functions with the single bar arrow, as in $X \to Y$.

3.2 Complete Metric Spaces

We are now in a position to define regular functions over an arbitrary metric space X. A function $x : \mathbb{Q}_\infty^+ \Rightarrow X$ is a **regular function** when

$$\forall \varepsilon_1 \varepsilon_2 : \mathbb{Q}^+ . B^X_{\varepsilon_1 + \varepsilon_2}(x(\varepsilon_1), x(\varepsilon_2)).$$

The function x is allowed to return anything when given ∞.

Two regular functions are equivalent $(x \asymp y)$ when their approximations are arbitrarily close to each other.

$$\forall \varepsilon_1 \varepsilon_2 : \mathbb{Q}^+ . B^X_{\varepsilon_1 + \varepsilon_2}(x(\varepsilon_1), y(\varepsilon_2))$$

Thus, a regular function is a function that is equivalent to itself under this relation.

Regular functions form a metric space [14], $\mathfrak{C}(X)$, where the ball relation $B^{\mathfrak{C}(X)}_\varepsilon(x, y)$ is

$$\forall \delta_1 \delta_2 : \mathbb{Q}^+ . B^X_{\delta_1 + \varepsilon + \delta_2}(x(\delta_1), y(\delta_2)).$$

This states that x and y are within ε of each other when their approximations are almost within ε of each other.

Completion is a Monad. The completion operator $\mathfrak{C}$ forms a monad in the category of metric spaces and uniformly continuous functions between them [14]. The injection of X into $\mathfrak{C}(X)$ is unit : $X \to \mathfrak{C}(X)$. The proof that a complete metric space is complete yields join : $\mathfrak{C}(\mathfrak{C}(X)) \to \mathfrak{C}(X)$. The function map : $(X \to Y) \Rightarrow \mathfrak{C}(X) \to \mathfrak{C}(Y)$ lifts uniformly continuous functions to the complete space. Finally, bind : $(X \to \mathfrak{C}(Y)) \Rightarrow \mathfrak{C}(X) \to \mathfrak{C}(Y)$ is defined in terms of map and join in the usual way.

$$\mathrm{unit}(a)(\varepsilon) := a$$
$$\mathrm{join}(x)(\varepsilon) := x\left(\frac{\varepsilon}{2}\right)\left(\frac{\varepsilon}{2}\right)$$
$$\mathrm{map}(f)(x)(\varepsilon) := f\left(x\left(\frac{\breve{\mu}_f(\varepsilon)}{2}\right)\right) \tag{1}$$
$$\mathrm{bind}(f) := \mathrm{join} \circ \mathrm{map}(f)$$

Here the function $\breve{\mu}_f : \mathbb{Q}_\infty^+ \Rightarrow \mathbb{Q}_\infty^+$ maps ∞ to ∞, and applies μ_f otherwise.

In my previous work, I used a simpler definition of map

$$\mathrm{map}'(f)(x)(\varepsilon) := f(x(\breve{\mu}_f(\varepsilon))). \tag{2}$$

Unfortunately, this definition requires the additional assumption that X be a prelength space [14]. Recently, I inferred from Richman's work [16] that map can be defined using equation 1 and works for all metric spaces if the modulus of continuity of map(f) is smaller than μ_f.

Despite the above, in the common case that X is a prelength space, the definition of map$'$ in equation 2 is more efficient, and map$'(f)$ has the same modulus of continuity as f. Because of this, I use map$'$ (and similarly bind$'$) throughout my work. I use map mostly for theoretical results.

Completion is a Strong Monad. Functions between two metric spaces form a metric space under the sup-norm. The ball relation between two functions $B_\varepsilon^{X \to Y}(f, g)$ is

$$\forall a. B_\varepsilon^Y(f(a), g(a))$$

Now the function map : $(X \to Y) \to \mathfrak{C}(X) \to \mathfrak{C}(Y)$ can be shown to be uniformly continuous [14]. By defining ap : $\mathfrak{C}(X \to Y) \to \mathfrak{C}(X) \to \mathfrak{C}(Y)$, higher arity maps such as map2 : $(X \to Y \to Z) \to \mathfrak{C}(X) \to \mathfrak{C}(Y) \to \mathfrak{C}(Z)$ can be constructed.

$$\mathrm{ap}(f)(x)(\varepsilon) := \mathrm{map}\left(f\left(\frac{\varepsilon}{2}\right)\right)(x)\left(\frac{\varepsilon}{2}\right)$$

$$\mathrm{map2}(f) := \mathrm{ap} \circ \mathrm{map}(f)$$

4 Real Numbers

Because the rational numbers $\mathbb{Q}$ are a metric space, the real numbers can be simply defined as the completion of $\mathbb{Q}$.

$$\mathbb{R} := \mathfrak{C}(\mathbb{Q})$$

Uniformly continuous operations on the real numbers are defined by lifting their rational counterparts with map or map2. This is how $+_\mathbb{R}$ and $-_\mathbb{R}$ are defined [14].

I find using monadic operators to define functions on $\mathbb{R}$ is easier than trying to define functions directly. It splits the problem into two parts. The first part is to define the the function over $\mathbb{Q}$, which is easier to work with because equality is decidable for $\mathbb{Q}$. The second part is to prove that the function is uniformly continuous.

4.1 Order

A real number x is non-negative when

$$\forall \varepsilon : \mathbb{Q}^+. - \varepsilon \leq_\mathbb{Q} x(\varepsilon).$$

The not-greater-than relation on real numbers, $x \leq_\mathbb{R} y$, means that $y - x$ is non-negative.

A real number x is positive when

$$\exists \varepsilon : \mathbb{Q}^+. \mathrm{unit}(\varepsilon) \leq_\mathbb{R} x$$

(recall that unit : $\mathbb{Q} \to \mathbb{R}$). One real number is less than another, $x <_\mathbb{R} y$, when $y - x$ is positive. Two real numbers are apart, $x \lessgtr y$, when $x < y \vee y < x$.

This definition of positivity differs from what would be analogous to Bishop and Bridges's definition, $\exists \varepsilon : \mathbb{Q}^+. \varepsilon <_\mathbb{Q} x(\varepsilon)$. Although the two definitions are equivalent, my definition above contains a rational number in $]0, x]$. This is exactly the information that will be needed to compute x^{-1} or $\ln(x)$ (Section 4.2). With Bishop and Bridges's definition, one must compute $x(\varepsilon) - \varepsilon$, which is a potentially expensive calculation.

4.2   Non-uniformly Continuous and Partial Functions

Unfortunately not all functions that we want to consider are uniformly continuous. One can deal with continuous functions by noting that they are uniformly continuous on some collection of closed sub-domains that cover the whole space. For example, $\lambda a : \mathbb{Q}.a^2$ is uniformly continuous on $[-c, c]$. Thus, a real number x can be squared by finding some domain $[-c, c]$ containing it and lifting $(\lambda a.(\max(\min(a, c), -c))^2$, which *is* uniformly continuous. In this case c can be chosen to be $|x(1)| + 1$. One can prove that the result is independent of the choice of c, so long as $x \in [-c, c]$.

Evaluating a non-uniformly continuous function is potentially a costly operation. The input x must be approximated twice. The first approximation finds a domain to operate in, and the second approximation is used to evaluate the function. In practice, I have found that one often has a suitable domain lying around for the particular problem at hand. If that is the case, then x only needs to be approximated once.

Partial functions with open domains are handled in the same way as non-uniformly continuous functions. For example, $\lambda x.x^{-1}$ is uniformly continuous on the domains $[c, \infty[$ and $] - \infty, -c]$ (where $0 < c$). One difference is that one cannot automatically find a domain containing x. One requires a proof that x is apart from 0. From such a proof, one can find a suitable domain containing x.

Partial functions with closed domains, such as $\lambda x.\sqrt{x}$, can be extended to continuous total functions. I extend the square root function to return 0 for negative values. If one wishes, one can then restrict the lifted function to only accept non-negative inputs.

4.3   Transcendental Functions

Transcendental functions are first defined from $\mathbb{Q}$ to $\mathbb{R}$. Once these functions are shown to be uniformly continuous (or otherwise using the techniques from the previous section), they are then lifted using bind to create functions from $\mathbb{R}$ to $\mathbb{R}$.

Most elementary functions can be defined on some sub-domain by an alternating decreasing series. Inputs outside this domain can often be dealt with by using range reduction. Range reduction uses elementary identities to reduce inputs from a wider to a narrower domain [14].

For example, the alternating series $\sum_{i=0}^{\infty}(-1)^i \frac{a^{2i+1}}{(2i+1)!}$ computes $\sin(a)$, and is decreasing when $a \in [-1, 1]$. For a outside this interval, range reduction is preformed by repeated application of the identity

$$\sin(a) \asymp 3\sin\left(\frac{a}{3}\right) - 4\sin^3\left(\frac{a}{3}\right).$$

The value of an infinite alternating series, is represented by a regular function that finds a partial sum having an error no more than ε. When an alternating series is decreasing, finding such a partial sum is easy because the last term also represents the error. One only needs to accumulate terms until a term becomes less than ε.

Coq will not accept a general recursive function that computes the above partial sum. It requires a proof of termination. This is done by computing an upper bound on the number of terms that will be needed. Strategies for doing this efficiently in Coq are discussed Section 5.2.

The elementary functions, sin, cos, and $\tan^{-1}$ are defined as described in my previous publication [14]. The implementation of ln has been improved by defining it in terms of $\tanh^{-1}$,

$$\ln\left(\frac{n}{d}\right) := 2\tanh^{-1}\left(\frac{n-d}{n+d}\right).$$

However, the input is still range reduced into $[\frac{1}{2}, 2]$ before using the above formula.

I have also implemented a function to sum sub-geometric series (a series where $|a_{n+1}| \leq r|a_n|$). The error of the partial sums of these series is easy to compute from the last term and r. I now use this function to compute the $\exp(a)$ function for $a \in]0, 1[$.

4.4   Compression

Without intervention, the numerators and denominators of rational numbers occurring in real number computations become too large for practical computation. To help prevent this, I defined a compression operation for real numbers.

$$\text{compress}(x)(\varepsilon) := \text{approx}_{\mathbb{Q}}\left(x\left(\frac{\varepsilon}{2}\right), \frac{\varepsilon}{2}\right)$$

where $\text{approx}_{\mathbb{Q}}(a, \delta)$ returns some rational number within δ of a. The idea is that $\text{approx}_{\mathbb{Q}}(a, \delta)$ quickly computes a rational number close to a but having a smaller numerator and denominator. In my implementation, I return $\frac{b}{2^n}$, where 2^n is the smallest power of 2 greater than the denominator of δ, and b is chosen appropriately so that the result is within δ of a.

The compress function is equivalent to the identity function on $\mathbb{R}$.

$$\text{compress}(x) \asymp x$$

By liberally inserting compress into one's expressions, one can often dramatically improve the efficiency of real number calculations. I am considering adding a call to compress with every use of map or bind so that the user does not need to add these calls themselves. Too many calls to compress can harm performance but perhaps not enough to cause worry.

5   Formalization in Coq

The theory of metric spaces and real numbers described in Sections 3 and 4 has been formalized in the Coq proof assistant. I developed functions and proofs simultaneously. I did not extract functions from constructive proofs, nor did I write functions entirely separately from their proofs of correctness. Proofs and functions are often mixed together, such as in the dependent records of metric spaces, uniformly continuous functions, and regular functions.

5.1   Efficient Proofs

A mixture of proofs and functions can still be efficient to evaluate by taking care to write the functional aspects efficiently and ensuring that the non-functional aspects are declared opaque. Declaring lemmas as opaque prevents call-by-value evaluation from normalizing irrelevant proofs.

I used Coq's Prop/Set distinction (two different universes of types) to assist in the separation of these concerns [4]. Types that have at most one member (extensionally) are proof-irrelevant and go into Prop. Lemmas having these types are declared opaque. Types that may have more than one member go into Set, and objects of such types are kept transparent. This criterion means that I use the Set based sum and dependent pair types for the constructive disjunction and constructive existential quantifier.

When proving a constructive existential goal, one has to deal with both Prop and Set during a proof. The existential lives in Set, but after supplying the witness, a Prop based proof obligation remains. The witness needs to be transparent, but the proof obligation should be opaque. It is best to try and separate these two parts into two different definitions, one transparent and one opaque. However, in some instances I make the entire development transparent, but I mark the proof obligation part with Coq's abstract tactic. The abstract tactic automatically defines an opaque lemma containing marked part of the proof and places this lemma into the proof object. Thus, the marked part is never evaluated.

5.2   Summing Series

One of the more challenging aspects of the formalization was computing the infinite series defined in Section 4.3 in an efficient manner. In order to convince Coq that the procedure of accumulating terms until the error becomes sufficiently small terminates, I provided Coq with an upper bound on the number of terms that would be required. I tried two different methods to accomplish this.

The first method computes an upper bound on the number of terms needed as a Peano natural number. The problem is that the call-by-value evaluation scheme used by Coq's virtual machine would first compute this value before computing the series. This upper bound is potentially extremely large, it is encoded in unary, and only a few terms may actually be needed in the computation. The solution to this problem was to create a lazy natural number using the standard trick of placing a function from the unit type inside the constructor.

The lambda expressions inside the lazy natural numbers delay the evaluation of the call-by-value scheme. With some care, only the number of constructors needed for the recursion are evaluated.

```
Inductive LazyNat : Set :=
| Lazy0 : LazyNat
| LazyS : (unit -> LazyNat) -> LazyNat.
```

Fig. 1. Inductive definition of lazy natural numbers

A second method, suggested by Benjamin Grégoire, is to compute the number of terms needed as a binary number. This prevents the term from becoming too big. It is possible to do recursion over the binary natural numbers such that two recursive calls are made with the output of one recursive call being threaded through the other. In this way, up to n recursive calls can be made even though only $\lg n$ constructors are provided by the witness of termination.

In the simplified example below, the function F is iterated up to n times. Continuation passing style is used to thread the recursive calls.

```
Variable A R : Type
Variable F : (A -> R) -> A -> R

Fixpoint iterate_pos (n:positive) (cont: A -> R) : A -> R :=
match n with
| xH => F cont
| xO n' => iterate_pos n' (fun a => iterate_pos n' cont a)
| xI n' => F (fun a => (iterate_pos n'
(fun a => iterate_pos n' cont a)) a)
end.
```

Fig. 2. The Coq function iterate_pos recurses F at up to n times, using continuation passing style

The η-expansion of the continuations in the above definition are important, otherwise the virtual machine would compute the value of the iterate_pos n' cont calls before reducing F. This is important because F may not utilize its recursive call depending on the value of a. In such a case, we do not want the recursive call to be evaluated.

5.3 π

A common definition of π is $4 \tan^{-1}(1)$. This is an inefficient way of computing π because the series for $\tan^{-1}(1)$ converges slowly. One can more efficiently compute π by calling $\tan^{-1}$ with smaller values [18]. I chose an optimized formula for π from a list [19]:

$$\pi := 176 \tan^{-1}\left(\frac{1}{57}\right) + 28 \tan^{-1}\left(\frac{1}{239}\right) - 48 \tan^{-1}\left(\frac{1}{682}\right) + 96 \tan^{-1}\left(\frac{1}{12943}\right)$$

This formula can easily be shown to be equivalent to $4 \tan^{-1}(1)$ by repeated application of the **arctangent sum law**:

$$\text{if } a, b \in]-1, 1[\text{ then } \tan^{-1}(a) + \tan^{-1}(b) \asymp \tan^{-1}\left(\frac{a+b}{1-ab}\right)$$

To apply the arctangent sum law, one needs to verify that a and b lie in $]-1, 1[$. To solve this, I wrote a Coq function to iterate the function $f(b) := \frac{a+b}{1-ab}$, and at each step verify that the result is in the interval $]-1, 1[$. This function, called ArcTan_multiple, has type

$$\forall a : \mathbb{Q}. -1 < a < 1 \Rightarrow \forall n. \top \vee \left(n \tan^{-1}(x) \asymp \tan^{-1}(f^{(n)}(0))\right)$$

It is easy to build a function of the above type that just proves $\top$ in all cases, but ArcTan_multiple tries to prove the non-trivial result if it can.

To apply this lemma I use a technique called reflection. The idea is to evaluate the ArcTan_multiple(a, r, n) into head normal form. This will yield either left(q) or right(p). If right(p) is returned then p is the proof we want.

My first attempt at building a tactic to implement this did not work well. I used Coq's eval hnf command to reduce my expression to head normal form. However, this command repeatedly calls simpl to expose a constructor instead of using the evaluation mechanism directly. The problem was that simpl does extra reductions that are not necessary to get head normal form, so using eval hnf was too time consuming.

Instead, I built a reflection lemma, called reflect_right, to assist in applying the ArcTan_multiple function:

$$\forall z : A \vee B.(\text{if } z \text{ then } \bot \text{ else } \top) \Rightarrow B$$

This simple lemma does case analysis on z. If z contains a proof of A, it returns a proof of $\bot \Rightarrow B$. If z contains a proof of B, it returns a proof of $\top \Rightarrow B$. To prove $n \tan^{-1}(a) \asymp \tan^{-1}(f^{(n)}(0))$, for the example $a := \frac{1}{57}$ and $n := 176$, one applies reflect_right composed with ArcTan_multiple to reduce the goal to

$$\text{if } (\text{ArcTan\_multiple } \frac{1}{57} * 176) \text{ then } \bot \text{ else } \top,$$

where $*$ is the trivial proof of $-1 < \frac{1}{57} < 1$. Then one normalizes this expression using *lazy evaluation* to either $\top$, if ArcTan_multiple succeeds, or $\bot$, if it fails.

5.4   Correctness

There are two ways to prove that functions are correct. One way is to prove that they satisfy some uniquely defining properties. The other way is to prove that the functions are equivalent to a given reference implementation. I have verified that my elementary functions are equivalent to the corresponding functions defined in the CoRN library [3]. The functions in the CoRN library can be seen to be correct from the large library of theorems available about them. The CoRN library contains many different characterizations of these functions and new characterizations can easily be developed.

The CoRN library defines a **real number structure** as a complete, ordered, Archimedean field. My first step was to prove that my operations form a real number structure. I first attempted to directly show that my real numbers satisfy all the axioms of a real number structure, but this approach was difficult. Instead, I created an isomorphism between my real numbers and the existing model of the real numbers developed by Niqui [6]. This was a much easier approach because Niqui's Cauchy sequence definition and my regular function definition are closely related. With this isomorphism in place, I proved my operations satisfied the axioms of a real number structure by passing through the isomorphism and using Niqui's existing lemmas. Niqui has also proved that all real number structures are isomorphic, so I can create an isomorphism between my real numbers and any other real number structure.

The next step was to define my elementary functions and prove that they are equivalent to the corresponding CoRN functions. These theorems are of the form $\Phi(f_{\mathrm{CoRN}}(x)) \asymp f(\Phi(x))$ where Φ is the isomorphism from CoRN's real numbers to my real numbers.

To aid in converting statements between different representations of real numbers, I have created a rewrite database that contains the correctness lemmas. By rewriting with this database, expressions can be automatically converted from CoRN's real numbers into my real numbers. This database can easily be extended with more functions in the future.

The CoRN library was more than just a specification; this library was useful throughout my development. For example, I was often able to prove that a differentiable function f is uniformly continuous with modulus $\lambda \varepsilon. \frac{\varepsilon}{M}$ when M is a bound on the derivative of f. I could prove this because the theory of derivatives had already been developed in CoRN. The CoRN library also helped me reduce the problem of proving the correctness of continuous functions on $\mathbb{R}$ to proving correctness only on $\mathbb{Q}$.

5.5   Solving Strict Inequalities Automatically

Whether a strict inequality holds between real numbers is semi-decidable. This question can be reduced to proving that some expression $e_0 : \mathbb{R}$ is positive. To prove e_0 is positive one must find an $\varepsilon : \mathbb{Q}^+$, such that $\mathrm{unit}(\varepsilon) \leq e_0$. I wrote a tactic to automate the search for such a witness. It starts with an initial $\delta : \mathbb{Q}^+$, and computes to see if $e_0(\delta) - \delta$ is positive. If it is positive, then $e_0(\delta) - \delta$ is such a witness; otherwise δ is halved and the process is repeated. If $e_0 \asymp 0$, then this process will never terminate. If $e_0 < 0$, then the tactic will notice that $e_0(\delta) + \delta$ is negative and terminate with an error indicating that e_0 is negative.

This tactic has been combined with the rewrite database of correctness lemmas to produce a tactic that solves strict inequalities of closed expressions over CoRN's real numbers. This allows users to work entirely with CoRN's real numbers. They need never be aware that my effective real numbers are running behind the scenes.

Recently Cezary Kaliszyk has proved that Coq's classical real numbers (from the standard library) form a CoRN real number structure, and he has shown that Coq's elementary functions are equivalent to CoRN's. Now strict inequalities composed from elementary functions over Coq's classical real numbers can automatically be solved.

The tactic currently only works for expressions composed from total functions. Partial functions with open domains pose a problem because proof objects witnessing, for example, that x is positive for $\ln(x)$ must be transparent for computation. However, proof objects for CoRN functions are opaque, and Coq's classical functions have no proof objects. The required proof objects are proofs of strict inequalities, so I am developing a tactic that recursively solves these strict inequalities and creates transparent proof objects. This will allow one prove strict inequalities over expressions that include partial functions such as ln and $\lambda x.x^{-1}$.

5.6 Setoids

Coq does not have quotient types. Setoids are used in place of quotient types. A **setoid** is a type associated with an equivalence relation on that type. A framework for working with setoids is built into Coq. Coq allows one to associate an equivalence relation with a type and register functions as morphisms by proving they are well-defined with respect to the given equivalence relations. Coq allows you substitute terms with other equivalent terms in expressions composed from morphisms. Coq automatically creates proof objects validating these substitutions.

Setoids have some advantages over quotient types. Some functions, most notably the function that approximates real numbers, are not well-defined with respect to the equivalence relation—two equivalent real numbers may compute different approximations. It is unclear how one would support these functions if a system with quotient types was used.

Support for setoids was invaluable for development; however, I encountered some difficulties when dealing with convertible types. The types `CR`, `Complete Q_as_MetricSpace`, and `cs_crr CRasCRing`, where `cs_crr` retrieves the carrier type, are all convertible. They are equivalent as far as the underlying type theory is concerned, but Coq's tactics work on the meta-level where these terms are distinguishable. The setoid system does not associate the equivalence relation on the real numbers with all of these various forms of the same type. Adding type annotations was not sufficient; they were simplified away by Coq. Instead, I used an identity function to force the types into a suitable form:

```
Definition ms_id (m:MetricSpace) (x:m) : m := x.
```

The setoid system is being reimplemented in the upcoming Coq 8.2 release. Therefore, some of these issues may no longer apply.

5.7 Timings

Table 1 shows examples of real number expressions that can be approximated. Approximations of these expressions were evaluated to within 10^{-20} on a 1.4 GHz ThinkPad X40 laptop using Coq's `vm_compute` command for computing with its virtual machine. These examples are taking from the "Many Digits" friendly competition problem set [13].

Table 1. Timings of approximations of various real number expressions

Coq Expression			
Mathematical Expression	**Time**	**Result**	**Error**
`(CRsqrt (compress (rational_exp (1))*compress (CRinv_pos (3#1) CRpi)))%CR`			
$\sqrt{\frac{e}{\pi}}$	1 sec	0.93019136710263285866	10^{-20}
`(sin (compress (CRpower_positive 3` `(translate (1#1) (compress (rational_exp (1))))))))%CR`			
$\sin((e+1)^3)$	25 sec	0.90949524105726624718	10^{-20}
`(exp (compress (exp (compress (rational_exp (1#2))))))%CR`			
$e^{e^{e^{\frac{1}{2}}}}$	146 sec	181.33130360854569351505	10^{-20}

6   Related Work

Julien is developing an implementation of real numbers in Coq using co-inductive streams of digits [11]. This representation allows common subexpressions to be easily shared because streams naturally memoize. Sharing does not work as well with my representation because real numbers are represented by functions. One would require additional structure to reuse approximations between subexpressions. Julien also uses the new machine integers implementation in Coq's virtual machine to make his computations even faster. It remains to be seen if using machine integers would provide a similar boost in my implementation.

Cruz-Filipe implemented CoRN's library of theorems and functions over the real numbers in Coq [2]. His implementation forms the reference specification of my work. Although his implementation is constructive, it was never designed for evaluation [5]. Many important definitions are opaque and efficiency of computation was not a concern during development. Cruz-Filipe showed that it is practical to develop a constructive theory of real analysis inside Coq. My work extends this result to show that it is also possible to develop a theory of real analysis that is practical to evaluate.

Muñoz and Lester implemented a system for approximating real number expressions in PVS [12]. Their system uses rational interval analysis for doing computation on monotone segments of transcendental functions. Unfortunately, this leads to some difficulties when reasoning at a local minimum or maximum, so their system cannot automatically prove $0 < \sin\left(\frac{\pi}{2}\right)$, for instance.

Harrison implemented a system to approximate real number expressions in HOL Light [9]. His system runs a tactic that externally computes an approximation to an expression and generates a proof that the approximation is correct. If such a technique were implemented for Coq, it would generate large proof objects. This is not an issue in HOL Light where proof objects are not kept.

Jones created a preliminary implementation of real numbers and complete metric spaces in LEGO [10]. She represented real numbers as a collection containing arbitrarily small intervals of rational numbers that all intersect. Complete metric spaces were similarly represented by using balls in place of intervals. Because the only way of getting an interval from the collection is by using the arbitrarily small interval property, her representation could have been simplified by removing the collection and let it implicitly be the image of a function that produces arbitrarily small intervals. This is similar to my work because one can interpret a regular function f as producing the interval $[f(\varepsilon) - \varepsilon, f(\varepsilon) + \varepsilon]$. Perhaps using functions that return intervals could improve computation by allowing one to see that an approximation maybe more accurate than requested.

My work is largely based on Bishop and Bridges's work [1]. Some definitions have been modified to make the resulting functions more efficient. My definition of a metric space is more general; it does not require that the distance function be computable. The original motivation for the ball relation was only to develop a theory of metric spaces that did not presuppose the existence of the real numbers; however, it allows me to form a metric space of functions. This metric space does not have a computable distance function in general and would not be a metric space according to Bishop and Bridge's definition.

7 Conclusion

We have seen a novel definition of a metric space using a ball relation. We have seen how to create an effective representation for complete metric spaces and seen that the completion operation forms a monad. Using this monad, we defined the real numbers and used the monad operations to define effective functions on the real numbers. This theory has been formalized in Coq and the elementary functions have been proved correct. Real number expressions can be approximated to any precision by evaluation inside Coq. Finally, a tactic was developed to automatically proof strict inequalities over closed real number expressions.

After completing the Haskell prototype and after writing up detailed paper proofs [14], it took about five months of work to complete the Coq formalization. This preparation allowed for a smooth formalization experience. Only a few minor errors were found in the paper proofs. These errors mostly consisted of failing to consider cases when ε may be too large, and they were easy to resolve.

My results show that one can implement constructive mathematics such that the resulting functionally can be efficiently executed. This may be seen as the beginning of the realization of Bishop's program to see constructive mathematics as programming language.

References

1. Bishop, E., Bridges, D.: Constructive Analysis. Grundlehren der mathematischen Wissenschaften, vol. 279. Springer, Heidelberg (1985)
2. Cruz-Filipe, L.: Constructive Real Analysis: a Type-Theoretical Formalization and Applications. PhD thesis, University of Nijmegen (April 2004)
3. Cruz-Filipe, L., Geuvers, H., Wiedijk, F.: C-CoRN: the constructive Coq repository at Nijmegen. In: Asperti, A., Bancerek, G., Trybulec, A. (eds.) MKM 2004. LNCS, vol. 3119, pp. 88–103. Springer, Heidelberg (2004)

4. Cruz-Filipe, L., Spitters, B.: Program extraction from large proof developments. In: Basin, D., Wolff, B. (eds.) TPHOLs 2003. LNCS, vol. 2758, pp. 205–220. Springer, Heidelberg (2003)
5. Cruz-Filipe, L., Letouzey, P.: A large-scale experiment in executing extracted programs. Electr. Notes Theor. Comput. Sci. 151(1), 75–91 (2006)
6. Geuvers, H., Niqui, M.: Constructive reals in Coq: Axioms and categoricity. In: Callaghan, P., Luo, Z., McKinna, J., Pollack, R. (eds.) TYPES 2000. LNCS, vol. 2277, pp. 79–95. Springer, Heidelberg (2002)
7. Gonthier, G.: A computer-checked proof of the four colour theorem. Technical report, Microsoft Research Cambridge (2005)
8. Hales, T.C.: A computer verification of the Kepler conjecture. In: Proceedings of the International Congress of Mathematicians, Beijing, vol. III, pp. 795–804. Higher Ed. Press (2002)
9. Harrison, J.: Theorem Proving with the Real Numbers. Springer, Heidelberg (1998)
10. Jones, C.: Completing the rationals and metric spaces in LEGO. In: The second annual Workshop on Logical environments, New York, NY, USA, pp. 297–316. Cambridge University Press, Cambridge (1993)
11. Julien, N.: Certified exact real arithmetic using co-induction in arbitrary integer base. In: Functional and Logic Programming Symposium (FLOPS). LNCS, Springer, Heidelberg (2008)
12. Muñoz, C., Lester, D.: Real number calculations and theorem proving. In: Hurd, J., Melham, T. (eds.) TPHOLs 2005. LNCS, vol. 3603, pp. 195–210. Springer, Heidelberg (2005)
13. Niqui, M., Wiedijk, F.: The "Many Digits" friendly competition (2005), http://www.cs.ru.nl/~milad/manydigits
14. O'Connor, R.: A monadic, functional implementation of real numbers. Mathematical. Structures in Comp. Sci. 17(1), 129–159 (2007)
15. Odlyzko, A.M., te Riele, H.J.J.: Disproof of the Mertens conjecture. J. Reine Angew. Math. 357, 138–160 (1985)
16. Richman, F.: Real numbers and other completions. Math. Log. Q. 54(1), 98–108 (2008)
17. The Coq Development Team. The Coq Proof Assistant Reference Manual – Version V8.0 (April 2004), http://coq.inria.fr
18. Weisstein, E.W.: Machin-like formulas. From MathWorld–A Wolfram Web Resource (January 2004), http://mathworld.wolfram.com/Machin-LikeFormulas.html
19. Williams, R.: Arctangent formulas for PI (December 2002), http://www.cacr.caltech.edu/~roy/upi/pi.formulas.html

Formalizing Soundness of Contextual Effects

Polyvios Pratikakis, Jeffrey S. Foster, Michael Hicks, and Iulian Neamtiu

University of Maryland, College Park, MD 20742

Abstract. A *contextual effect* system generalizes standard type and effect systems: where a standard effect system computes the effect of an expression e, a contextual effect system additionally computes the *prior* and *future* effect of e, which characterize the behavior of computation prior to and following, respectively, the evaluation of e. This paper describes the formalization and proof of soundness of contextual effects, which we mechanized using the Coq proof assistant. Contextual effect soundness is an unusual property because the prior and future effect of a term e depends not on e itself (or its evaluation), but rather on the evaluation of the context in which e appears. Therefore, to state and prove soundness we must "match up" a subterm in the original typing derivation with the possibly-many evaluations of that subterm during the evaluation of the program, in a way that is robust under substitution. We do this using a novel typed operational semantics. We conjecture that our approach could prove useful for reasoning about other properties of derivations that rely on the context in which that derivation appears.

1 Introduction

Type and effect systems are used to reason about a program's computational effects [5,8,11]. Such systems have various applications in program analysis, e.g., to compute the set of memory accesses, I/O calls, function calls or new threads that occur in any given part of the program. Generally speaking, a type and effect system proves judgments of the form $\varepsilon; \Gamma \vdash e : \tau$ where ε is the effect of expression e. Recently, we proposed generalizing such systems to track what we call *contextual effects*, which capture the effects of the context in which an expression occurs [7]. In our contextual effect system, judgments have the form $\Phi; \Gamma \vdash e : \tau$, where Φ is a tuple $[\alpha; \varepsilon; \omega]$ containing ε, the standard effect of e, and α and ω, the effects of the program evaluation prior to and after computing e, respectively.

Our prior work explored the utility of contextual effects by studying two applications, one related to dynamic software updating correctness, and the other to analysis of multi-threaded programs. This paper presents the formalization and proof of soundness of contextual effects, which we have mechanized using the Coq proof assistant [2]. Intuitively, for all subexpressions e of a given program e_p, a contextual effect $[\alpha; \varepsilon; \omega]$ is sound for e if (1) α contains the actual, run-time effect of evaluating e_p prior to evaluating e, (2) ε contains the run-time effect of evaluating e itself, and (3) ω contains the run-time effect of evaluating the remainder of e_p after e's evaluation has finished. (Discussed in Section 2.)

O. Ait Mohamed, C. Muñoz, and S. Tahar (Eds.): TPHOLs 2008, LNCS 5170, pp. 262–277, 2008.

There are two main challenges with formalizing this intuition to prove that our contextual effect system is sound. First, we must find a way to define what constitute the *actual* prior and future effects of e when it is evaluated as part of e_p. Interestingly, these effects cannot be computed compositionally (i.e., by considering the subterms of e), as they depend on the relative position of the evaluation of e within the evaluation of e_p, and not on the evaluation of e itself. Moreover, the future effect of e models the evaluation after e has reduced to a value. In a small-step semantics, specifying the future effect by finding the end of e's computation would be possible but awkward. Thus we opt for a big-step operational semantics, in which we can easily and naturally define the prior, standard, and future effect of every subterm in a derivation. (Section 3)

The second challenge, and the main novelty of our proof, is specifying how to match up the contextual effect Φ of e, as determined by the *original* typing derivation of $\Phi_p; \Gamma \vdash e_p : \tau_p$, with the run-time effects of e recorded in the evaluation derivation. The difficulty here is that, due to substitution, e may appear many times and in different forms in the evaluation of e_p. In particular, a value containing e may be passed to a function $\lambda x.e'$ such that x occurs several times in e', and thus after evaluating the application, e will be duplicated. Moreover, variables within e itself could be substituted away by other reductions. Thus we cannot just syntactically match a subterm e of the original program e_p with its corresponding terms in the evaluation derivation.

To solve this problem, we define a *typed operational semantics* in which each subderivation is annotated with two typing derivations, one for the term under consideration and one for its final value. Subterms in the original program e_p are annotated with subderivations of the original typing derivation $\Phi_p; \Gamma \vdash e_p : \tau_p$. As subterms are duplicated and have substitutions applied to them, our semantics propagates the typing derivations in the natural way to the new terms. In particular, if Φ is the contextual effect of subterm e of e_p, then all of the terms derived from e will also have contextual effect Φ in the typed operational semantics. Given this semantics, we can now express soundness formally, namely that in every subderivation of the typed evaluation of a program, the contextual effect Φ in its typing contains the run-time prior, standard, and future effects of its computation. (Section 4)

We mechanized our proof using the Coq proof assistant, starting from the framework developed by Aydemir et al [1]. We found the mechanization process worthwhile, because our proof structure, while conceptually clear, required getting a lot of details right. Most notably, typing derivations are nested inside of evaluation derivations in the typed operational semantics, and thus the proofs of each case of the lemmas are somewhat messy. Using a proof assistant made it easy to ensure we had not missed anything. We found that, modulo some typos, our paper proof was correct, though the mechanization required that we precisely define the meaning of "subderivation." (Section 5)

We believe that our approach to proving soundness of contextual effects could be useful for other systems, in particular ones in which properties of subderivations depend on their position within the larger derivation in which they appear.

2   Background: Contextual Effects

This section reviews our type and effect system, and largely follows our previous presentation [7]. Readers familiar with the system can safely skip this section.

2.1   Language

Figure 1 presents our source language, a simple calculus with expressions that consist of values v (integers, functions or pointers), variables and call-by-value function application. Our language also includes updateable references, created with $\mathsf{ref}^L\ e$, along with dereference and assignment. We annotate each syntactic occurrence of ref with a label L, which serves as the abstract name for the locations allocated at that program point. Evaluating $\mathsf{ref}^L\ e$ creates a pointer r_L, where r is a fresh name in the heap and L is the declared label. Dereferencing or assigning to r_L during evaluation has effect $\{L\}$. Note that pointers r_L do not appear in the syntax of the program, but only during its evaluation. For simplicity we do not model recursive functions directly, but they can be encoded using references. Also, due to space constraints we omit let and if. They are included in the mechanized proof; supporting them is straightforward.

An *effect*, written α, ε, or ω, is a possibly-empty set of labels, and may be 1, the set of all labels. A *contextual effect*, written Φ, is a tuple $[\alpha; \varepsilon; \omega]$. If e' is a subexpression of e, and e' has contextual effect $[\alpha; \varepsilon; \omega]$, then

- The *current effect* ε is the effect of evaluating e' itself.
- The *prior effect* α is the effect of evaluating e until we begin evaluating e'.
- The *future effect* ω is the effect of the remainder of the evaluation of e after e' is fully evaluated.

Thus ε is the effect of e' itself, $\alpha \cup \omega$ is the effect of the context in which e' appears, and therefore $\alpha \cup \varepsilon \cup \omega$ is the effect of evaluating e.

To make contextual effects easier to work with, we introduce some shorthand. We write Φ^α, Φ^ε, and Φ^ω for the prior, current, and future effect components, respectively, of Φ. We also write $\Phi_\emptyset$ for the empty effect $[1; \emptyset; 1]$—by subsumption, discussed below, an expression with this effect may appear in any context. For brevity, whenever it is clear we will refer to contextual effects simply as *effects*.

Expressions	$e ::= v \mid x \mid e\ e \mid \mathsf{ref}^L\ e \mid\ !e \mid e := e$
Values	$v ::= n \mid \lambda x.e \mid r_L$
Effects	$\alpha, \varepsilon, \omega ::= \emptyset \mid 1 \mid \{L\} \mid \varepsilon \cup \varepsilon$
Contextual Effects	$\Phi ::= [\alpha; \varepsilon; \omega]$
Types	$\tau ::= int \mid ref^\varepsilon\ \tau \mid \tau \xrightarrow{\Phi} \tau$
Environments	$\Gamma ::= \cdot \mid (\Gamma, x \mapsto \tau) \mid (\Gamma, r \mapsto \tau)$
Labels	L

Fig. 1. Syntax

2.2 Typing

Figure 2 presents our contextual type and effect system. The rules prove judgments of the form $\Phi; \Gamma \vdash e : \tau$, meaning in type environment Γ, expression e has type τ and contextual effect Φ.

Types τ, listed in Figure 1, include the integer type int; reference types $ref^\varepsilon \tau$, which denote a reference to memory location of type τ where the reference itself is annotated with a label $L \in \varepsilon$; and function types $\tau \longrightarrow^\Phi \tau'$, where τ and τ' are the domain and range types, respectively, and the function has contextual effect Φ. Environments Γ, defined in Figure 1, are maps from variable names or (unlabeled) pointers to types.

The first two rules, (TINT) and (TVAR), assign the expected types and the empty effect, since values have no effect. Rule (TLAM) types the function body e and annotates the function's type with the effect of e. The expression as a whole has no effect, since the function produces no run-time effects until it is actually called. Rule (TAPP) types function application, which combines Φ_1, the effect

$$(\text{TINT}) \frac{}{\Phi_\emptyset; \Gamma \vdash n : int} \qquad (\text{TVAR}) \frac{\Gamma(x) = \tau}{\Phi_\emptyset; \Gamma \vdash x : \tau}$$

$$(\text{TLAM}) \frac{\Phi; \Gamma, x : \tau' \vdash e : \tau}{\Phi_\emptyset; \Gamma \vdash \lambda x.e : \tau' \longrightarrow^\Phi \tau} \qquad (\text{TAPP}) \frac{\begin{array}{c}\Phi_1; \Gamma \vdash e_1 : \tau_1 \longrightarrow^{\Phi_f} \tau_2 \\ \Phi_2; \Gamma \vdash e_2 : \tau_1 \\ \Phi_1 \rhd \Phi_2 \rhd \Phi_f \hookrightarrow \Phi\end{array}}{\Phi; \Gamma \vdash e_1\, e_2 : \tau_2}$$

$$(\text{TREF}) \frac{\Phi; \Gamma \vdash e : \tau}{\Phi; \Gamma \vdash \mathsf{ref}^L\, e : ref^{\{L\}} \tau} \qquad (\text{TDEREF}) \frac{\begin{array}{c}\Phi_1; \Gamma \vdash e : ref^\varepsilon \tau \\ \Phi_2^\varepsilon = \varepsilon \quad \Phi_1 \rhd \Phi_2 \hookrightarrow \Phi\end{array}}{\Phi; \Gamma \vdash !e : \tau}$$

$$(\text{TASSIGN}) \frac{\Phi_1; \Gamma \vdash e_1 : ref^\varepsilon \tau \quad \Phi_2; \Gamma \vdash e_2 : \tau \quad \Phi_3^\varepsilon = \varepsilon \quad \Phi_1 \rhd \Phi_2 \rhd \Phi_3 \hookrightarrow \Phi}{\Phi; \Gamma \vdash e_1 := e_2 : \tau}$$

$$(\text{TLOC}) \frac{\Gamma(r) = \tau}{\Phi_\emptyset; \Gamma \vdash r_L : ref^{\{L\}} \tau} \qquad (\text{TSUB}) \frac{\begin{array}{c}\Phi'; \Gamma \vdash e : \tau' \\ \tau' \leq \tau \quad \Phi' \leq \Phi\end{array}}{\Phi; \Gamma \vdash e : \tau}$$

$$(\text{XFLOW-CTXT}) \frac{\Phi_1 = [\alpha_1; \varepsilon_1; (\varepsilon_2 \cup \omega_2)] \quad \Phi_2 = [(\varepsilon_1 \cup \alpha_1); \varepsilon_2; \omega_2] \quad \Phi = [\alpha_1; (\varepsilon_1 \cup \varepsilon_2); \omega_2]}{\Phi_1 \rhd \Phi_2 \hookrightarrow \Phi}$$

$$(\text{SINT}) \frac{}{int \leq int} \qquad (\text{SREF}) \frac{\tau \leq \tau' \quad \tau' \leq \tau \quad \varepsilon \subseteq \varepsilon'}{ref^\varepsilon \tau \leq ref^{\varepsilon'} \tau'}$$

$$(\text{SFUN}) \frac{\tau_1' \leq \tau_1 \quad \tau_2 \leq \tau_2' \quad \Phi \leq \Phi'}{\tau_1 \longrightarrow^\Phi \tau_2 \leq \tau_1' \longrightarrow^{\Phi'} \tau_2'} \qquad (\text{SCTXT}) \frac{\alpha_2 \subseteq \alpha_1 \quad \varepsilon_1 \subseteq \varepsilon_2 \quad \omega_2 \subseteq \omega_1}{[\alpha_1; \varepsilon_1; \omega_1] \leq [\alpha_2; \varepsilon_2; \omega_2]}$$

Fig. 2. Typing

of e_1, with Φ_2, the effect of e_2, and Φ_f, the effect of the function. We specify the sequencing of effects with the combinator $\Phi_1 \triangleright \Phi_2 \hookrightarrow \Phi$, defined by (XFLOW-CTXT). Since e_1 evaluates before e_2, this rule requires that the future effect of e_1 be $\varepsilon_2 \cup \omega_2$, i.e., everything that happens during the evaluation of e_2, captured by ε_2, plus everything that happens after, captured by ω_2. Similarly, the past effect of e_2 must be $\varepsilon_1 \cup \alpha_1$, since e_2 is evaluated just after e_1. Lastly, the effect Φ of the entire expression has α_1 as its prior effect, since e_1 is evaluated first; ω_2 as its future effect, since e_2 is evaluated last; and $\varepsilon_1 \cup \varepsilon_2$ as its current effect, since both e_1 and e_2 are evaluated. We write $\Phi_1 \triangleright \Phi_2 \triangleright \Phi_3 \hookrightarrow \Phi$ as shorthand for $(\Phi_1 \triangleright \Phi_2 \hookrightarrow \Phi') \wedge (\Phi' \triangleright \Phi_3 \hookrightarrow \Phi)$.

(TREF) types memory allocation, which has no effect but places the annotation L into a singleton effect $\{L\}$ on the output type. This singleton effect can be increased as necessary by using subsumption. (TDEREF) types the dereference of a memory location of type $ref^\varepsilon \tau$. In a standard effect system, the effect of $!\,e$ is the effect of e plus the effect ε of accessing the pointed-to memory. Here, the effect of e is captured by Φ_1, and because the dereference occurs after e is evaluated, (TDEREF) puts Φ_1 in sequence just before some Φ_2 such that Φ_2's current effect is ε. Therefore by (XFLOW-CTXT), Φ^ε is $\Phi_1^\varepsilon \cup \varepsilon$, and e's future effect Φ_1^ω must include ε and the future effect of Φ_2. On the other hand, Φ_2^ω is unconstrained by this rule, but it will be constrained by the context, assuming the dereference is followed by another expression. (TASSIGN) is similar to (TDEREF), combining the effects Φ_1 and Φ_2 of its subexpressions with a Φ_3 whose current effect is ε. (TLOC) gives a pointer r_L the type of a reference to the type of r in Γ.

Finally, (TSUB) introduces subsumption on types and effects. The judgments $\tau' \leq \tau$ and $\Phi' \leq \Phi$ are defined at the bottom of Figure 2. (SINT), (SREF), and (SFUN) are standard, with the usual co- and contravariance where appropriate. (SCTXT) defines subsumption on effects, which is covariant in the current effect, as expected, and contravariant in both the prior and future effects. To understand the contravariance, first consider an expression e with future effect ω_1. Since ω_1 should contain (i.e., be a superset of) the locations that may be accessed in the future, we can use e in any context that accesses *at most* locations in ω_1. Similarly, since past effects should contain the locations that were accessed in the past, we can use e in any context that accessed at most locations in α_1.

3   Operational Semantics

As discussed in the introduction, to establish the soundness of the static semantics we must address two concerns. First, we must give an operational semantics that specifies the run-time contextual effects of each subterm e appearing in the evaluation of a term e_p. Second, we must find a way to match up subterms e that arise in the evaluation of e_p with the corresponding terms e' in the unevaluated e_p, to see whether the effects ascribed to the original terms e' by the type system approximate the actual effects of the subterms e. This section defines an operational semantics that addresses the first concern, and the next section augments it to address the second concern, allowing us to prove our system sound.

3.1   The Problem of Future Effects

Consider an expression e appearing in program e_p. We write $e_p = C[e]$ for a context C, to make this relationship more clear. Using a small-step operational semantics, we can intuitively view the contextual effects of e as follows:

$$\underbrace{C[e] \to \cdots \to C'[e]}_{\text{prior effect } \alpha} \overbrace{\to C'[e'] \to \cdots \to C'[v]}^{\text{evaluation of } e} \underbrace{\to \cdots \to v_p}_{\text{future effect } \omega}$$

with the middle brace labeled *standard effect ε*.

(The evaluation of e_p could contain several evaluations of e, each of which could differ from e according to previous substitutions of e's free variables, but we ignore these difficulties for now and consider them in the next section.)

For this evaluation, the actual, run-time prior effect α of e is the effect of the evaluation that occurs before e starts evaluating, the actual standard effect ε of e is the effect of the evaluation of e to a value v, and the actual future effect ω of e is the effect of the remainder of the computation. For every expression in the program, there exist similar partitions of the evaluation to define the appropriate contextual effects.

However, while this picture is conceptually clear, formalizing contextual effects, particularly future effects, is awkward in small-step semantics. Suppose we have some contextual effect Φ associated with subterm e in the context $C'[e]$ above. Then Φ^ω, the future effect of subterm e, models everything that happens after we evaluate to $C'[v]$—but that happens some arbitrary number of steps after we begin evaluating $C'[e]$, making it difficult to associate with the subterm e. We could solve this problem by inserting "brackets" into the semantics to identify the end of a subterm's evaluation, but that adds complication, especially since there are many different subterms whose contextual effects we wish to track and prove sound.

Our solution to this problem is to use big-step semantics, since in big-step semantics, each subderivation is a full evaluation. This lets us easily identify both the beginning and the end of each sub-evaluation in the derivation tree, and gives us a natural specification of contextual effects.

3.2   Big-Step Semantics

Figure 3 shows key rules in a big-step operational semantics for our language. Reductions operate on *configurations* $\langle \alpha, \omega, H, e \rangle$, where α and ω are the sets of locations accessed before and after that point in the evaluation, respectively; H is the heap (a map from locations r to values); and e is the expression to be evaluated. Evaluations have the form

$$\langle \alpha, \omega, H, e \rangle \longrightarrow_\varepsilon \langle \alpha', \omega', H', R \rangle$$

where ε is the effect of evaluating e and R is the result of reduction, either a value v or **err**, indicating evaluation failed. Intuitively, as evaluation proceeds, labels move from the future effect ω to the past effect α.

$$[\text{ID}]\frac{}{\langle \alpha, \omega, H, v\rangle \longrightarrow_\emptyset \langle \alpha, \omega, H, v\rangle} \qquad \boxed{\text{Heaps } H ::= \emptyset \mid H, r \mapsto v}$$

$$[\text{REF}]\frac{\langle \alpha, \omega, H, e\rangle \longrightarrow_\varepsilon \langle \alpha', \omega', H', v\rangle \quad r \notin \text{dom}(H')}{\langle \alpha, \omega, H, \text{ref}^L e\rangle \longrightarrow_\varepsilon \langle \alpha', \omega', (H', r \mapsto v), r_L\rangle}$$

$$[\text{DEREF}]\frac{\langle \alpha, \omega, H, e\rangle \longrightarrow_\varepsilon \langle \alpha', \omega' \cup \{L\}, H', r_L\rangle \quad r \in \text{dom}(H')}{\langle \alpha, \omega, H, !\, e\rangle \longrightarrow_{\varepsilon \cup \{L\}} \langle \alpha' \cup \{L\}, \omega', H', H'(r)\rangle}$$

$$[\text{ASSIGN}]\frac{\begin{array}{c}\langle \alpha, \omega, H, e_1\rangle \longrightarrow_{\varepsilon_1} \langle \alpha_1, \omega_1, H_1, r_L\rangle \\ \langle \alpha_1, \omega_1, H_1, e_2\rangle \longrightarrow_{\varepsilon_2} \langle \alpha_2, \omega_2 \cup \{L\}, (H_2, r \mapsto v'), v\rangle\end{array}}{\langle \alpha, \omega, H, e_1 := e_2\rangle \longrightarrow_{\varepsilon_1 \cup \varepsilon_2 \cup \{L\}} \langle \alpha_2 \cup \{L\}, \omega_2, (H_2, r \mapsto v), v\rangle}$$

$$[\text{CALL}]\frac{\begin{array}{c}\langle \alpha, \omega, H, e_1\rangle \longrightarrow_{\varepsilon_1} \langle \alpha_1, \omega_1, H_1, \lambda x.e\rangle \\ \langle \alpha_1, \omega_1, H_1, e_2\rangle \longrightarrow_{\varepsilon_2} \langle \alpha_2, \omega_2, H_2, v_2\rangle \\ \langle \alpha_2, \omega_2, H_2, e[x \mapsto v_2]\rangle \longrightarrow_{\varepsilon_3} \langle \alpha', \omega', H', v\rangle\end{array}}{\langle \alpha, \omega, H, e_1\, e_2\rangle \longrightarrow_{\varepsilon_1 \cup \varepsilon_2 \cup \varepsilon_3} \langle \alpha', \omega', H', v\rangle}$$

$$[\text{CALL-W}]\frac{\langle \alpha, \omega, H, e_1\rangle \longrightarrow_{\varepsilon_1} \langle \alpha', \omega', H', v\rangle \quad v \neq \lambda x.e}{\langle \alpha, \omega, H, e_1\, e_2\rangle \longrightarrow_\emptyset \langle \alpha, \omega, H, \mathbf{err}\rangle}$$

$$[\text{DEREF-H-W}]\frac{\langle \alpha, \omega, H, e\rangle \longrightarrow_\varepsilon \langle \alpha', \omega', H', r_L\rangle \quad r \notin \text{dom}(H')}{\langle \alpha, \omega, H, !\, e\rangle \longrightarrow_\emptyset \langle \alpha, \omega, H, \mathbf{err}\rangle}$$

$$[\text{DEREF-L-W}]\frac{\langle \alpha, \omega, H, e\rangle \longrightarrow_\varepsilon \langle \alpha', \omega', H', r_L\rangle \quad r \in \text{dom}(H') \quad L \notin \omega'}{\langle \alpha, \omega, H, !\, e\rangle \longrightarrow_\emptyset \langle \alpha, \omega, H, \mathbf{err}\rangle}$$

Fig. 3. Operational Semantics

With respect to the definitions of Section 3.1, the prior effect α in Section 3.1 corresponds to α here, and the future effect ω in Section 3.1 corresponds to ω' here. The future effect ω before the evaluation of e contains both the future and the standard effect of e, i.e., $\omega = \omega' \cup \varepsilon$. Similarly, the past effect α' after the evaluation of e contains the past effect α and the effect of e, i.e., $\alpha' = \alpha \cup \varepsilon$. We prove below that our semantics preserves this property.

The reduction rules are straightforward. [ID] reduces a value to itself without changing the state or the effects. [REF] generates a fresh location r, which is bound in the heap to v and evaluates to r_L. [DEREF] reads the location r in the heap and adds L to the standard evaluation effect. This rule requires that the future effect after evaluating e have the form $\omega' \cup \{L\}$, i.e., L must be in the future effect after evaluating e, but prior to dereferencing the result. Then L is added to α' in the output configuration of the rule. Notice that $\omega' \cup \{L\}$ is a standard union, hence L may also be in ω', which allows the same location to be accessed multiple times. Also note that we require L to be in the future effect just after the evaluation of e, but do not require that it be in ω. However, this will actually hold—below we prove that $\omega = \omega' \cup \{L\} \cup \varepsilon$, and in general when the semantics takes a step, effects move from the future to the past.

[ASSIGN] behaves similarly to [DEREF]. [CALL] evaluates the first expression to a function, the second expression to a value, and then the function body with the formal argument replaced by the actual argument. Our semantics also includes rules [CALL-W], [DEREF-H-W] and [DEREF-L-W] that produce **err** when the program tries to access a location that is not in the future effect of the input, or when values are used at the wrong type. Our system includes similar error rules for assignment (not shown).

3.3   Standard Effect Soundness

We can now prove standard effect soundness. First, we prove an *adequacy* property of our semantics that helps ensure they make sense:

Lemma 1 (Adequacy of Semantics). *If* $\langle \alpha, \omega, H, e \rangle \longrightarrow_\varepsilon \langle \alpha', \omega', H', v \rangle$*, then* $\alpha' = \alpha \cup \varepsilon$ *and* $\omega = \omega' \cup \varepsilon$*.*

This lemma formalizes our intuition that labels move from the future to prior effect during evaluation.

We can then prove that the static Φ^ε associated to a term by our type and effect system soundly approximates the actual effect ε of an expression. We ignore actual effects α and ω by setting them to 1. We say heap H is well-typed under Γ, written $\Gamma \vdash H$, if $\mathrm{dom}(\Gamma) = \mathrm{dom}(H)$ and for every $r \in \mathrm{dom}(H)$, we have $\Phi_\emptyset; \Gamma \vdash H(r) : \Gamma(r)$. The standard effect soundness lemma is:

Theorem 1 (Standard Effect Soundness). *If*

1. $\Phi; \Gamma \vdash e : \tau$*,*
2. $\Gamma \vdash H$ *and*
3. $\langle 1, 1, H, e \rangle \longrightarrow_\varepsilon \langle 1, 1, H', R \rangle$

then there is a Γ' such that:

1. *R is a value v for which $\Phi_\emptyset; (\Gamma', \Gamma) \vdash v : \tau$,*
2. *$(\Gamma', \Gamma) \vdash H'$ and*
3. *$\varepsilon \subseteq \Phi^\varepsilon$.*

Here (Γ', Γ) is the concatenation of environments Γ' and Γ. The proof of this theorem is by induction on the evaluation derivation, and follows traditional type-and-effect system proofs, adapted for our semantics.

Next, we prove that if the program evaluates to a value, then there is a *canonical evaluation* in which the program evaluates to the same value, but starting with an empty α and ending with an empty ω. This will produce an evaluation derivation with the *most precise* α and ω values for every configuration, which we can then prove we soundly approximate using our type and effect system.

Lemma 2 (Canonical Evaluation). *If* $\langle 1, 1, H, e \rangle \longrightarrow_\varepsilon \langle 1, 1, H', v \rangle$ *then there exists a derivation* $\langle \emptyset, \varepsilon, H, e \rangle \longrightarrow_\varepsilon \langle \varepsilon, \emptyset, H', v \rangle$*.*

4 Contextual Effect Soundness

Now we turn to proving contextual effect soundness. We aim to show that the prior and future effect of some subterm e of a program e_p approximate the evaluation of e_p before and after, respectively, the evaluation of e. Suppose for the moment that e_p contains no function applications. As a result, an evaluation derivation D_p of e_p according to the operational semantics in Figure 3 will be isomorphic to a typing derivation T_p of e_p according to the rules in Figure 2. In this situation, soundness for contextual effects is easy to define. For any subterm e of e_p, we have an evaluation derivation D and a typing derivation T:

$$D :: \langle \alpha, \omega, H, e \rangle \longrightarrow_\varepsilon \langle \alpha', \omega', H', v \rangle \qquad T :: \Phi; \Gamma \vdash e : \tau$$

where D is a subderivation of D_p and T is a subderivation of T_p. Then the prior and future effects computed by our contextual effect system are sound if $\alpha \subseteq \Phi^\alpha$ (the effect of the evaluation before e is contained in Φ^α) and $\omega' \subseteq \Phi^\omega$ (the effect of the evaluation after v is contained in Φ^ω).

For example, consider the evaluation of $!(\mathsf{ref}^L\ n)$.

$$
\text{(DEREF)} \cfrac{
\text{(REF)} \cfrac{
\text{(ID)} \cfrac{}{\langle \emptyset, \emptyset \cup \{L\}, H, n \rangle \longrightarrow \langle \emptyset, \emptyset \cup \{L\}, H, n \rangle}
}{\langle \emptyset, \emptyset \cup \{L\}, H, \mathsf{ref}^L\ n \rangle \longrightarrow \langle \emptyset, \emptyset \cup \{L\}, (H, r_L \mapsto n), r_L \rangle}
}{\langle \emptyset, \emptyset \cup \{L\}, H, !(\mathsf{ref}^L\ n) \rangle \longrightarrow_{\{L\}} \langle \emptyset \cup \{L\}, \emptyset, (H, r_L \mapsto n), n \rangle}
$$

Here is the typing derivation (where we have rolled a use of (TSUB) into (TINT)):

$$
\text{(TDEREF)} \cfrac{
\text{(TREF)} \cfrac{
\text{(TINT')} \cfrac{}{[\emptyset; \emptyset; \{L\}]; \cdot \vdash n : int}
}{[\emptyset; \emptyset; \{L\}]; \cdot \vdash \mathsf{ref}^L\ n : ref^L\ int}
\quad [\emptyset; \{L\}; \emptyset]^\varepsilon = \{L\} \quad [\emptyset; \emptyset; \{L\}] \rhd [\emptyset; \{L\}; \emptyset] \hookrightarrow [\emptyset; \{L\}; \emptyset]
}{[\emptyset; \{L\}; \emptyset]; \cdot \vdash !(\mathsf{ref}^L\ n) : int}
$$

We can see that these derivations are isomorphic, and thus it is easy to read the contextual effect from the typing derivation for $\mathsf{ref}^L\ n$ and to match it up with the actual effect of the corresponding subderivation of the evaluation derivation.

Unfortunately, function applications add significant complication because D_p and T_p are no longer isomorphic. Indeed, a subterm e of the original program e_p may appear multiple times in D_p, possibly with substitutions applied to it. For example, consider the term $(\lambda x. ! x; ! x)\ \mathsf{ref}^L\ n$ (where we introduce the sequencing operator ; with the obvious semantics, for brevity), typed as:

$$
\text{(TAPP)} \cfrac{
\text{(TLAM)} \cfrac{\Phi_f; \Gamma, x : ref^{\{L\}}\ int \vdash\ ! x; ! x : int}{\Phi_\emptyset; \Gamma \vdash \lambda x. ! x; ! x : ref^{\{L\}}\ int \longrightarrow^{\Phi_f} int \quad (T_1)}
\quad
\begin{array}{c}\Phi_2; \Gamma \vdash \mathsf{ref}^L\ n : ref^{\{L\}}\ int \quad (T_2) \\ \Phi_\emptyset \rhd \Phi_2 \rhd \Phi_f \hookrightarrow \Phi \end{array}
}{\Phi; \Gamma \vdash (\lambda x. ! x; ! x)\ \mathsf{ref}^L\ n : int}
$$

The evaluation derivation has the following structure:

$$\langle \emptyset, \emptyset \cup \{L\}, H, (\lambda x.\,!\,x;\,!\,x) \rangle \longrightarrow \langle \emptyset, \emptyset \cup \{L\}, H, (\lambda x.\,!\,x;\,!\,x) \rangle \quad (1)$$
$$\langle \emptyset, \emptyset \cup \{L\}, H, \mathsf{ref}^L\ n \rangle \longrightarrow \langle \emptyset, \emptyset \cup \{L\}, H', r_L \rangle \quad (2)$$

$$(\text{CALL})\ \frac{\langle \emptyset, \emptyset \cup \{L\}, H', (!\,x;\,!\,x)[x \mapsto r_L] \rangle \longrightarrow_{\{L\}} \langle \emptyset \cup \{L\}, \emptyset, H', n \rangle \quad (3)}{\langle \emptyset, \emptyset \cup \{L\}, H, (\lambda x.\,!\,x;\,!\,x)\ \mathsf{ref}^L\ n \rangle \longrightarrow_{\{L\}} \langle \emptyset \cup \{L\}, \emptyset, H', n \rangle}$$

where $H' = (H, r_L \mapsto n)$. Subderivations *(1)* and *(2)* correspond to the two sub-derivations (T_1) and (T_2) of (TAPP), but there is no analogue for subderivation *(3)*, which captures the actual evaluation of the function. Clearly this relates to the function's effect Φ_f, but how exactly is not structurally apparent from the derivation. Returning to our example, we must match up the effect in the typing derivation for $!\,x$, which is part of the typing of the function $(\lambda x.\,!\,x;\,!\,x)$, with evaluation of $!\,r_L$ that occurs when the function evaluates in subderivation *(3)*.

To do this, we instrument the big-step semantics from Figure 3 with typing derivations, and define exactly how to associate a typing derivation with each derived subterm in an evaluation derivation. The key property of the resulting *typed operational semantics* is that the contextual effect Φ associated with a subterm e in the original typing derivation T_p is also associated with all terms derived from e via copying or substitution. In the example, the relevant typing subderivation for $!\,x$ in T_p will be copied and substituted according to the evaluation so that it can be matched with $!\,r_L$ in subderivation *(3)*.

4.1   Typed Operational Semantics

In our typed operational semantics, evaluations have the form:

$$\langle T, \alpha, \omega, H, e \rangle \longrightarrow_\varepsilon \langle T', \alpha', \omega', H', v \rangle$$

where T is a typing derivation for the expression e, and T' is a typing derivation for v:

$$T :: \Phi; \Gamma \vdash e : \tau \qquad T' :: \Phi_\emptyset; (\Gamma', \Gamma) \vdash v : \tau$$

Note that we include T' in our rules mostly to emphasize that v is well-typed with the same type as e. The only information from T' we need that is not present in T is the new environment (Γ', Γ), which may contain the types of pointers newly allocated in the heap during the evaluation of e. Also, the environments Γ and Γ' only refer to heap locations, since e and v have no free variables and could always be typed under the empty environment.

Figure 4 presents the typed evaluation rules. New hypotheses are highlighted with a gray background. While these rules look complicated, they are actually quite easy to construct. We begin with the original rules in Figure 3, add a typing derivation to each configuration, and then specify appropriate hypotheses about each typing derivation to connect up the derivation of the whole term with the derivation of each of the subterms. We discuss this process for each of the rules.

[ID-A] is the same as [ID], except we introduce typing derivations T_v and T'_v for the left- and right-hand sides of the evaluation, respectively. T_v may be any

$$[\text{ID-A}]\frac{T_v :: \Phi; \Gamma \vdash v : \tau \qquad T_v' :: \Phi_\emptyset; \Gamma \vdash v : \tau}{\langle T_v, \alpha, \omega, H, v \rangle \longrightarrow_\emptyset \langle T_v', \alpha, \omega, H, v \rangle}$$

$$[\text{REF-A}]\frac{\begin{array}{c}\langle T', \alpha, \omega, H, e \rangle \longrightarrow_\varepsilon \langle T_v, \alpha', \omega', H', v \rangle \qquad r \notin \text{dom}(H) \\ T :: \Phi; \Gamma \vdash \mathsf{ref}^L e : \mathit{ref}^\varepsilon \tau \qquad T' :: \Phi'; \Gamma \vdash e : \tau \\ T_v :: \Phi_\emptyset; \Gamma' \vdash v : \tau \qquad T_r :: \Phi_\emptyset; (\Gamma', r \mapsto \tau) \vdash r_L : \mathit{ref}^\varepsilon \tau \qquad \Phi' \leq \Phi\end{array}}{\langle T, \alpha, \omega, H, \mathsf{ref}^L e \rangle \longrightarrow_\varepsilon \langle T_r, \alpha', \omega', (H', r \mapsto v), r_L \rangle}$$

$$[\text{DEREF-A}]\frac{\begin{array}{c}\langle T', \alpha, \omega, H, e \rangle \longrightarrow_\varepsilon \langle T_r, \alpha', \omega' \cup \{L\}, H', r_L \rangle \qquad r \in \text{dom}(H') \\ T :: \Phi; \Gamma \vdash \,!e : \tau \qquad\qquad T' :: \Phi_1; \Gamma \vdash e : \mathit{ref}^{\varepsilon'} \tau' \\ T_r :: \Phi_\emptyset; \Gamma' \vdash r_L : \mathit{ref}^{\varepsilon'} \tau' \qquad T_v :: \Phi_\emptyset; \Gamma' \vdash H'(r) : \tau \\ \Phi' \leq \Phi \qquad \tau' \leq \tau \qquad \Phi_1 \triangleright [\alpha_1; \varepsilon'; \omega_1] \hookrightarrow \Phi'\end{array}}{\langle T, \alpha, \omega, H, !e \rangle \longrightarrow_{\varepsilon \cup \{L\}} \langle T_v, \alpha' \cup \{L\}, \omega', H', H'(r) \rangle}$$

$$[\text{ASSIGN-A}]\frac{\begin{array}{c}\langle T_1, \alpha, \omega, H, e_1 \rangle \longrightarrow_{\varepsilon_1} \langle T_r, \alpha_1, \omega_1, H_1, r_L \rangle \\ \langle T_2, \alpha_1, \omega_1, H_1, e_2 \rangle \longrightarrow_{\varepsilon_2} \langle T_v, \alpha_2, \omega_2 \cup \{L\}, (H_2, r \mapsto v'), v \rangle \\ T :: \Phi; \Gamma \vdash e_1 := e_2 : \tau \qquad T_1 :: \Phi_1; \Gamma \vdash e_1 : \mathit{ref}^\varepsilon \tau' \\ T_r :: \Phi_\emptyset; \Gamma_1 \vdash r_L : \mathit{ref}^\varepsilon \tau' \qquad T_2 :: \Phi_2; \Gamma_1 \vdash e_2 : \tau' \\ T_v :: \Phi_\emptyset; \Gamma_2 \vdash v : \tau' \qquad T_v' :: \Phi_\emptyset; \Gamma_2 \vdash v : \tau \\ \Phi' \leq \Phi \qquad \tau' \leq \tau \qquad \Phi_1 \triangleright \Phi_2 \triangleright [\alpha_3; \varepsilon; \omega_3] \hookrightarrow \Phi'\end{array}}{\langle T, \alpha, \omega, H, e_1 := e_2 \rangle \longrightarrow_{\varepsilon_1 \cup \varepsilon_2 \cup \{L\}} \langle T_v', \alpha_2 \cup \{L\}, \omega_2, (H_2, r \mapsto v), v \rangle}$$

$$[\text{CALL-A}]\frac{\begin{array}{c}\langle T_1, \alpha, \omega, H, e_1 \rangle \longrightarrow_{\varepsilon_1} \langle T_f, \alpha_1, \omega_1, H_1, \lambda x.e \rangle \\ \langle T_2, \alpha_1, \omega_1, H_1, e_2 \rangle \longrightarrow_{\varepsilon_2} \langle T_{v_2}, \alpha_2, \omega_2, H_2, v_2 \rangle \\ \langle T_3, \alpha_2, \omega_2, H_2, e[v_2 \mapsto x] \rangle \longrightarrow_{\varepsilon_3} \langle T_v, \alpha', \omega', H', v \rangle \\ T :: \Phi; \Gamma \vdash e_1 e_2 : \tau \qquad T_1 :: \Phi_1; \Gamma \vdash e_1 : \tau_1 \longrightarrow^{\Phi_f} \tau_2 \\ T_f :: \Phi_\emptyset; \Gamma_1 \vdash \lambda x.e : \tau_1 \longrightarrow^{\Phi_f} \tau_2 \qquad T_2 :: \Phi_2; \Gamma_1 \vdash e_2 : \tau_1 \\ T_{v_2} :: \Phi_\emptyset; \Gamma_2 \vdash v_2 : \tau_1 \qquad T_3 :: \Phi_f; \Gamma_2 \vdash e[x \mapsto v_2] : \tau \\ T_v :: \Phi_\emptyset; \Gamma_3 \vdash v : \tau \qquad \Phi_1 \triangleright \Phi_2 \triangleright \Phi_f \hookrightarrow \Phi' \qquad \Phi' \leq \Phi\end{array}}{\langle T, \alpha, \omega, H, e_1 e_2 \rangle \longrightarrow_{\varepsilon_1 \cup \varepsilon_2 \cup \varepsilon_3} \langle T_v, \alpha', \omega', H', v \rangle}$$

Fig. 4. Typed operational semantics

typing derivation that assigns a type to v. Here, and in the other rules in the typed operational semantics, we allow subsumption in the typing derivations on the left-hand side of a reduction. Thus T_v may type the value v under some effect Φ that is not $\Phi_\emptyset$. The output typing derivation T_v' is the same as T_v, except it uses the effect $\Phi_\emptyset$ (recall the only information we use from T_v' is the new environment, which is this case is unchanged from T_v).

[REF-A] is a more complicated case. Here the typing derivation T must (by observation of the rules in Figure 2) assign $\mathsf{ref}^L e$ a type $\mathit{ref}^\varepsilon \tau$ and some effect

Φ. By inversion, then, we know that T must in fact assign the subterm e the type τ as witnessed by some typing derivation T', which we use in the typed evaluation of e. We allow $\Phi' \leq \Phi$ to account for subsumption applied to the term $\mathsf{ref}^L\ e$. Note that this rule does not specify how to construct T' from T. Later on, we will prove that if there is a valid standard reduction of a well-typed term, then there is a valid typed reduction of the same term. Continuing with the rule, our semantics assigns some typing derivation T_v to v. Then the output typing derivation T_r should assign a type to r_L. Hence we take the environment Γ' from T_v, which contains types for locations in the heap allocated thus far, and extend it with a new binding for r of the correct type.

[DEREF-A] follows the same pattern as above. Given the initial typing derivation T of the term $!\,e$, we assume there exists a typing derivation T' of the appropriate shape for subterm e. Reducing e yields a new typing derivation T_r, and the final typing derivation T_v assigns the type τ to the value $H'(r)$ returned by the dereference. As above, we add subtyping constraints $\Phi' \leq \Phi$ and $\tau' \leq \tau$ to account for subsumption of the term $!\,e$. The most interesting feature of this rule is the last constraint, $\Phi_1 \rhd [\alpha_1; \varepsilon'; \omega_1] \hookrightarrow \Phi'$, which states that the effect $\Phi \geq \Phi'$ of the whole expression $!\,e$ (from typing derivation T) must contain the effect Φ_1 of e followed by some contextual effect containing standard effect ε'. Again, we will prove below that it is always possible to construct a typed derivation that satisfies this constraint, intuitively because [DEREF] from Figure 2 enforces exactly the same constraint. [ASSIGN-A] is similar to [DEREF].

[CALL-A] is the most complex of the four rules, but the approach is exactly the same as above. Starting with typing derivation T for the function application, we require that there exist typing derivations T_1 and T_2 for e_1 and e_2, where the type of e_2 is the domain type of e_1. Furthermore, T_f and T_{v_2} assign the same types as T_1 and T_2, respectively. Then by the substitution lemma, we know there exists a typing derivation T_3 that assigns type τ to the function body e in which the formal x is mapped to the actual v_2. The output typing derivation T_v assigns v the same type τ as T_3 assigns to the function body. We finish the rule with the usual effect sequencing and subtyping constraints.

4.2 Soundness

The semantics in Figure 4 precisely associate a typing derivation—and most importantly, a contextual effect—with each subterm in an evaluation derivation. We prove soundness in two steps. First, we argue that given a typing derivation of a program and an evaluation derivation according to the rules in Figure 3, we can always construct a typed evaluation derivation.

Lemma 3 (Typed evaluation derivations exist). *If* $T :: \Phi; \Gamma \vdash e : \tau$ *and* $D :: \langle \alpha, \omega, H, e \rangle \longrightarrow_\varepsilon \langle \alpha'\omega', H', v \rangle$ *where* $\Gamma \vdash H$, *then there exists* T_v *such that*

$$\langle T, \alpha, \omega, H, e \rangle \longrightarrow_\varepsilon \langle T_v, \alpha', \omega', H', v \rangle$$

The proof is by induction on the evaluation derivation D. For each case, we show we can always construct a typed evaluation by performing inversion on the typing derivation T, using T's premises to apply the corresponding typed operational semantics rule. Due to subsumption, we cannot perform direct inversion on T. Instead, we used a number of inversion lemmas (not shown) that generalize the premises of the syntax-driven typing rule that applies to e, for any number of following [TSUB] applications.

Next, we prove that if we have a typed evaluation derivation, then the contextual effects assigned in the derivation soundly model the actual run-time effects. Since contextual effects are non-compositional, we reason about the soundness of contextual effects in a derivation in relation to its context inside a larger derivation. To do that, we use $E_1 \in E_2$ to denote that E_1 is a subderivation of E_2. We define the subderivation relation inductively on evaluation derivations in the typed operational semantics, with base cases corresponding to each evaluation rule, and one inductive case for transitivity. For example, given an application of [CALL-A] (uninteresting premises omitted):

$$\cdots$$

$$\frac{\begin{array}{l} E_1 :: \langle T_1, \alpha, \omega, H, e_1 \rangle \longrightarrow_{\varepsilon_1} \langle T_f, \alpha_1, \omega_1, H_1, \lambda x.e \rangle \\ E_2 :: \langle T_2, \alpha_1, \omega_1, H_1, e_2 \rangle \longrightarrow_{\varepsilon_2} \langle T_{v_2}, \alpha_2, \omega_2, H_2, v_2 \rangle \\ E_3 :: \langle T_3, \alpha_2, \omega_2, H_2, e[v_2 \mapsto x] \rangle \longrightarrow_{\varepsilon_3} \langle T_v, \alpha', \omega', H', v \rangle \end{array}}{E :: \langle T, \alpha, \omega, H, e_1\ e_2 \rangle \longrightarrow_{\varepsilon_1 \cup \varepsilon_2 \cup \varepsilon_3} \langle T', \alpha', \omega', H, v \rangle}$$

we have $E_1 \in E$, $E_2 \in E$ and $E_3 \in E$. The subderivation relationship is also transitive, i.e., if $E_1 \in E_2$ and $E_2 \in E_3$ then $E_1 \in E_3$.

The following lemma states that if E_2 is an evaluation derivation whose contextual effects are sound (premises 2, 5, and 6) and E_1 is a subderivation of E_2 (premise 3), then the effects of E_1 are sound (conclusions 2 and 3).

Lemma 4 (Soundness of sub-derivation contextual effects). *If*

1. $E_1 :: \langle T_1, \alpha_1, \omega_1, H_1, e_1 \rangle \longrightarrow_{\varepsilon_1} \langle T_{v_1}, \alpha'_1, \omega'_1, H'_1, v_1 \rangle$ *with* $T_1 :: \Phi_1; \Gamma_1 \vdash e_1 : \tau_1$,
2. $E_2 :: \langle T_2, \alpha_2, \omega_2, H_2, e_2 \rangle \longrightarrow_{\varepsilon_2} \langle T_{v_2}, \alpha'_2, \omega'_2, H'_2, v_2 \rangle$ *with* $T_2 :: \Phi_2; \Gamma_2 \vdash e_2 : \tau_2$,
3. $E_1 \in E_2$
4. $\Gamma_2 \vdash H_2$
5. $\alpha_2 \subseteq \Phi_2^\alpha$
6. $\omega_2 \subseteq \Phi_2^\omega$

then

1. $\Gamma_1 \vdash H_1$
2. $\alpha_1 \subseteq \Phi_1^\alpha$
3. $\omega_1 \subseteq \Phi_1^\omega$

The proof is by induction on $E_1 \in E_2$. The work occurs in the base cases of the $\in$ relation, and the transitivity case trivially applies induction.

The statement of Lemma 4 may seem odd: we assume a derivation's effects are sound and then prove the soundness of the effects of its subderivation(s). Nevertheless, this technique is efficacious. If E_2 is the topmost derivation (for the whole program) then the lemma can be trivially applied for E_2 and any of its subderivations, as α_2 and ω'_2 will be $\emptyset$, and thus trivially approximated by the effects defined in Φ_2. Given this, and the fact (from Lemma 3) that typed derivations always exist, we can easily state and prove contextual effect soundness.

Theorem 2 (Contextual Effect Soundness). *Given a program e_p with no free variables, a typing derivation T and a (standard) evaluation D according to the rules in Figure 3, we can construct a typed evaluation derivation*

$$E :: \langle T, \emptyset, \varepsilon_p, \emptyset, e_p \rangle \longrightarrow_{\varepsilon_p} \langle T_v, \varepsilon_p, \emptyset, H, v \rangle$$

such that for every subderivation E' of E:

$$E' :: \langle T', \alpha, \omega, H, e \rangle \longrightarrow_{\varepsilon} \langle T_v, \alpha', \omega', H', v \rangle$$

with $T' :: \Phi; \Gamma \vdash e : \tau$, it is always the case that $\alpha \subseteq \Phi^\alpha$, $\varepsilon \subseteq \Phi^\varepsilon$, and $\omega' \subseteq \Phi^\omega$.

This theorem follows as a corollary of Lemma 2, Lemma 3 and Lemma 4, since the initial heap and Γ are empty, and the whole program is typed under $[\emptyset; \varepsilon; \emptyset]$, where ε soundly approximates the effect of the whole program by Theorem 1.

The full (paper) proof can be found in a technical report [6].

5   Mechanization

We encoded the above formalization and soundness proof using the Coq proof assistant. The source code for the formalization and the proof scripts can be found at `http://www.cs.umd.edu/projects/PL/contextual/contextual-coq.tgz`. We were pleased that the mechanization of the system largely followed the paper proof, with only a few minor differences.

First, we used the framework developed by Aydemir et al. [1] for modeling bound and named variables, whereas the paper proof assumes alpha equivalence of all terms and does not reason about capturing and renaming.

Second, Lemma 4 states a property of all subderivations of a derivation. On paper, we had left the definition of subderivation informal, whereas we had to formally define it in Coq. This was straightforward if tedious. In Coq we defined $E \in E'$, described earlier, as an inductive relation, with one case for each premise of each evaluation rule.

While our mechanized proof is similar to our paper proof, it does have some awkwardness. Our encoding of typed operational semantics is dependent on typing derivations, and the encoding of the subderivation relation is dependent on typed evaluations. This causes the definitions of typed evaluations and subderivations to be dependent on large sets of variables, which decreases readability. We were unable to use Coq's system for implicit variables to address this issue, due to its current limitations.

In total, the formalization and proof scripts for the contextual effect system takes 5,503 lines of Coq, of which we wrote 2,692 lines and the remaining 2,811 lines came from Aydemir et al [1]. It took the first author approximately ten days to encode the definitions and lemmas and do the proofs, starting from minimal

Coq experience, limited to attending a tutorial at POPL 2008. It took roughly equal time and effort to construct the encodings as to do the actual proofs. In the process of performing the proofs, we discovered some typographical errors in the paper proof, and we found some cases where we had failed to account for possible subsumption in the type and effect system. Perhaps the biggest insight we gained was that to prove Lemma 4, we needed to do induction on the subderivation relation, rather than on the derivation itself.

6   Related Work

Our original paper on contextual effects [7] presented the same type system and operational semantics shown in Sections 2 and 3, but placed scant emphasis on the details of the proof of soundness in favor of describing novel applications. Indeed, we felt that the proof technique described in the published paper was unnecessarily unintuitive and complicated, and that led us to ultimately discover the technique presented in this paper. To our knowledge, ours is the first mechanized proof of a property of typing and evaluation derivations that depends on the positions of subderivations in the super-derivation tree.

Type and effect systems [5,8,11] are widely used to statically enforce restrictions, check properties, or in static analysis to infer the behavior of computations [4,9,3,10,12]. Some more detailed comparisons with these systems can be found in our previous publication [7]. Talpin and Jouvelot [11] use a big-step operational semantics to prove standard effect soundness. In their system, operational semantics are not annotated with effects. Instead, the soundness property is that the static effect, unioned with a static description of the starting heap, describes the heap at the end of the computation. In addition to addressing contextual effects, our operational semantics can also be used as a definition of the *actual* effect (prior, standard, or future) of the computation, regardless of the static system used to infer or check effects. The soundness property for standard effects by Talpin and Jouvelot immediately follows for our system from Theorem 1.

7   Conclusions

This paper presents the proof of soundness for contextual effects [7]. We have mechanized and verified the proof using the Coq proof assistant.

Contextual effect soundness is interesting because the soundness of the effect of e depends on the *position* of e's evaluation within the evaluation derivation of the whole program e_p. That is, the prior and future effects of e depend not on the evaluation of e itself, but rather on the evaluation of e_p prior to, and after, evaluating e, respectively. Adding further complication, a subterm e within the original program, for which the contextual effect is computed by the type and effect system, may change during the evaluation of e_p. In particular, it may be duplicated or modified due to substitutions. To match up these modified terms with the term in the original typing derivation, we employ a novel *typed operational semantics* that correlates the relevant portion of the typing derivation with the evaluation of every subexpression in the program. In mechanizing our proof, we discovered a missing definition (subderivations) in our formal system, and we gained much more assurance that our proof, which had to carefully coordinate the many parts of typed evaluation derivations, was correct.

We conjecture that our proof technique can be used to reason about other non-compositional properties that span a derivation, such as the freshness of a name, or computations that depend on context.

Acknowledgments. The authors thank the Penn PLClub for holding the Coq tutorial at POPL 2008, which got us interested in this topic. We thank the anonymous reviewers for their useful feedback. This work was funded in part by NSF grants CNS-0346989, CCF-0541036, CCF-0430118, and the University of Maryland Partnership with the Laboratory for Telecommunications Sciences.

References

1. Aydemir, B., Charguéraud, A., Pierce, B.C., Pollack, R., Weirich, S.: Engineering formal metatheory. In: POPL (2008)
2. The Coq proof assistant., `http://coq.inria.fr`
3. Hicks, M., Foster, J.S., Pratikakis, P.: Lock Inference for Atomic Sections. In: TRANSACT (2006)
4. Igarashi, A., Kobayashi, N.: Resource Usage Analysis. In: POPL (2002)
5. Lucassen, J.M.: Types and Effects: Towards the Integration of Functional and Imperative Programming. PhD thesis, MIT Laboratory for Computer Science, MIT/LCS/TR-408 (August 1987)
6. Neamtiu, I., Hicks, M., Foster, J.S., Pratikakis, P.: Contextual Effects for Version-Consistent Dynamic Software Updating and Safe Concurrent Programming. Technical Report CS-TR-4920, Dept. of Computer Science, University of Maryland (November 2007)
7. Neamtiu, I., Hicks, M., Foster, J.S., Pratikakis, P.: Contextual effects for version-consistent dynamic software updating and safe concurrent programming. In: POPL (2008)
8. Nielson, F., Nielson, H.R., Hankin, C.: Principles of Program Analysis. Springer, Heidelberg (1999)
9. Skalka, C., Smith, S., Horn, D.V.: Types and trace effects of higher order programs. Journal of Functional Programming (July 2007)
10. Smith, F., Walker, D., Morrisett, G.: Alias types. In: Smolka, G. (ed.) ESOP 2000 and ETAPS 2000. LNCS, vol. 1782. Springer, Heidelberg (2000)
11. Talpin, J.-P., Jouvelot, P.: The type and effect discipline. Inf. Comput. 111(2), 245–296 (1994)
12. Walker, D., Crary, K., Morrisett, G.: Typed memory management in a calculus of capabilities. In: TOPLAS, July 2000, vol. 24(4), pp. 701–771 (2000)

First-Class Type Classes

Matthieu Sozeau[1] and Nicolas Oury[2]

[1] Univ. Paris Sud, CNRS, Laboratoire LRI, UMR 8623, Orsay, F-91405
INRIA Saclay, ProVal, Parc Orsay Université, F-91893
sozeau@lri.fr
[2] University of Nottingham
npo@cs.nott.ac.uk

Abstract. Type Classes have met a large success in HASKELL and IS-ABELLE, as a solution for sharing notations by overloading and for specifying with abstract structures by quantification on contexts. However, both systems are limited by second-class implementations of these constructs, and these limitations are only overcomed by ad-hoc extensions to the respective systems. We propose an embedding of type classes into a dependent type theory that is first-class and supports some of the most popular extensions right away. The implementation is correspondingly cheap, general and integrates well inside the system, as we have experimented in COQ. We show how it can be used to help structured programming and proving by way of examples.

1 Introduction

Since its introduction in programming languages [1], *overloading* has met an important success and is one of the core features of object–oriented languages. Overloading allows to use a common name for different objects which are *instances* of the same type schema and to automatically select an *instance* given a particular type. In the functional programming community, overloading has mainly been introduced by way of *type classes*, making ad-hoc polymorphism less ad hoc [17].

A *type class* is a set of functions specified for a parametric type but defined only for some types. For example, in the HASKELL language, it is possible to define the class **Eq** of types that have an equality operator as:

class **Eq** a where (==) :: $a \rightarrow a \rightarrow$ **Bool**

It is then possible to define the behavior of == for types whose equality is decidable, using an instance declaration:

instance **Eq Bool** where x == y = if x then y else not y

This feature has been widely used as it fulfills two important goals. First, it allows the programmer to have a uniform way of naming the same function over different types of data. In our example, the *method* == can be used to compare any type of data, as long as this type has been declared an instance of the **Eq**

O. Ait Mohamed, C. Muñoz, and S. Tahar (Eds.): TPHOLs 2008, LNCS 5170, pp. 278–293, 2008.

class. It also adds to the usual Damas–Milner [4] parametric polymorphism a form of ad-hoc polymorphism. E.g, one can constrain a polymorphic parameter to have an instance of a type class.

=/= :: **Eq** $a \Rightarrow a \to a \to$ **Bool**
x =/= y = not $(x$ == $y)$

Morally, we can use == on x and y because we have supposed that a has an implementation of == using the type class *constraint* '**Eq** $a \Rightarrow$'. This construct is the equivalent of a functor argument specification in module systems, where we can require arbitrary structure on abstract types.

Strangely enough, overloading has not met such a big success in the domain of computer assisted theorem proving as it has in the programming language community. Yet, overloading is very present in non formal—pen and paper—mathematical developments, as it helps to solve two problems:

– It allows the use of the same operator or function name on different types. For example, + can be used for addition of both natural and rational numbers. This is close to the usage made of type classes in functional programming. This can be extended to proofs as well, overloading the reflexive name for every reflexivity proof for example (see §3.4).
– It shortens some notations when it is easy to recover the whole meaning from the context. For example, after proving $(M, \varepsilon, \cdot)$ is a *monoid*, a proof will often mention the monoid M, leaving the reader implicitly inferring the neutral element and the law of the monoid. If the context does not make clear which structure is used, it is always possible to be more precise (an extension known as named instances [9] in HASKELL). Here the name M is overloaded to represent both the carrier and the (possibly numerous) structures built on it.

These conventions are widely used, because they allow to write proofs, specifications and programs that are easier to read and to understand.

The COQ proof assistant can be considered as both a programming language and a system for computer assisted theorem proving. Hence, it can doubly benefit from a powerful form of overloading.

In this paper, we introduce *type classes* for COQ and we show how it fulfills all the goals we have sketched. We also explain a very simple and cheap implementation, that combines existing building blocks: *dependent records*, *implicit arguments*, and *proof automation*. This simple setup, combined with the expressive power of dependent types subsumes some of the most popular extensions to HASKELL type classes [2,8].

More precisely, in this paper,

– we first recall some basic notions of type theory and some more specific notions we are relying on (§2) ;
– then, we define *type classes* and *instances* before explaining the basic design and implementation decisions we made (§3) ;
– we show how we handle *superclasses* and *substructures* in section 4 ;

- we present two detailed examples (§5): a library on monads and a development of category theory ;
- we discuss the implementation and possible extensions to the system (§6) ;
- finally, we present related and future work (§7).

Though this paper focuses mainly on the COQ proof assistant, the ideas developed here could easily be used in a similar way in any language based on Type Theory and providing both dependent records and implicit arguments.

2   Preliminaries

A complete introduction to Type Theory and the COQ proof assistant is out of the scope of this article. The interested reader may refer to the reference manual [16] for an introduction to both. Anyway, we recall here a few notations and concepts used in the following sections.

2.1   Dependent Types

COQ's type system is a member of the family of Type Theories. In a few words, it is a lambda-calculus equipped with *dependent types*. It means that types can depend upon values of the language. This allows to refine the type of a piece of data depending on its content and the type of a function depending on its behavior. For example, it is possible, using COQ, to define the type of the lists of a given size n and to define a function *head* that can only be applied to lists of a strictly positive size.

Indeed, the type can specify the behavior of a function so precisely, that a type can be used to express a *mathematical property*. The corresponding program is then the *proof* of this property. Hence, the COQ system serves both as a programming language and a proof assistant, being based on a single unifying formalism.

2.2   Dependent Records

Among the different types that can be defined in COQ, we need to explain more precisely *dependent records*, as they serve as the basis of our implementation of type classes. A dependent record is similar to a record in a programming language, aggregating different fields, each possessing a *label*, a *type* and a *value*. Dependent records are a slight generalization of this as the type of a field may refer to the value of a preceding field in the record.

For example, for a type A and a property P on A, one can express the property *"there exist an element a in a type A such that (P a) holds"* using a dependent record whose:

- First field is named *witness* and has type A.
- Second field is named *proof* and is of type P *witness*. The type of this second field *depends* on the value of the first field *witness*.

Another common use of dependent records, that we are going to revisit in the following, is to encode *mathematical structures* gluing some *types*, *operators* and *properties*. For example, the type of monoids can be defined using a dependent record. In COQ syntax:

```
Record Monoid := {
  carrier : Type ;
  neutral : carrier ; op : carrier → carrier → carrier ;
  neutral_left : ∀ x, op neutral x = x ; ⋯ }
```

We can observe that the neutral element and the operation's types refer to carrier, the first field of our record, which is a type. Along with the type and operations we can also put the monoid laws. When we build a record object of type **Monoid** we will have to provide proof objects correspondingly. Another possibility for declaring the **Monoid** data type would be to have the carrier type as a parameter instead of a field, making the definition more polymorphic:

```
Record Monoid (carrier : Type) := {
  neutral : carrier ; op : carrier → carrier → carrier ; ⋯ }
```

One can always abstract fields into parameters this way. We will see in section §4 that the two presentations, while looking roughly equivalent, are quite different when we want to specify sharing between structures.

2.3   Implicit Arguments

Sometimes, because of type dependency, it is easy to infer the value of a function argument by the context where it is used. For example, to apply the function id of type $\forall A : \text{Type}, A \rightarrow A$ one has to first give a type X and then an element x of this type because of the explicit polymorphism inherent to the COQ system. However, we can easily deduce X from the type of x and hence find the *value* of the first argument automatically if the second argument is given. This is exactly what the implicit argument system does. It permits to write shortened applications and use unification to find the values of the implicit arguments. In the case of id, we can declare the first argument as implicit and then apply id as if it was a unary function, making the following type-check:

```
Definition foo : bool := id true.
```

One can always explicitly specify an implicit argument at application time, to disambiguate or simply help the type-checker using the syntax:

```
Definition foo : bool := id (A:=bool) true.
```

This is very useful when the user himself specifies which are the implicit arguments, knowing that they will be resolved in some contexts (due to typing constraints giving enough information for example) but not in others. Both mechanisms of resolution and disambiguation will be used by the implementation of type classes. Implicit arguments permit to recover reasonable (close to ML) syntax in a polymorphic setting even if the system with non-annotated terms does not have a complete inference algorithm.

3   First Steps

We now present the embedding of type classes in our type theory beginning with a raw translation and then incrementally refine it. We use the syntax $\overrightarrow{x_n : t_n}$ to denote the ordered typing context $x_1 : t_1, \cdots, x_n : t_n$. Following the HASKELL convention, we use capitalized names for type classes.

3.1   Declaring Classes and Instances

We extend the syntax of CoQ commands to declare classes and instances. We can declare a class **Id** on the types τ_1 to τ_n having members $f_1 \cdots f_m$ and an instance of **Id** using the following syntax:

$$\text{Class } \textbf{Id } (\alpha_1 : \tau_1) \cdots (\alpha_n : \tau_n) := \qquad \text{Instance } \textbf{Id } t_1 \cdots t_n :=$$
$$\qquad f_1 : \phi_1 ; \qquad\qquad\qquad\qquad f_1 := b_1 ;$$
$$\qquad \vdots \qquad\qquad\qquad\qquad\qquad \vdots$$
$$\qquad f_m : \phi_m. \qquad\qquad\qquad\qquad f_m := b_m.$$

Translation. We immediately benefit from using a powerful dependently-typed language like CoQ. To handle these new declarations, we do not need to extend the underlying formalism as in HASKELL or ISABELLE but can directly use existing constructs of the language.

A newly declared class is automatically translated into a record type having the index types as *parameters* and the same members. When the user declares a new instance of such a type class, the system creates a record object of the type of the corresponding class.

In the following explanations, we denote by $\Gamma_{\textbf{Id}} \triangleq \overrightarrow{\alpha_n : \tau_n}$ the context of parameters of the class **Id** and by $\Delta_{\textbf{Id}} \triangleq \overrightarrow{f_m : \phi_m}$ its context of definitions. Quite intuitively, when declaring an instance $\overrightarrow{t_n}$ should be an instance of $\Gamma_{\textbf{Id}}$ and $\overrightarrow{b_m}$ an instance of $\Delta_{\textbf{Id}}[\overrightarrow{t_n}]$.

Up to now, we just did wrapping around the Record commands, but the declarations are not treated exactly the same. In particular, the projections from a type class are declared with particular implicit parameters.

3.2   Method Calls

When the programmer calls a type class method f_i, she does not have to specify which instance she is referring to: that's the whole point of overloading. However, the particular instance that is being used is bound to appear inside the final term. Indeed, class methods are encoded, in the final proof term, as applications of some projection of a record object representing the instance $f_i : \forall \Gamma_{\textbf{Id}}, \textbf{Id } \overrightarrow{\alpha_i} \to \phi_i$. To be able to hide this from the user we declare the instance of the class **Id** as an implicit argument of the projection, as well as all the arguments corresponding to $\Gamma_{\textbf{Id}}$.

For example, the method call x == y where x, y : **bool** expands to the application (eq) (?B:Type) (?e:**Eq** ?B) x y, where ?$A : \tau$ denotes unification variables corresponding to implicit arguments, whose value we hope to discover during type checking. In this case $?B$ will be instantiated by **bool** leaving an **Eq bool** constraint.

Indeed, when we write a class method we expect to find the instance automatically by resolution of implicit arguments. However, it will not be possible to infer the instance argument using only unification as is done for regular arguments: we have to do a kind of proof search (e.g., here we are left with a **Eq** $?B$ constraint). So, after type-checking a term, we do constraint solving on the unification variables corresponding to implicit arguments and are left with the ones corresponding to type class constraints.

3.3   Instance Search

These remaining constraints are solved using a special purpose tactic that tries to build an instance from a set of declared instances and a type class constraint. Indeed, each time an instance is declared, we add a lemma to a database corresponding to its type. For example, when an instance of the class **Eq** for booleans is declared, a lemma of type **Eq bool** is created. In fact, any definition whose type has a conclusion of the form **Eq** τ can be used as an instance of **Eq**.

When searching for an instance of a type class, a simple proof search algorithm is performed, using the lemmas of the database as the only possible hints to solve the goal. Following our example, the remaining constraint **Eq bool** will be resolved if we have (at least) one corresponding lemma in our database. We will discuss the resolution tactic in section 6. Note that we can see the result of the proof search interactively in COQ by looking at the code, and specify explicitly which instance we want if needed using the $(A:=t)$ syntax.

3.4   Quantification

Finally, we need a way to parameterize a definition by an arbitrary instance of a type class. This is done using a regular dependent product which quantifies on implementations of the record type.

For convenience, we introduce a new binding notation $[\ \textsf{Id}\ \overrightarrow{t_n}\ ]$ to add a type class constraint in the environment. This boils down to specifying a binding $(instance : \textsf{Id}\ \overrightarrow{t_n})$ in this case, but we will refine this construct soon to attain the ease of use of the HASKELL class binder. Now, all the ingredients are there to write our first example:

```
Definition neq (A : Type) [ Eq A ] (x y : A) : bool := not (x == y).
```

The process goes like this: when we interpret the term, we transform the $[\ \textbf{Eq}\ A\ ]$ binding into an $(eqa : \textbf{Eq}\ A)$ binding, and we add unification variables for the first two arguments of $(==) : \forall A : \textsf{Type}, \textbf{Eq}\ A \rightarrow A \rightarrow A \rightarrow \textbf{bool}$. After type-checking, we are left with a constraint **Eq** A in the environment $[\ A : \textsf{Type},\ eqa : \textbf{Eq}\ A, x\ y : A\ ]$, which is easily resolved using the eqa assumption. At this point, we can substitute the result of unification into the term and give it to the kernel for validation.

Implicit generalization. When writing quantifications on type classes, we will often find ourselves having to bind some parameters just before the type class constraint, as for the A argument above. This suggests that it would be convenient to see type class binders as constructs binding their arguments. Indeed, when we give variables as part of the arguments of a class binder for **C**, we are effectively constraining the type of the variable as it is known from the Γ_C context. Hence we can consider omitting the external quantification on this variable and implicitly quantify on it as part of the class binder. No names need be generated and type inference is not more complicated. For example we can rewrite neq:

Definition neq [**Eq** A] $(x\ y : A)$: **bool** := negb $(x == y)$.

This corresponds to the way type classes are introduced in HASKELL in prenex positions, effectively constraining the implicitly quantified type variables. If one wants to specify the type of A it is also possible using the notation $(A : \mathsf{Type})$ in the class binder. The arguments of class binders are regular contexts.

Hence the syntax also supports instantiated constraints like [**Eq** (**list** A)]. More formally, to interpret a list of binders we proceed recursively on the list to build products or lambda abstractions. For a type class binder [**C** $\overrightarrow{t_n : \tau_n}$] we collect the set of free variables in the global environment extended by the terms $\overrightarrow{t_n}$, we extend the environment with them and the class binder and recursively interpret the rest of the binders in this environment.

General contexts. As hinted in the previous section, all the contexts we are manipulating are arbitrary. In particular this means that the parameters of a type class are not restricted to be types of kind **Type** (a strong limitation of ISABELLE's implementation). Any type constructor is allowed as a parameter, hence the standard **Monad** (m : **Type** $\rightarrow$ **Type**) class causes no problem in our setting (see §5.1). In fact, as we are in a dependently-typed setting, one can even imagine having classes parameterized by terms like:

Class **Reflexive** $(A : \mathsf{Type})$ $(R : \mathsf{relation}\ A) :=$ reflexive : $\forall\ x : A,\ R\ x\ x$.

We can instantiate this class for any A using Leibniz equality:

Instance **Reflexive** A (eq A) := reflexive x := refl_equal x.

Note that we are using implicit generalization again to quantify on A here. Now, if we happen to have a goal of the form R t t where t : T we can just apply the reflexive method and the type class mechanism will automatically do a proof search for a **Reflexive** T R instance in the environment.

4   Superclasses as Parameters, Substructures as Instances

We have presented a way to introduce and to use type classes and instances. The design relies on the strength of the underlying logical system. This strength makes it straightforward to extend this simple implementation with two key structuring concepts: inheritance and composition of structures.

4.1   Superclasses as Parameters

Inheritance permits to specify hierarchies of structures. The idea is to allow a class declaration to include type class constraints, known as superclasses in the HASKELL terminology.

This kind of superclasses happens a lot in the formalization of mathematics, when the definition of a structure on an object makes sense only if the object already supports another structure. For example, talking about the class of functors from C to D only makes sense if C and D are categories or defining a group G may suppose that G is already a monoid. This can be expressed by superclass constraints:

> Class [C : **Category**, D : **Category**] $\Rightarrow$ **Functor** := ...
> Class [G : **Monoid**] $\Rightarrow$ **Group** := ...

As often when dealing with parameterization, there are two ways to translate inheritance in our system: either by making the superstructures *parameters* of the type class or regular *members*. However, in the latter case it becomes harder to specify relations between structures because we have to introduce equalities between record fields. For example, we may wish to talk about two **Functor**s F and G between two categories C and D to define the concept of an adjunction. If we have the categories on which a **Functor** is defined as members src and dst of the **Functor** class, then we must ensure that F and G's fields are coherent using equality hypotheses:

> Definition adjunction $(F : \textbf{Functor})$ $(G : \textbf{Functor})$,
>   src F = dst $G \rightarrow$ dst F = src G ...

This gets very awkward because the equalities will be needed to go from one type to another in the definitions, obfuscating the term with coercions, and the user will also have to pass these equalities explicitly.

The other possibility, known as Pebble-style structuring [13], encourages the use of parameters instead. In this case, superstructures are explicitly specified as part of the structure type. Consider the following:

> Class [C : **Category** *obj hom*, D : **Category** *obj' hom'*] $\Rightarrow$ **Functor** := ...

This declaration is translated into the record type:

> **Functor** $(obj : \textbf{Type})$ $(hom : obj \rightarrow obj \rightarrow \textbf{Type})$ $(C : \textbf{Category}\ obj\ hom)$
> $(obj' : \textbf{Type})$ $(hom' : obj' \rightarrow obj' \rightarrow \textbf{Type})$ $(D : \textbf{Category}\ obj'\ hom') := ...$

We depart from HASKELL's syntax where the parameters of a superclass have to be repeated as part of the type class parameters: they are automatically considered as parameters of the subclass in our case, in the order they appear in the superclass binders.

Note that we directly gave names to the type class binders instead of letting the system guess them. We will use these names C and D to specify sharing of structures using the regular dependent product, e.g.:

`Definition` adjunction [C : **Category** *obj hom*, D : **Category** *obj' hom'*,
  F : **Functor** *obj hom* C *obj' hom'* D,
  G : **Functor** *obj' hom'* D *obj hom* C] := . . .

The verbosity problem of using Pebble-style definitions is quite obvious in this example. However we can easily solve it using the existing implicit arguments technology and bind F and G using (F : **Functor** C D) (G : **Functor** D C) outside the brackets. One can also use the binding modifier '!' inside brackets to turn back to the implicit arguments parsing of COQ, plus implicit generalization.

This is great if we want to talk about superclasses explicitly, but what about the use of type classes for programming, where we do not necessarily want to specify hierarchies? In this case, we want to specify the parameters of the class only, and not the implementations of superclasses. We just need to extend the type-checking process of our regular binder to implicitly generalize superclasses. For example, when one writes [**Functor** *obj hom obj' hom'*], we implicitly generalize not only by the free variables appearing in the arguments but also by implementations C and D of the two category superclasses (the example is developed in section §5.2). This way we recover the ability to quantify on a subclass without talking about its superclasses but we still have them handy if needed. They can be bound using the (A:=t) syntax as usual.

4.2 Substructures as Instances

Superclasses are used to formalize *is-a* relations between structures: it allows, for example, to express easily that a group *is a* monoid with some additional operators and laws.

However, it is often convenient to describe a structure by its components. For example, one can define a ring by a carrier set S that *has* two components: a group structure and a monoid structure on S. In this situation, it is convenient that an instance declaration of a ring automatically declares instances for its components: an instance for the group and an instance for the monoid.

We introduce the syntax :> for methods which are themselves type classes. This adds instance declarations for the corresponding record projections, allowing to automatically use overloaded methods on the substructures when working with the composite structure.

Remark. This is very similar to the coercion mechanism available in COQ [14], with which one can say that a structure is a subtype of another substructure. The source-level type-checking algorithm is extended to use this subtyping relation in addition to the usual conversion rule. It puts the corresponding projections into the term to witness the subtyping, much in the same way we fill unification variables with instances to build a completed term.

5   Examples

We now present two examples of type classes for programming and proving: an excerpt from a monad library and a development of some basic concepts of category theory. Both complete developments [1] can be processed by the latest version of COQ which implements all the features we have described.

5.1 Monads

We define the **Monad** type class over a type constructor η having operations unit and bind respecting the usual monad laws (we only write one for conciseness). We use the syntax $\{a\}$ for specifying implicit arguments in our definitions.

Class **Monad** (η : Type $\rightarrow$ Type) :=
    unit : $\forall$ $\{\alpha\}$, $\alpha \rightarrow \eta\ \alpha$;
    bind : $\forall$ $\{\alpha\ \beta\}$, $\eta\ \alpha \rightarrow (\alpha \rightarrow \eta\ \beta) \rightarrow \eta\ \beta$;
    bind_unit_right : $\forall$ α $(x : \eta\ \alpha)$, bind x unit $= x$.

We recover standard notations for monads using COQ's notation system, e.g.:

Infix ">>=" := bind (at *level* 55).

We are now ready to begin a section to develop common combinators on a given **Monad** on η. Every definition in the section will become an overloaded method on this monad *mon*.

Section *Monad_Defs*.
    Context [*mon* : **Monad** η].

The following functions are directly translated from HASKELL's prelude. Definitions are completely straightforward and concise.

    Definition join $\{\alpha\}$ $(x : \eta\ (\eta\ \alpha))$: $\eta\ \alpha$:= x >>= id.

    Definition liftM $\{\alpha\ \beta\}$ $(f : \alpha \rightarrow \beta)$ $(x : \eta\ \alpha)$: $\eta\ \beta$:= $a \leftarrow x$;; return $(f\ a)$.

We can use the overloading support to write more concise proof scripts too.

    Lemma do_return_eta : $\forall$ α $(u : \eta\ \alpha)$, $x \leftarrow u$;; return $x = u$.
    Proof. intros α u.
      rewrite $\leftarrow$ (bind_unit_right u) at 2.
      rewrite (eta_expansion (unit $(\alpha{:=}\alpha)$)).
      reflexivity.
    Qed.

Type classes are transparently supported in all constructions, notably fixpoints.

    Fixpoint sequence $\{\alpha\}$ $(l : $**list**$\ (\eta\ \alpha))$: $\eta\ ($**list**$\ \alpha)$:=
      match l with
       | nil $\Rightarrow$ return nil
       | $hd :: tl \Rightarrow x \leftarrow hd$;; $r \leftarrow$ sequence tl ;; return $(x :: r)$
      end.

They work just as well when using higher-order combinators.

    Definition mapM $\{\alpha\ \beta\}$ $(f : \alpha \rightarrow \eta\ \beta)$: **list** $\alpha \rightarrow \eta\ ($**list** $\beta)$:=
      sequence $\cdot$ map f.

---

[1] http://www.lri.fr/~sozeau/research/coq/classes.en.html

5.2   Category Theory

Our second example has a bit more mathematical flavor: we make a few constructions of category theory, culminating in a definition of the map functor on lists.

A **Category** is a set of objects and morphisms with a distinguished id morphism and a compose law satisfiying the monoid laws. We use a setoid (a set with an equivalence relation) for the morphisms because we may want to redefine equality on them, e.g. to use extensional equality for functions. The $\equiv$ notation refers to the overloaded equivalence method of **Setoid**.

Class **Category** $(obj$: Type$)$ $(hom : obj \rightarrow obj \rightarrow$ Type$)$:=
  morphisms :> $\forall$ a b, **Setoid** $(hom$ a $b)$;
  id : $\forall$ a, hom a a;
  compose : $\forall$ a b c, hom a b $\rightarrow$ hom b c $\rightarrow$ hom a c;
  id_unit_left : $\forall$ a b $(f : hom$ a $b)$, compose f (id b) $\equiv f$;
  id_unit_right : $\forall$ a b $(f : hom$ a $b)$, compose (id a) $f \equiv f$;
  assoc : $\forall$ a b c d $(f : hom$ a $b)$ $(g : hom$ b $c)$ $(h : hom$ c $d)$,
    compose f (compose g h) $\equiv$ compose (compose f g) h.

We define a notation for the overloaded composition operator.

Notation " x $\cdot$ y " := (compose y x) (at *level 40, left associativity*).

A **Terminal** object is an object to which every object has a unique morphism (modulo the equivalence on morphisms).

Class [C : **Category** obj hom] $\Rightarrow$ **Terminal** $(one : obj)$:=
  bang : $\forall$ x, hom x one ;
  unique : $\forall$ x $(f$ $g : hom$ x $one)$, $f \equiv g$.

Two objects are isomorphic if the morphisms between them are unique and inverses.

Definition isomorphic [**Category** obj hom] a b : _ :=
  $\{$ $f : hom$ a b & $\{$ $g : hom$ b a | $f \cdot y \equiv$ id $b \wedge g \cdot f \equiv$ id a $\}$ $\}$.

Using these definition, we can do a proof on abstract instances of the type classes.

Lemma terminal_isomorphic
  [C : **Category** obj hom, ! **Terminal** C x, ! **Terminal** C y] : isomorphic x y.
Proof.
  intros ; red.
  exists (bang x). exists (bang $(one:=x)$ y).
  split.
    apply unique.
    apply (unique $(one:=x)$).
Qed.

We can define a **Functor** between two categories with its two components, on objects and morphisms. We keep them as parameters because they are an essential part of the definition and we want to be able to specify them later.

Class [C : **Category** obj hom, D : **Category** obj' hom'] $\Rightarrow$
 Functor $(Fobj : obj \rightarrow obj')$
         $(Fmor : \forall\ a\ b,\ hom\ a\ b \rightarrow hom'\ (Fobj\ a)\ (Fobj\ b)) :=$
 preserve_ident : $\forall\ a,\ Fmor\ a\ a\ (\text{id}\ a) \equiv \text{id}\ (Fobj\ a)$;
 preserve_assoc : $\forall\ a\ b\ c\ (f : hom\ a\ b)\ (g : hom\ b\ c)$,
    $Fmor\ a\ c\ (\text{compose}\ f\ g) \equiv Fmor\ b\ c\ g \cdot Fmor\ a\ b\ f.$

Let's build the **Functor** instance for the map combinator on lists. We will be working in the category of Coq types and functions. The arrow homset takes Types A and B to the $A \rightarrow B$ function space.

Definition arrow $A\ B := A \rightarrow B.$

Here we build a setoid instance for functions which relates the ones which are pointwise equal, i.e: functional extensionality. We do not show the straightforward proofs accompanying each Instance declaration in the following. For example, we have to show that pointwise equality is an equivalence here.

Instance arrow_setoid : **Setoid** (arrow $a\ b$) $:=$
 equiv $f\ g := \forall\ x,\ f\ x = g\ x.$

We define the category TYPE of Coq types and functions.

Instance TYPE : **Category** Type arrow $:=$
 morphisms $a\ b :=$ arrow_setoid ;
 id $a\ x := x$;
 compose $a\ b\ c\ f\ g := g \cdot f.$

It is then possible to create a **Functor** instance for map on **list**s.

Instance list_map_functor : **Functor** TYPE TYPE **list** map.

6   Discussion

The previous sections contain a description of the overall design and implementation of type classes and a few examples demonstrating their effective use in the current version of Coq. Our overall technical contribution is a realistic implementation of type classes in a dependently-typed setting. We have also introduced convenient syntactic constructions to build, use and reason about type classes. We now summarize our implementation choices and explore some further alleys in the design space of type classes for dependently–typed languages.

6.1   Overview

All the mechanisms we have described are implemented in the current version of Coq, as a rather thin layer on top of the implicit arguments system, the record implementation, and the type-checking algorithm. Each of these components had to be slightly enhanced to handle the features needed by type classes. We added the possibility of specifying (maximally inserted) implicit arguments in toplevel definitions. We adapted the record declaration front-end to handle implicit arguments in Class and Instance definitions properly and permit giving only a subset of an instance's fields explicitly, leaving the rest to be proved interactively. We also have a binding of type classes with Russell [15] that uses its enriched type system and allows to handle missing fields as obligations.

6.2   Naming

We allow for a lot of names to be omitted by the user that we really need to have programmatically. To create fresh names is difficult and error-prone, and we tried to use as much information that the user gives as possible during name generation. We need more experiments with the system to know where we should require names to be given to get reasonably robust code and specifications.

One thing that we learned doing bigger examples is that naming instances is very useful, even if we don't know beforehand if we will need to refer to them explicitly to disambiguate terms. Disambiguation is rarely needed in practice but it is very helpful. We always allow it, so that one can specify a particular super-class instance when specifying a class binder or a specific type class parameter for any method if needed.

6.3   Computing

A common concern when using heavily parameterized definitions is how efficiently the code can be run. Indeed, when we define type classes having multiple superclasses or parameters and call a method it gets applied to a large number of arguments. The good thing is that both the interpreted evaluation functions and the virtual machine [5] of Coq are optimized to handle large applications. Also, we avoid trouble with proofs preventing reductions in terms that happen when using other styles of formalizations.

6.4   Searching

The current tactic that searches for instances is a port of the eauto tactic that does bounded breadth- or depth-first search using a set of applicable lemmas. The tactic is applied to all the constraints at once to find a solution satisfying all of them, which requires a backtracking algorithm. Our first experiments suggest that this is sufficient for handling type classes in the simply-typed version, à la Isabelle. However, with multiple parameters and arbitrary instances the problem becomes undecidable and we can only do our best efforts to solve as much problems as possible. We envisage a system to give more control over this part of the system, allowing to experiment with new ways of using type classes in a controlled environment.

An especially important related issue is the current non-deterministic choice of instances. The problem is not as acute as in Haskell because we have an *interactive* system and disambiguation is always possible. Also, in some cases (e.g., when using reflexive to search for a reflexivity proof) we simply do not care what instance is found, applying the principle of proof-irrelevance. However, in a programming environment, one may want to know if ambiguity arises due to multiple instances for the same constraint appearing in the environment. This is undecidable with arbitrary instances, so one would also want to restrict the shape of allowed instances of some classes to ensure decidability of this test. We have not yet explored how to do this in CoQ, but we think this could be internalized in the system using a reflection technique. We hope we could transport existing results on type inference for HASKELL type classes in this setting.

Finally, we did not study the relation between type-inference and instance search formally as this has been done in the other systems. In our current setting, the two systems could be given separate treatment, as we do instance search only after the whole term is type-checked and unification of type variables is finished.

7   Related Work

The first family of related work concerns type classes in functional programming. Type classes were introduced and extended as part of the HASKELL language. Some extensions has been proposed to make the type classes system more powerful, while still retaining good properties of type inference. Most of them solve problems that do not arise when introducing type classes in a dependently-typed language like CoQ.

- The use of dependent records gives a straightforward support of multi-parameter, dependent type classes.
- The possibility to mix more liberally types and values could give us *associated types* with classes [2] (and with them functional dependencies [8]) for free, although we have not worked out the necessary changes on the system yet.
- Named instances [9] are a direct consequence of having a first-class encoding.

Some work has recently been done to extend HASKELL with some partial support for types depending on types or values like Generalized Algebraic Data Types. These extensions often make use of type classes to write programs at the level of types. Our approach, using the full power of dependent types, allows to program types and values in the same functional language and is therefore much less contrived.

Another embedding of type classes using an implicit argument mechanism was done in the SCALA object-oriented programming language [12] recently.

Isabelle. HASKELL-style type classes are also an integral part of the ISABELLE proof assistant [19]. Wenzel [18] has studied type classes in the ISABELLE proof assistant, and recently F. Haftmann and him have given [6] a constructive explanation of the original axiomatic type classes in terms of locales [10]. This explanation is not needed in our case because the evidence passing is done directly in the core language. We always have the kernel's type-checker to tell us if the elaboration mechanism did something wrong at any point. However, the inference is clearly formalized in ISABELLE whereas we have no similar result.

The relative (compared to CoQ) lack of power of the higher-order logic in ISABELLE/HOL makes it impossible to quantify on type constructors directly, using ordinary constructs of the language. This prevents using the system to implement our library on monads for example. To overcome this problem, one has to work in an extension of HOL with Scott's Logic of Computable Functions. It is then possible to construct axiomatic type classes for type constructors using a domain-theoretic construction [7].

Dependent Records. Ample litterature exists on how to extend dependent type theories with records, e.g [3] gives a good starting point. We stress that our implementation did absolutely not change CoQ's kernel and the associated type theory, and we would only benefit from more fully-featured records. We leave a thorough comparison with the Canonical Structure mechanism of CoQ for future work.

7.1   Future Work

Classes and Modules. We could customize the existing extraction mechanism [11] from CoQ to HASKELL to handle type classes specially. However, this translation is partial as our type system is more powerful. It would be interesting to study this correspondence and also the connection between type classes and modules in the CoQ system itself.

Examples. Some examples have already been developed and we have seen some improvements in the elegance and clarity of the formalization of mathematics and the ease of programming. Some further work needs to be done there.

On top of classes. The type class system gives a basis for communication between users and the system, by adhering to a common structuring principle. We are exploring ways to use this mechanism, for example by developing a new setoid rewriting tactic based on an internalization of signatures in the class system.

8   Conclusion

We have presented a type class system for the CoQ proof assistant. This system is useful both for developing elegant programs and concise mathematical formalizations on abstract structures. Yet, the implementation is relatively simple and benefits from dependent types and pre–existing technologies. It could easily be ported on other systems based on dependent types.

References

1. Birtwistle, G.M., Dahl, O.-J., Myhrhaug, B., Nygaard, K.: Simula Begin. Studentlitteratur (Lund, Sweden), Bratt Institut fuer neues Lernen (Goch, FRG), Chartwell-Bratt Ltd (Kent, England) (1979)
2. Chakravarty, M.M.T., Keller, G., Jones, S.L.P., Marlow, S.: Associated types with class. In: Palsberg, J., Abadi, M. (eds.) POPL, pp. 1–13. ACM Press, New York (2005)

3. Coquand, T., Pollack, R., Takeyama, M.: A logical framework with dependently typed records. In: Hofmann, M.O. (ed.) TLCA 2003. LNCS, vol. 2701, pp. 105–119. Springer, Heidelberg (2003)
4. Damas, L., Milner, R.: Principal type schemes for functional programs. In: POPL, Albuquerque, New, Mexico, pp. 207–212 (1982)
5. Grégoire, B., Leroy, X.: A compiled implementation of strong reduction. In: ICFP 2002, pp. 235–246. ACM Press, New York (2002)
6. Haftmann, F., Wenzel, M.: Constructive Type Classes in Isabelle. In: Altenkirch, T., McBride, C. (eds.) TYPES 2006. LNCS, vol. 4502, pp. 160–174. Springer, Heidelberg (2007)
7. Huffman, B., Matthews, J., White, P.: Axiomatic Constructor Classes in Isabelle/HOLCF. In: Hurd, J., Melham, T. (eds.) TPHOLs 2005. LNCS, vol. 3603, pp. 147–162. Springer, Heidelberg (2005)
8. Jones, M.P.: Type classes with functional dependencies. In: Smolka, G. (ed.) ESOP 2000 and ETAPS 2000. LNCS, vol. 1782, pp. 230–244. Springer, Heidelberg (2000)
9. Kahl, W., Scheffczyk, J.: Named instances for haskell type classes. In: Hinze, R. (ed.) A Comparative Study of Very Large Data Bases. LNCS, vol. 59. Springer, Heidelberg (2001)
10. Kammüller, F., Wenzel, M., Paulson, L.C.: Locales - A Sectioning Concept for Isabelle. In: Bertot, Y., Dowek, G., Hirschowitz, A., Paulin, C., Théry, L. (eds.) TPHOLs 1999. LNCS, vol. 1690, pp. 149–166. Springer, Heidelberg (1999)
11. Letouzey, P.: Programmation fonctionnelle certifie – L'extraction de programmes dans l'assistant Coq. PhD thesis, Universit Paris-Sud (July 2004)
12. Moors, A., Piessens, F., Odersky, M.: Generics of a higher kind. In: ECOOP 2008 (submitted, 2008)
13. Pollack, R.: Dependently typed records for representing mathematical structure. In: Aagaard, M.D., Harrison, J. (eds.) TPHOLs 2000. LNCS, vol. 1869, pp. 462–479. Springer, Heidelberg (2000)
14. Saïbi, A.: Typing algorithm in type theory with inheritance. In: POPL, La Sorbonne, Paris, France, January 15-17, 1997, pp. 292–301. ACM Press, New York (1997)
15. Sozeau, M.: Subset coercions in Coq. In: Altenkirch, T., McBride, C. (eds.) TYPES 2006. LNCS, vol. 4502, pp. 237–252. Springer, Heidelberg (2007)
16. The Coq Development Team. The Coq Proof Assistant Reference Manual – Version V8.1 (July 2006), http://coq.inria.fr
17. Wadler, P., Blott, S.: How to make ad-hoc polymorphism less ad hoc. In: POPL, Austin, Texas, pp. 60–76 (1989)
18. Wenzel, M.: Type classes and overloading in higher-order logic. In: Gunter, E.L., Felty, A.P. (eds.) TPHOLs 1997. LNCS, vol. 1275, pp. 307–322. Springer, Heidelberg (1997)
19. Wenzel, M., Paulson, L.: Isabelle/isar. In: Wiedijk, F. (ed.) The Seventeen Provers of the World. LNCS (LNAI), vol. 3600, pp. 41–49. Springer, Heidelberg (2006)

Formalizing a Framework for Dynamic Slicing of Program Dependence Graphs in Isabelle/HOL

Daniel Wasserrab and Andreas Lochbihler*

Universität Karlsruhe,
{wasserra,lochbihl}@ipd.info.uni-karlsruhe.de

Abstract. Slicing is a widely-used technique with applications in e.g. compiler technology and software security. Thus verification of algorithms in these areas is often based on the correctness of slicing, which should ideally be proven independent of concrete programming languages and with the help of well-known verifying techniques such as proof assistants. As a first step in this direction, this contribution presents a framework for dynamic slicing based on control flow and program dependence graphs and machine checked in Isabelle/HOL. Abstracting from concrete syntax we base the framework on a graph representation of the program fulfilling certain structural and well-formedness properties.

1 Introduction

Slicing is a widely-used technique with applications in e.g. compiler technology, debugging and software security. Thus, many algorithms in these areas rely on the different variants of slicing being correct. Suppose there was a tool with which to prove slicing correct. Clearly, this would immensely faciliate verification of such algorithms. Ideally, such a tool would not be restricted to a specific programming language or syntax and utilize well-known verification techniques such as proof assistants. In this contribution, to tackle a subtask of this idea, we present a framework for dynamic slicing based on control flow and program dependence graphs and machine-checked in Isabelle/HOL.

In aiming for versatility, our approach rests upon a special graph representation in the style of control flow graphs (CFG) and not on concrete syntax. If such a representation fulfills certain basic structural and well-formedness properties, we call it trace control flow graph (TCFG) and associate a path of edges to every program trace. For this graph, which can even be infinite, we define the program dependence graph (PDG) using the standard notions of control and data dependence. Then we compute the backward slice and obtain the sliced path from the original TCFG path by invalidating (i.e. replacing them with an operation doing nothing) all operations triggered by nodes that are not in the backward slice. Our main theorem shows that executing the remaining operations on a sliced path yields the same result as performing those on the initial path w.r.t. the variables used at the target node for every suitable input to the program trace.

In the second part, we also present how to embed a simple While language (without procedures) in the framework. On the one hand, we illustrate this way both how to construct a trace control flow graph for a semantics and how to validate the required

---

* This work was supported by DFG grant Sn11/10-1.

O. Ait Mohamed, C. Muñoz, and S. Tahar (Eds.): TPHOLs 2008, LNCS 5170, pp. 294–309, 2008.

well-formedness properties using the programming language semantics. On the other hand, this demonstrates that these properties are indeed sensible and chosen well.

1.1 Slicing

To collect all program points that can influence a certain statement in a program a program analysis called *slicing* was defined by Weiser [18]. There are many approaches to how to accomplish this task, for an overview, see Tip [15] or Krinke [8]. Our formalization uses the graph-based approach in relying on the dynamic equivalents of control flow (CFG) and program dependence graphs (PDG). CFGs consist of nodes denoting the program statements and edges that represent the order, in which these statements are executed. Two relations on these nodes, called data and control dependence, determine the edges, which connect the CFG nodes to constitute the PDG. A *backward slice* of a node is then defined as the set of all nodes on which the given node is transitively data and control dependent. This set is a conservative approximation of all program points that can influence this statement. Our approach is dynamic as we can have infinite graphs (e.g. because of method inlining) and we only apply slicing to paths which are executable in a certain state, not to the graph as a whole.

1.2 Isabelle

The formalization is written completely in Isabelle/HOL [12], including all lemmas and theorems, i.e. every single proof is machine-checked. To be as generic as possible with respect to possible languages/CFGs instantiating the framework, we extensively use the Locales concept in Isabelle [3]. This encapsulation enables one to write generic functions and predicates, limited only by imposing certain constraints on them.

Isabelle also provides the means to automatically generate LaTeX documents (such as this one) based on Isabelle input files by automatically replacing references to definitions and lemmas in the LaTeX text with the respective pretty-printed and typeset formulae. This avoids typing errors introduced via the transfer from Isabelle to LaTeX files and shows the convenient, human readable syntax of the formalization.

1.3 Notation

Types include the basic types of truth values, natural numbers and integers, which are called *bool*, *nat*, and *int* respectively. The space of total functions is denoted by $\Rightarrow$. Type variables are written $'a$, $'b$, etc. $t::\tau$ means that HOL term t has HOL type τ.

Pairs come with the two projection functions $fst :: 'a \times 'b \Rightarrow 'a$ and $snd :: 'a \times 'b \Rightarrow 'b$. We identify tuples with pairs nested to the right: (a, b, c) is identical to $(a, (b, c))$ and $'a \times 'b \times 'c$ is identical to $'a \times ('b \times 'c)$.

Sets (type $'a$ *set*) follow the usual mathematical convention.

Lists (type $'a$ *list*) come with the empty list [], the infix constructor $\cdot$, the infix @ that appends two lists, and the conversion function *set* from lists to sets. Variable names ending in "s" usually stand for lists and $|xs|$ is the length of xs. If $i < |xs|$ then $xs_{[i]}$ denotes the i-th element of xs. The standard function *map*, which applies a function to every element in a list, is also available.

datatype $'a\ option = None \mid Some\ 'a$ adjoins a new element *None* to a type $'a$. All existing elements in type $'a$ are also in $'a\ option$, but are prefixed by *Some*. Hence *bool option* has the values *Some True*, *Some False* and *None*.

Case distinctions on data types use guards, where every guard must be followed by a data type constructor. E.g. *case x of Some y* $\Rightarrow$ *f y* $\mid$ *None* $\Rightarrow$ *g* means that if x is some y then the result is $f\ y$ where f may refer to value y, and if x is *None*, then the result is g.

Partial functions are modeled as functions of type $'a \Rightarrow 'b\ option$, where *None* represents undefinedness and $f\ x = Some\ y$ means x is mapped to y. Instead of $'a \Rightarrow 'b\ option$ we write $'a \rightharpoonup 'b$ and call such functions **maps**.

Function update is defined as follows: $f(a := b) \equiv \lambda x.\ \textit{if } x = a \textit{ then } b \textit{ else } f\ x$, where $f :: 'a \Rightarrow 'b$ and $a :: 'a$ and $b :: 'b$.

2   The Framework

Our framework is generic, i.e. we do not restrict ourselves to a specific programming language (in fact, not even to a certain programming paradigm such as imperative or object oriented programming). To construct a PDG, on which to perform dynamic slicing, we instantiate the framework with a so called *trace control flow graph*, representing code from any source code language. E.g. this trace CFG can be obtained by inlining the CFG for every procedure at the respective calling site in a certain program, thus being potentially infinite (e.g. if we have a recursive function). The constraints on this trace CFG has to fulfill are described in detail in the next section.

2.1   The Input Trace Control Flow Graph

The trace CFG (called TCFG in the following) consists of nodes of type $'node$ and edges of type $'edge$, with an edge a being in the set of edges if it fulfills some property *valid-edge a*, a parameter of the instantiating language. Using the functions *sourcenode*, *targetnode* and *kind*, we can determine the source node, target node and edge kind of an edge, respectively. An edge kind describes the action taken when traversing this edge, so we have two different edge kinds of type $'state\ edge\text{-}kind$, both parameterized with a state (or *context* as it is called in other programming languages) type variable $'state$: updating the current state with a function $f :: 'state \Rightarrow 'state$, written $\Uparrow f$, or assuring that predicate $Q :: 'state \Rightarrow bool$ in the current state is fulfilled, written $(Q)_{\sqrt{}}$. We define a function *transfer* such that traversing an edge in a state s then means that we apply this function to the corresponding edge kind, thus calculating $f\ s$, if we have an update edge $\Uparrow f$, otherwise leaving the state unchanged. Function *pred* determines if predicate Q of a predicate edge $(Q)_{\sqrt{}}$ in a certain state s is fulfilled, returning *True* for update edges.

A node n is in the node set of a TCFG, if it fulfills the property *valid-node n* with the following definition: *valid-node n* $\equiv \exists a.\ valid\text{-}edge\ a \wedge (n = sourcenode\ a \vee n = targetnode\ a)$. Furthermore we need two special nodes (*-Entry-*), which may only have outgoing edges, and (*-Exit-*), having incoming edges only. Also there is a special edge which has (*-Entry-*) as source and (*-Exit-*) as target node, the respective edge kind being $(\lambda s.\ False)_{\sqrt{}}$, a predicate which can never be fulfilled. The last restrictions on the TCFG are that we do not allow multi edges nor self loops, so if the source and target nodes of two

$$\frac{}{Def\,(\text{-}Entry\text{-}) = \emptyset \wedge Use\,(\text{-}Entry\text{-}) = \emptyset} \qquad \frac{}{Def\,(\text{-}Exit\text{-}) = \emptyset \wedge Use\,(\text{-}Exit\text{-}) = \emptyset}$$

$$\frac{valid\text{-}edge\ a \qquad V \notin Def\,(sourcenode\ a)}{state\text{-}val\,(transfer\,(kind\ a)\ s)\ V = state\text{-}val\ s\ V} \qquad \frac{valid\text{-}edge\ a \qquad kind\ a = (Q)_{\surd}}{Def\,(sourcenode\ a) = \emptyset}$$

$$\frac{valid\text{-}edge\ a \qquad \forall\,V \in Use\,(sourcenode\ a).\ state\text{-}val\ s\ V = state\text{-}val\ s'\ V}{\forall\,V \in Def\,(sourcenode\ a).\ state\text{-}val\,(transfer\,(kind\ a)\ s)\ V = state\text{-}val\,(transfer\,(kind\ a)\ s')\ V}$$

$$\frac{valid\text{-}edge\ a \qquad pred\,(kind\ a)\ s \qquad \forall\,V \in Use\,(sourcenode\ a).\ state\text{-}val\ s\ V = state\text{-}val\ s'\ V}{pred\,(kind\ a)\ s'}$$

Fig. 1. The well-formedness properties of a trace CFG

valid edges coincide, so do the two edges, and the source and target node of no valid edge may be equal.

Based on these constraints, we can now specify a property *inner-node n*, which defines the subset of valid nodes not being (*-Entry-*) or (*-Exit-*). We also allow to combine edges in paths, written $n -as \rightarrow * \ n'$, meaning that node n can reach n' via edges $as::'edge\ list$. These paths are constructed via the following two rules:

$$\frac{valid\text{-}node\ n}{n -[]\rightarrow * \ n} \qquad \frac{n'' -as \rightarrow * \ n' \qquad valid\text{-}edge\ a \qquad sourcenode\ a = n \qquad targetnode\ a = n''}{n -a \cdot as \rightarrow * \ n'}$$

Moreover we define *sourcenodes*, *targetnodes* and *kinds* as mappings of the respective functions to edge lists using *map*. Using these, we can show that the source node of an edge in a TCFG path edge list matches the target node of the edge preceding it, and the target node of an edge matches the source node of the edge succeeding it.

We also lift *transfer* and *pred* to lists of edge kinds via

$$transfers\ []\ s = s \qquad transfers\,(e \cdot es)\ s = transfers\ es\,(transfer\ e\ s)$$
$$preds\ []\ s = True \qquad preds\,(e \cdot es)\ s = pred\ e\ s \wedge preds\ es\,(transfer\ e\ s)$$

After having defined the structural properties of the TCFG, we furthermore need some well-formedness properties for its edges and the *Def* and *Use* sets of the source nodes of its edges, which collect the defined and used variables in this node, respectively, and a function *state-val s V* returning the value currently in variable V in state s (the formal rules are shown in Fig. 1):

- *Def* and *Use* sets of (*-Entry-*) and (*-Exit-*) are empty,
- traversing an edge leaves all variables which are not defined in the source node of this edge unchanged,
- the source node of a predicate edge does not define any variables,
- if two states agree on all variables in the *Use* set of the source node of an edge, after traversing this edge the two states agree on all variables in the *Def* set of this node, so different values in the variables not in the *Use* set cannot influence the values of the variables in the *Def* set,
- if two states agree on all variables in the *Use* set of the source node of a predicate edge and this predicate is valid in one state, it is also valid in the other one.

If we also have a semantics of the language – where $\langle c,s \rangle \Rightarrow \langle c',s' \rangle$ means that evaluating statement c in state s results in fully evaluated statement c' and final state s' – and a mapping from a node n to its corresponding statement c via n *identifies* c, we have another well-formedness property (called *semantically well-formed*):

$$\frac{n \text{ identifies } c \qquad \langle c,s \rangle \Rightarrow \langle c',s' \rangle}{\exists n' \text{ as. } n -as \rightarrow* n' \wedge transfers \ (kinds \ as) \ s = s' \wedge preds \ (kinds \ as) \ s \wedge n' \text{ identifies } c'}$$

This property states that if the complete evaluation of statement c in state s results in a state s' and node n corresponds to statement c, then there is a path in the TCFG beginning at n, on which, taking s as initial state, all predicates in predicate edges hold and the traversal of the path edge kinds also yields state s'.

2.2 Constructing the Program Dependence Graph

Though we instantiate the framework with a possibly infinite trace CFG, we call the data structure constructed in the following program dependence graph (PDG), as we are using the standard definitions of control and data dependence described e.g. in [15, Sec. 2]. For control dependence, the trace CFG must contain all possible paths and not only a special one, i.e. a trace.

Control Dependence. First we formalize the notion of *postdomination*. A node n' postdominates a node n, if both are valid nodes and n' must lie on every path from n to (-Exit-). Note: If no path from n to (-Exit-) exists (e.g. because of unstructured control flow), every valid node postdominates n.

Definition 1. *(Postdomination)*

$$n' \text{ postdominates } n \equiv valid\text{-}node \ n \wedge valid\text{-}node \ n' \wedge$$
$$(\forall \text{ as. } n -as \rightarrow* (\text{-}Exit\text{-}) \longrightarrow n' \in set \ (sourcenodes \ as))$$

Using this notion, control dependence is straightforward (see [19], Sec. 3.3). A node n' is control dependent on a node n via edges as, iff there is a path from n to n' using edges as and n' postdominates the targetnode of the first edge of as but does not postdominate the targetnode of another valid edge a' leaving n. The notion $n' \notin set \ (sourcenodes \ as)$ states that as is a minimal path without loops starting at n'.

Definition 2. *(Control Dependence)*

$$n \text{ controls } n' \text{ via } as \equiv n' \notin set \ (sourcenodes \ as) \wedge n -as \rightarrow* n' \wedge (\exists a \ a' \ as'. \ as = a \cdot as' \wedge$$
$$sourcenode \ a = n \wedge n' \text{ postdominates } targetnode \ a \wedge valid\text{-}edge \ a' \wedge$$
$$sourcenode \ a' = n \wedge \neg \ n' \text{ postdominates } targetnode \ a')$$

In Lem. 1 we prove that this definition is equivalent to the one given in [15, Sec. 2], which says that node n' is control dependent on node n via edges as, iff there is a path from n to n' using edges as and n' postdominates every node on this path except n – since every node in this path except n is a target node of an edge of this edge list as, we can rewrite this proposition using *targetnodes* – and n' does not postdominate n. Property $n' \notin set \ (sourcenodes \ as)$ again guarantees minimal paths without loops starting at n' and we have to explicitly forbid n being the (-Exit-) node.

Lemma 1. *(Control Dependence Variant)*

$$n \text{ controls } n' \text{ via } as \equiv n -as \rightarrow * n' \wedge \neg n' \text{ postdominates } n \wedge$$
$$(\forall n'' \in set \ (targetnodes \ as). \ n' \text{ postdominates } n'') \wedge$$
$$n' \notin set \ (sourcenodes \ as) \wedge n \neq (\text{-}Exit\text{-})$$

The next lemma shows that every inner node n, which is reachable via a path from ($\text{-}Entry\text{-}$) and has a path leading to ($\text{-}Exit\text{-}$), has a node n' on which it is control dependent. Note that there may be more than one such node if control flow is unstructured.

Lemma 2. *(Control Dependence Predecessor)*

$$\frac{inner\text{-}node \ n \qquad (\text{-}Entry\text{-}) -as \rightarrow * n \qquad n -as' \rightarrow * (\text{-}Exit\text{-})}{\exists n' \ as. \ n' \text{ controls } n \text{ via } as}$$

Data Dependence. Node n' is data dependent on a node n, iff there is a variable V which gets defined at n and used at n' and there is a path from n to n' using a nonempty list of edges as such that no node in the path redefines V.

Definition 3. *(Data Dependence)*

$$n \text{ influences } V \text{ in } n' \text{ via } as \equiv V \in Def \ n \wedge V \in Use \ n' \wedge n -as \rightarrow * n' \wedge$$
$$(\exists a' \ as'. \ as = a' \cdot as' \wedge (\forall n'' \in set \ (sourcenodes \ as'). \ V \notin Def \ n''))$$

This definition forbids data dependences from a node to itself as well as source node n occurring twice on path as.

The PDG. A PDG consists of edges of two different types: control dependence edges $n -as \rightarrow_{cd} n'$ and data dependence edges $n -\{V\}as \rightarrow_{dd} n'$. The definitions are straightforward, using the definitions above:

Definition 4. *(PDG Edges)*

$$\frac{n \text{ controls } n' \text{ via } as}{n -as \rightarrow_{cd} n'} \qquad \frac{n \text{ influences } V \text{ in } n' \text{ via } as}{n -\{V\}as \rightarrow_{dd} n'}$$

A path in the PDG using edges as, written $n -as \rightarrow_d * n'$, is constructed by concatenating PDG edges (if the respective source and target nodes match), as being the concatenation of the CFG edge lists of all PDG edges.

2.3 Dynamic Backward Slicing with Respect to a Node

In the standard understanding of slicing in dependence graph-based approaches, the *slicing criterion* corresponds to a node; thus, we call this node *slicing node*. In the following, we compute a dynamic backward slice of a node n' and show that the slice fulfills the fundamental property of dynamic slicing in Theorem 1, i.e. for all variables used in n' traversing the sliced path returns the same result as traversing the respective non-sliced TCFG path. Since we formalise dynamic backward slices as the backward traversal of PDG edges and do not model def-def dependence edges, we only regard variables in the *Use* set of the slicing node, not those in the *Def* set.

Computing a Dynamic Path Slice. The only relevant information the slice of a path must provide is if a certain edge gets included in it or not – as their respective edge kind carries the transition information in this framework. Thus we can model a path slice as a bit vector, i.e. a *bool list*, of the same length as the edge list *as* of the path, being *True* at position i iff the edge at position i of edge list *as* has to be considered. An edge has to be considered if its source node has a PDG path to the slicing node n' with an edge list corresponding to the according suffix of *as*. Function *slice-path as* computes this bit vector by traversing the edge list *as*. Note that the last node of the reduced path being the slicing node n' is invariant throughout this computation:

Definition 5. *(Dynamic Path Slice)*

$$slice\text{-}path\ [] \equiv []$$
$$slice\text{-}path\ (a{\cdot}as) \equiv let\ n' = last(targetnodes\ (a{\cdot}as))\ in$$
$$(sourcenode\ a\ -a{\cdot}as \rightarrow_d * n'){\cdot}slice\text{-}path\ as$$

The fact that we only consider PDG paths via the executed CFG edge list makes this slicing dynamic, static slicing would consider all possible dependences, i.e. also dependences via other CFG edge lists.

Bit vectors can be compared via the less-or-equal relation $\preceq_b$, where $bs \preceq_b bs'$ holds if $|bs| = |bs'|$ and bs' is *True* at least at those elements where bs is *True*. Thus we can say that bs contains less or equal information on the former edge list than bs'. The maximal elements for relation $\preceq_b$, are the bit vectors which are *True* at every entry.

To obtain the edge kind list of the sliced path, we have to combine the bit vector with the initial edge list of the path. If the bit vector is *True* for a certain edge, we copy the respective edge kind in the list, otherwise, if the edge kind is an update edge, we include $\Uparrow id$ which does not alter the state, if it is a predicate edge, we include $(\lambda s.\ True)_{\sqrt{}}$ being *True* in any state. Again the calculation is done via iteration over the edge list:

Definition 6. *(Edge Kind List for a Sliced Path)*

$$select\text{-}edge\text{-}kinds\ []\ [] \equiv []$$
$$select\text{-}edge\text{-}kinds\ (a{\cdot}as)\ (b{\cdot}bs) \equiv (if\ b\ then\ kind\ a$$
$$elsc\ (case\ kind\ a\ of\ \Uparrow f \Rightarrow \Uparrow id\ |\ (Q)_{\sqrt{}} \Rightarrow (\lambda s.\ True)_{\sqrt{}})))$$
$${\cdot}select\text{-}edge\text{-}kinds\ as\ bs$$
$$slice\text{-}kinds\ as \equiv select\text{-}edge\text{-}kinds\ as\ (slice\text{-}path\ as)$$

See Fig. 2 for an example of a path, the bit vector representing its sliced path and the edge kind list of the respective sliced path. In this and all following examples, we number edges and use a simplified language where we replaced the update functions with easier to understand variable assignments, writing predicates as boolean expressions over variables in brackets without the $\sqrt{}$. Data dependences are indicated with solid arrows and annotated with the respective variable name, control dependence edges with dashed arrows. All the variables on the right hand side of an assignment as well as those used in a predicate are in the *Use* set of the source node of the respective edge, the variables on the left hand side of an assignment constitute its *Def* set.

Dependent Live Variables. Our next aim is to prove that dynamic path slicing is indeed correct, i.e. traversing the edge kinds in the sliced path returns the same result as

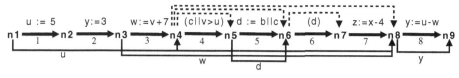

Assumption: *Use* **n9** = { x, y }
Bit vector representing the sliced path: [*True, False, True, True, False, False, False, True*]
Sliced path edge kind list: [u:=5, **id**, w:=v+7, (c||v>u), **id**, **True**, **id**, y:=u-w]

Fig. 2. Example of how to calculate the edge kinds of a sliced path with slice node **n9**

traversing those on the original path for all variables used in the slicing node and if all predicates on the original path are satisfiable, so are they in the sliced path. An auxiliary definition to reach this goal is the notion of *dep-live-vars n′ as*, a collection of variables whose altering can change the value of a variable used in node n'. In fact, we compute a kind of *Live Variables Analysis* as described in [11], restricted to only one path and ignoring those nodes, on which the parameter node is not (transitively) dependent.

But collecting just the variables is not enough for our goal, we furthermore need information about via which TCFG edges we can reach the node where this variable was used (and not redefined in between) starting from our current position in the path. Hence it is possible to have the same variable multiple times in the set, but the respective edge list component differs. *dep-live-vars* has two parameters, first the node n' for which this calculation is made, and second the edge list *as* we have already traversed previous the iterations (note that this traversal happens from right to left). Example: $(V, as') \in$ *dep-live-vars n′ as* states that a node where V is used can be reached via edges as' from the node from which edges *as* lead to n' and no node on as' redefines V. The formal rules for *dep-live-vars*:

Definition 7. *(Dependent Live Variables)*

$$\frac{V \in Use\ n'}{(V, []) \in dep\text{-}live\text{-}vars\ n'\ []} \qquad \frac{V \in Use\ (sourcenode\ a) \qquad sourcenode\ a - a \cdot as' \rightarrow_{cd} n'}{(V, []) \in dep\text{-}live\text{-}vars\ n'\ (a \cdot as')}$$

$$\frac{V \in Use\ (sourcenode\ a) \qquad sourcenode\ a - a \cdot as' \rightarrow_{cd} n'' \qquad n'' - as'' \rightarrow_d * n'}{(V, []) \in dep\text{-}live\text{-}vars\ n'\ (a \cdot (as' @ as''))}$$

$$\frac{(V, as') \in dep\text{-}live\text{-}vars\ n'\ as \qquad V' \in Use\ (sourcenode\ a)}{n' = last\ (targetnodes\ (a \cdot as)) \qquad sourcenode\ a - \{V\} a \cdot as' \rightarrow_{dd} last(targetnodes\ (a \cdot as'))}{(V', []) \in dep\text{-}live\text{-}vars\ n'\ (a \cdot as)}$$

$$\frac{(V, as') \in dep\text{-}live\text{-}vars\ n'\ as}{n' = last\ (targetnodes\ (a \cdot as)) \qquad \neg\ sourcenode\ a - \{V\} a \cdot as' \rightarrow_{dd} last(targetnodes\ (a \cdot as'))}{(V, a \cdot as') \in dep\text{-}live\text{-}vars\ n'\ (a \cdot as)}$$

An easy corollary of this definition is that for all $(V, as') \in$ *dep-live-vars n′ as*, list as' is a prefix of list *as*. For an example of a calculation of the dependent live variables, see Fig. 3. Note that there is no PDG path from **n5** to **n9** (as there is no dependency edge leaving **n9**), so we can ignore the variables used at this node.

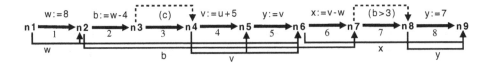

Calculation of *dep-live-vars* for node n9 (if x and y are in its *Use* set):

$as = \dots$	elements in set *dep-live-vars* n9 *as*
$[]$	$\{(x,[]),(y,[])\}$
$[8]$	$\{(x,[8])\}$
$[7,8]$	$\{(x,[7,8]),(b,[])\}$
$[6,7,8]$	$\{(b,[6]),(v,[]),(w,[])\}$
$[5,6,7,8]$	$\{(b,[5,6]),(v,[5]),(w,[5])\}$
$[4,5,6,7,8]$	$\{(b,[4,5,6]),(u,[]),(w,[4,5])\}$
$[3,4,5,6,7,8]$	$\{(b,[3,4,5,6]),(u,[3]),(w,[3,4,5]),(c,[])\}$
$[2,3,4,5,6,7,8]$	$\{(w,[]),(u,[2,3]),(w,[2,3,4,5]),(c,[2])\}$
$[1,2,3,4,5,6,7,8]$	$\{(u,[1,2,3]),(c,[1,2])\}$

Fig. 3. Example for the calculation of Dependent Live Variables for node **n9**

Lemma 3. *(Dependent Live Variables and Use sets)*
If $(V, as') \in$ *dep-live-vars* $n'\ as$ and $n -as\rightarrow* n'$ then
$V \in Use\ n' \wedge (\forall n''{\in}set\ (sourcenodes\ as).\ V \notin Def\ n'') \wedge as = as'$ or
$(\exists nx\ as''.\ as = as'\ @\ as'' \wedge n -as'\rightarrow* nx \wedge nx -as''\rightarrow_d * n' \wedge V \in Use\ nx \wedge$
$(\forall n''{\in}set\ (sourcenodes\ as').\ V \notin Def\ n''))$.

This lemma, which is proved by induction on the dependent live variables rules, now leads us directly to an important statement coded in Corollary 1: Suppose we have a variable V with edge list component as' in the *dep-live-vars* set of node n' and path as and a TCFG path from the target node of a valid edge a leading to n' also using edges as. If now V gets defined at the source node of edge a, there is a PDG path from *sourcenode* a via edges as to n' with a leading data dependency edge for variable V:

Corollary 1. *(Dependent Live Variables and PDG paths)*
If $(V, as') \in$ *dep-live-vars* $n'\ as$ and *targetnode* $a -as\rightarrow* n'$ and
$V \in Def\ (sourcenode\ a)$ and *valid-edge* a then
$(\exists nx\ as''.\ as = as'@as'' \wedge sourcenode\ a -\{V\}a{\cdot}as'\rightarrow_{dd} nx \wedge nx -as''\rightarrow_d * n')$

Machine Checked Dynamic Slicing. Now, before proving the desired fundamental property of path slicing, we show a more general lemma. We have a TCFG path n $-as\rightarrow* n'$, n' not being (*-Exit-*), and two bit vectors $bs \preceq_b bs'$, the first one the result of *slice-path as*. *es* and *es'* are their respective edge kind lists obtained using *select-edge-kinds* on edges as. Furthermore we have two states s and s' agreeing on all variables in the *dep-live-vars* set for the slicing node n' on edge list as. If now all predicates hold while traversing edge kinds *es'* with starting state s', so do all predicates while traversing edge kinds *es* with starting state s and the value of any variable V in the *Use* set of slicing node n' agrees in the states yielded by both traversals.

Lemma 4. *(Generalized Fundamental Property of Dynamic Path Slicing)*

$$\frac{n -as\rightarrow* n' \qquad bs \preceq_b bs'}{slice\text{-}path\ as = bs \quad select\text{-}edge\text{-}kinds\ as\ bs = es \quad select\text{-}edge\text{-}kinds\ as\ bs' = es'} \\ \frac{\forall V\ xs.\ (V, xs) \in dep\text{-}live\text{-}vars\ n'\ as \longrightarrow state\text{-}val\ s\ V = state\text{-}val\ s'\ V \qquad preds\ es'\ s'}{(\forall V \in Use\ n'.\ state\text{-}val\ (transfers\ es\ s)\ V = state\text{-}val\ (transfers\ es'\ s')\ V) \wedge preds\ es\ s}$$

We prove this lemma by induction on bs, where the base case ($bs = []$ and thus $bs' = []$) is trivial. In the induction step we know that since bs is nonempty as consists of a (valid) leading edge a' and the tail list as'. The proof then does the following case analysis: if traversing edge list as changes one of the values in the *Use* set of n' compared to traversing just the tail edge list as', the source node n of the leading edge a' has to define a variable in the dependent live variables set of n' reached via as' – otherwise no such influence would be possible. This implies by Corollary 1 that there is a PDG path from n to n', so edge a' must be part of the slice, i.e. the first element of bs must be *True* and so is the first element of bs' by definition of $\preceq_b$. If however traversing edge list as gives the same values in the *Use* set of n' as traversing just the tail edge list as', the traversal of the leading edge a' does not matter for the variables used in slicing node n'. This proposition, combined with the induction hypothesis and the well-formedness properties of the TCFG, leads to the proof of this lemma.

Replacing bs' with the maximal bit vector w.r.t. $\preceq_b$ of the matching size, using the definition of *slice-kinds* (see Def. 6) and instantiating s and s' with the same state s, the fundamental property is now an easy consequence:

Theorem 1. *(Fundamental Property of Dynamic Path Slicing)*

$$\frac{n -as\rightarrow* n' \qquad preds\ (kinds\ as)\ s}{(\forall V \in Use\ n'.\ state\text{-}val\ (transfers\ (slice\text{-}kinds\ as)\ s)\ V = state\text{-}val\ (transfers\ (kinds\ as)\ s)\ V) \\ \wedge preds\ (slice\text{-}kinds\ as)\ s}$$

Provided that the TCFG is also semantically well-formed, we easily extend this theorem to the semantics: a statement c evaluated in state s returns residual statement c' and state s'. Then there exists a path between the corresponding nodes of the statements and after traversing the sliced version of this path, all variables that are used in the target node have the same value as in state s', the result of the semantics evaluation.

Theorem 2. *(Dynamic Path Slicing and Semantics)*

$$\frac{n\ identifies\ c \qquad \langle c,s\rangle \Rightarrow \langle c',s'\rangle}{\exists n'\ as.\ n -as\rightarrow* n' \wedge preds\ (slice\text{-}kinds\ as)\ s \wedge n'\ identifies\ c' \wedge \\ (\forall V \in Use\ n'.\ state\text{-}val\ (transfers\ (slice\text{-}kinds\ as)\ s)\ V = state\text{-}val\ s'\ V)}$$

3   Instantiation of the Framework with a Simple While-Language

We demonstrate that the framework is applicable and the well-formedness conditions are sensible by showing how to embed a simple imperative While-language (without procedures) called WHILE. In this section we use the basic definitions shown in Sec. 2.1, instantiating the type variables accordingly.

The Language. Our language features two value types $Intg::val$ and $Bool::val$, which represent integer and boolean (i.e. $true$ and $false$) values. Expressions consist of constant values, variables and binary operators. We support five different statements of type cmd: the no-op statement $Skip$, assignment of expression e to a variable V, written $V{:}{=}e$, sequential composition of statements $;;$, conditionals if (b) c_1 $else$ c_2 and while loops $while$ (b) c'. Defining the state is easy: it is just a simple mapping from variables to values $var \rightharpoonup val$. The partial function $[e]s$ returns $Some$ v, if expression e evaluates to value v in state s, $None$, if e cannot be evaluated in state s (e.g. in the case of non well-formed programs).

The Control Flow Graph. As WHILE does not provide procedures, its trace CFG is equal to its CFG. Nodes in this CFG are of the type w-$node$, which incorporates inner nodes $(\text{-}l\,\text{-})$ bearing a label l of type nat, and the special nodes $(\text{-}Entry\text{-})$ and $(\text{-}Exit\text{-})$. $n \oplus i$ adds i to the label number of n and returns a new node bearing this number as label, if n is an inner node, otherwise leaving $(\text{-}Entry\text{-})$ and $(\text{-}Exit\text{-})$ unchanged. $\#{:}c$ denotes the number of inner nodes we need for a CFG of statement c.

WHILE CFG edges have the type w-$edge = w$-$node \times state\ edge$-$kind \times w$-$node$. A CFG edge valid for program $prog$ is written $prog \vdash n -et\rightarrow n'$ and consists of a description of the program $prog$ of type cmd, i.e. the statement for which this CFG is generated, a source node n, an edge kind et of type $state\ edge$-$kind$ and a target node n'. Basically, the CFG is constructed via first constructing recursively the CFGs for the substatements, then combining these graphs into a single one, eventually adjusting the labels so that they remain unique. Also, we add one additional node after every variable assignment node and one after every $while$ node (reachable via the edge, where the loop predicate is false). We will describe the motivation for this later.

Definition 8. *(WHILE CFG Edges)*
Basic rules:
$$prog \vdash (\text{-}Entry\text{-}) -(\lambda s.\ False)_{\surd} \rightarrow (\text{-}Exit\text{-}) \qquad prog \vdash (\text{-}Entry\text{-}) -(\lambda s.\ True)_{\surd} \rightarrow (\text{-}0\,\text{-})$$

Skip: $Skip \vdash (\text{-}0\,\text{-}) -\Uparrow id \rightarrow (\ Exit\text{-})$

V:=e:
$$V{:}{=}e \vdash (\text{-}0\,\text{-}) -\Uparrow \lambda s.\ s(V := [e]s) \rightarrow (\text{-}1\,\text{-}) \qquad V{:}{=}e \vdash (\text{-}1\,\text{-}) -\Uparrow id \rightarrow (\text{-}Exit\text{-})$$

$c_1\,;;c_2$:
$$\frac{c_1 \vdash n -et\rightarrow n' \qquad n' \neq (\text{-}Exit\text{-})}{c_1 \,;;\, c_2 \vdash n -et\rightarrow n'} \qquad \frac{c_1 \vdash n -et\rightarrow (\text{-}Exit\text{-}) \qquad n \neq (\text{-}Entry\text{-})}{c_1 \,;;\, c_2 \vdash n -et\rightarrow (\text{-}0\,\text{-}) \oplus \#{:}c_1}$$

$$\frac{c_2 \vdash n -et\rightarrow n' \qquad n \neq (\text{-}Entry\text{-})}{c_1 \,;;\, c_2 \vdash n \oplus \#{:}c_1 -et\rightarrow n' \oplus \#{:}c_1}$$

if (b) c_1 $else$ c_2 : if (b) c_1 $else$ $c_2 \vdash (\text{-}0\,\text{-}) -(\lambda s.\ [b]s = Some\ true)_{\surd} \rightarrow (\text{-}0\,\text{-}) \oplus 1$

$\qquad if$ (b) c_1 $else$ $c_2 \vdash (\text{-}0\,\text{-}) -(\lambda s.\ [b]s = Some\ false)_{\surd} \rightarrow (\text{-}0\,\text{-}) \oplus (\#{:}c_1 + 1)$

$$\frac{c_1 \vdash n -et\rightarrow n' \qquad n \neq (\text{-}Entry\text{-})}{if\ (b)\ c_1\ else\ c_2 \vdash n \oplus 1 -et\rightarrow n' \oplus 1}$$

$$\frac{c_2 \vdash n -et\rightarrow n' \qquad n \neq (\text{-}Entry\text{-})}{if\ (b)\ c_1\ else\ c_2 \vdash n \oplus (\#{:}c_1 + 1) -et\rightarrow n' \oplus (\#{:}c_1 + 1)}$$

4.

Skip: $(\text{-}Entry\text{-}) \xrightarrow{\text{True}} (\text{-}0\text{-}) \xrightarrow{\text{id}} (\text{-}Exit\text{-})$

if (y<0) (while (x<0) (x:=x+1;; y:=x)) else Skip: (y>=0) $(\text{-}7\text{-})$ id

5. $(\text{-}Entry\text{-}) \xrightarrow{\text{True}} (\text{-}0\text{-}) \xrightarrow{(y<0)} (\text{-}1\text{-}) \xrightarrow{(x>=0)} (\text{-}2\text{-}) \xrightarrow{\text{id}} (\text{-}Exit\text{-})$

(x<0)

$(\text{-}3\text{-}) \xrightarrow{\text{x:=x+1}} (\text{-}4\text{-}) \xrightarrow{\text{id}} (\text{-}5\text{-}) \xrightarrow{\text{y:=x}} (\text{-}6\text{-})$

3. while (x<0) (x:=x+1;; y:=x): $(\text{-}Entry\text{-}) \xrightarrow{\text{True}} (\text{-}0\text{-}) \xrightarrow{(x>=0)} (\text{-}1\text{-}) \xrightarrow{\text{id}} (\text{-}Exit\text{-})$

(x<0)

$(\text{-}2\text{-}) \xrightarrow{\text{x:=x+1}} (\text{-}3\text{-}) \xrightarrow{\text{id}} (\text{-}4\text{-}) \xrightarrow{\text{y:=x}} (\text{-}5\text{-})$

2. x:=x+1;; y:=x: $(\text{-}Entry\text{-}) \xrightarrow{\text{True}} (\text{-}0\text{-}) \xrightarrow{\text{x:=x+1}} (\text{-}1\text{-}) \xrightarrow{\text{id}} (\text{-}2\text{-}) \xrightarrow{\text{y:=x}} (\text{-}3\text{-}) \xrightarrow{\text{id}} (\text{-}Exit\text{-})$

1. x:=x+1: $(\text{-}Entry\text{-}) \xrightarrow{\text{True}} (\text{-}0\text{-}) \xrightarrow{\text{x:=x+1}} (\text{-}1\text{-})$ $(\text{-}0\text{-}) \xrightarrow{\text{y:=x}} (\text{-}1\text{-}) \xrightarrow{\text{id}} (\text{-}Exit\text{-})$

$(\text{-}Exit\text{-})$ y:=x: $(\text{-}Entry\text{-})$

Fig. 4. Construction of the CFG for *if* $(y<0)$ $(while$ $(x<0)$ $(x:=x+1;;y:=x))$ *else Skip*

while (b) c'**:** *while* (b) $c' \vdash (\text{-}0\text{-}) - (\lambda s.\ [\![b]\!]s = Some\ true)_{\sqrt{}} \to (\text{-}0\text{-}) \oplus 2$

while (b) $c' \vdash (\text{-}0\text{-}) - (\lambda s.\ [\![b]\!]s = Some\ false)_{\sqrt{}} \to (\text{-}1\text{-})$

while (b) $c' \vdash (\text{-}1\text{-}) - \Uparrow id \to (\text{-}Exit\text{-})$ $\dfrac{c' \vdash n - et \to (\text{-}Exit\text{-})\qquad n \neq (\text{-}Entry\text{-})}{while\ (b)\ c' \vdash n \oplus 2 - et \to (\text{-}0\text{-})}$

$\dfrac{c' \vdash n - et \to n'\qquad n \neq (\text{-}Entry\text{-})\qquad n' \neq (\text{-}Exit\text{-})}{while\ (b)\ c' \vdash n \oplus 2 - et \to n' \oplus 2}$

In Fig. 4, we show schematically the single steps 1.-5. of the recursive CFG construction for the statement *if* $(y<0)$ $(while$ $(x<0)$ $(x:=x+1;;y:=x))$ *else Skip*. Now we have to prove that these definitions fulfill the well-formedness criteria given in Sec. 2.1. Hence we define *sourcenode*, *targetnode* and *kind* accordingly, say that a is a valid edge iff for a program *prog prog* $\vdash$ *sourcenode* a $-kind$ $a \to$ *targetnode* a holds and that n is a valid node iff it is the source or target node of a valid edge. Also we define the *Use* set as the set of all variables in the expression on the right hand side of a variable assignment and those occuring in a condition or loop predicate. The *Def* set contains only the variable that eventually gets assigned. Using these, the basic and well-formedness properties of the WHILE CFG are easily shown via rule induction on the CFG edge rules.

The Semantics. The rules for a small step or structural operational semantics then look as expected – only the *while*-rule has been split in two cases instead of a translation into a conditional –, $\langle c, s \rangle \to \langle c', s' \rangle$ stating that reduction of statement c in state s results in remainder statement c' and state s':

Definition 9. *(Small Step Semantics of* WHILE*)*

$\langle V:=e, s \rangle \to \langle Skip, s(V := [\![e]\!]s) \rangle$ $\dfrac{\langle c, s \rangle \to \langle c', s' \rangle}{\langle c;; c_2, s \rangle \to \langle c';; c_2, s' \rangle}$ $\langle Skip;; c_2, s \rangle \to \langle c_2, s \rangle$

$$\frac{[\![b]\!]s = Some\ true}{\langle if\ (b)\ c_1\ else\ c_2\ ,s\rangle \rightarrow \langle c_1\ ,s\rangle} \qquad \frac{[\![b]\!]s = Some\ false}{\langle if\ (b)\ c_1\ else\ c_2\ ,s\rangle \rightarrow \langle c_2\ ,s\rangle}$$

$$\frac{[\![b]\!]s = Some\ true}{\langle while\ (b)\ c,s\rangle \rightarrow \langle c;;\ while\ (b)\ c,s\rangle} \qquad \frac{[\![b]\!]s = Some\ false}{\langle while\ (b)\ c,s\rangle \rightarrow \langle Skip,s\rangle}$$

To prove that the WHILE CFG in combination with this semantics is semantically well-formed, we need some kind of simulation of reducing statements in the semantics via following the CFG edges. Thus, we need a one-to-one mapping from the current position in the CFG to the current state of reduction in the semantics, i.e. from node n to statement c, which we called n *identifies* c in Sec. 2.1. Predicate *labels prog n c* accomplishes this task in program *prog*, defined by recursively computing the predicate for substatements and adjusting the labels if necessary (taking care of the additional nodes for $V:=e$ and *while*, being in fact *Skip* nodes).

Definition 10. *(Mapping from Nodes to Statements)*

$$labels\ c\ 0\ c \qquad\qquad labels\ (V:=e)\ 1\ Skip$$

$$\frac{labels\ c_1\ l\ c}{labels\ (c_1\ ;;\ c_2\ )\ l\ (c;;\ c_2\ )} \qquad \frac{labels\ c_2\ l\ c}{labels\ (c_1\ ;;\ c_2\ )\ (l + \#:c_1\ )\ c}$$

$$\frac{labels\ c_1\ l\ c}{labels\ (if\ (b)\ c_1\ else\ c_2\ )\ (l + 1)\ c} \qquad \frac{labels\ c_2\ l\ c}{labels\ (if\ (b)\ c_1\ else\ c_2\ )\ (l + \#:c_1\ + 1)\ c}$$

$$\frac{labels\ c'\ l\ c}{labels\ (while\ (b)\ c')\ (l + 2)\ (c;;\ while\ (b)\ c')} \qquad labels\ (while\ (b)\ c')\ 1\ Skip$$

Trying to define that traversing the CFG edges conforms to reducing the semantics will fail as the semantics is not strong enough, since it reduces (and thus destroys) the initial statement and does not contain any information on node labels. So we need another means for reducing statements, called *label semantics*, which contains both statement and label information and also remembers the initial statement. A step in this semantics is written $prog \vdash \langle c,s,l\rangle \rightsquigarrow \langle c',s',l'\rangle$, meaning that in program *prog* (the initial statement), statement c in state s with label l reduces to a remainder statement c' in state s' with label l'. The label semantics rules can be divided into two groups: first, we have seven rules corresponding to the seven rules of the standard semantics in Def. 9, with identical rules for $V:=e$, $if\ (b)\ c_1\ else\ c_2$ and $while\ (b)\ c'$ (only label information added). The rules for reducing a sequential composition need more premises to guarantee that the reduction provides the correct label for the right hand side of the composition:

$$\frac{labels\ (c_1\ ;;\ c_2\ )\ l\ (Skip;;\ c_2\ ) \qquad labels\ (c_1\ ;;\ c_2\ )\ \#:c_1\ c_2 \qquad l < \#:c_1}{c_1\ ;;\ c_2\ \vdash \langle Skip;;\ c_2\ ,s,l\rangle \rightsquigarrow \langle c_2\ ,s,\#:c_1\ \rangle}$$

$$\frac{labels\ (while\ (b)\ c')\ l\ (Skip;;\ while\ (b)\ c')}{while\ (b)\ c' \vdash \langle Skip;;\ while\ (b)\ c',s,l\rangle \rightsquigarrow \langle while\ (b)\ c',s,0\rangle}$$

Second, we have five more rules to take care of valid semantics steps in substatements also being valid in composite statements (i.e. ;;, $if\ (b)\ c_1\ else\ c_2$ and $while\ (b)\ c'$). We show two examples of these rules:

$$\frac{prog \vdash \langle c,s,l\rangle \rightsquigarrow \langle c',s',l'\rangle}{c_1\ ;;\ prog \vdash \langle c,s,l + \#:c_1\ \rangle \rightsquigarrow \langle c',s',l' + \#:c_1\ \rangle}$$

$$\frac{cx;;\ while\ (b)\ cx \vdash \langle c,s,l\rangle \rightsquigarrow \langle c',s',l'\rangle \qquad l < \#:cx \qquad l' < \#:cx}{while\ (b)\ cx \vdash \langle c,s,l + 2\rangle \rightsquigarrow \langle c',s',l' + 2\rangle}$$

Using *labels* we now show that a step in the label semantics simulates a step in the small step semantics (proven via induction on c):

Lemma 5. *(Label Semantics Simulates Semantics)*

$$\frac{labels\ prog\ l\ c \qquad \langle c,s \rangle \rightarrow \langle c',s' \rangle}{\exists\, l'.\ prog \vdash \langle c,s,l \rangle \rightsquigarrow \langle c',s',l' \rangle \wedge labels\ prog\ l'\ c'}$$

The label semantics now provides all information we need to show that traversing a CFG edge simulates a reducing step in it:

Lemma 6. *(CFG Edge Simulates Label Semantics)*

$$\frac{prog \vdash \langle c,s,l \rangle \rightsquigarrow \langle c',s',l' \rangle}{\exists\, et.\ prog \vdash (\text{-}\,l\,\text{-}) -et\rightarrow (\text{-}\,l'\,\text{-}) \wedge transfer\ et\ s = s' \wedge pred\ et\ s}$$

When proving this proposition by rule induction on the label semantics rules, the need for the additional *Skip* nodes after variable assignments and while loops gets clear: as the reduction of a variable assignment or while loop with invalid predicate leads to a *Skip* statement, we also need a CFG edge from the respective nodes to a node corresponding to a *Skip* statement. As the outgoing edge of this node is by construction of the CFG a no-op edge (function *id* is applied to the state, see the second rule for $V:=e$ and the third rule for *while* (b) c' in Def. 8), this adjustment is valid.

Showing that the WHILE CFG is semantically well-formed w.r.t. its semantics is just an easy consequence of Lem. 5 and 6, lifted to the corresponding transitive closures.

4   Related Work

Our slicing formalization approach is related to the work by Agrawal and Horgan [1], primarily to Approach 3, since nodes can occur multiple times in path, which is our equivalent to their *execution history*. From such an execution history, they build the *Dynamic Dependence Graph*, but this graph does not correspond to the PDG computed here as the latter contains all paths, not only a selected one. Although data dependence can be computed for a single path in isolation, we need this additional information about all possible paths for computing the control dependence relation. Agrawal and Horgan also use (in their case static) PDG information to determine the control dependences in their execution history. Having all possible traces in a TCFG is only a formalization trick that is of course not applicable in algorithms really computing dynamic slices because TCFGs are potentially infinite .

The quite natural idea of replacing parts irrelevant for slicing with operations doing nothing can also be found in the work by Amtoft [2]. He uses a *code map* for mapping CFG nodes to the corresponding program statements. Slicing then modifies this code map for all nodes not in the slice, mapping statement nodes to **skip** (equivalent to our $\Uparrow id$), predicate nodes to $true?$ or $false?$. Encoding the predicate directly in the edges, we can restrict ourselves to $(\lambda s.\ True)_{\sqrt{}}$ for these edges in the dynamic case.

Jhala and Majumdar [6] also use the notion of *path slicing*, but they focus on eliminating all edges not relevant for the (un)reachability of the target location. The resulting (path) slices have two main properties: First, if the slice is infeasible, so is the original path, and second, if the slice is feasible, then the target location is reachable. Similar to our approach, they do not focus on one special path or trace, but also take alternative paths into consideration, e.g. to find feasible variants of the path under consideration.

Ranganath et al. [13] discuss interesting questions that arise if some nodes in the CFG cannot reach the Exit node. Our formalization is not immune to this problem as unstructured control flow can lead to this undesired situation. At the moment we can only guarantee that paths are sliced correctly for graphs without such nodes, otherwise certain control dependences may not be recognized, i.e. slices may be too small. However the main theorems remain correct, but the slices do not necessarily conform the standard understanding of correct slices. Since we also want to apply our framework to unstructured control flow, future efforts will be made to tackle this problem.

There exist some formal approaches to program slicing and its correctness. Reps and Yang showed in [14] the correctness of static PDG based slicing, restricted to a simple While language without procedures. The approach of Gouranton and Le Métayer [5] is similar to ours as they present a language independent framework and use it to show the correctness of dynamic slicing. Instead of using graph structures, they base their work on natural semantics. The slicing itself uses annotations where program points with annotation *False* are treated as *Skip*, the same strategy we pursue with our notion of bit vectors. They also present the embedding of three different languages in their framework: an imperative, a logic programming and a functional one. In [16], Ward and Zedan model slicing as a program transformation, thereby abstracting from specific representations. A program transformation in their sense is any operation on a program which generates a semantically equivalent program. The aim of their work is to provide a unified mathematical framework for sequential programs. We think that our approach is closer to the intuitive understanding of PDG based slicing than the last two methods as we use the standard notions of control flow, and control and data dependence. Also, both works rely on pen-and-paper proofs whereas our framework is fully machine-checked.

The notion of control flow graphs can be found in various works on verification using theorem provers, e.g. see [10] and [4], using control flow implicitly, i.e. as a relation, not as a real graph structure. Lammich and Müller-Olm [9] define a parallel flow graph similar in structure to our control flow graph (but formalising procedures and parallelism), which is not restricted to a certain language either. While our work uses flow graphs to construct dependence graphs and to prove certain properties of them, they focus on the correctness of analyses on the flow level.

5   Conclusion and Future Work

We have presented a generic framework for dynamic slicing using a PDG based approach and proven it correct in the proof assistant Isabelle/HOL. This framework is independent from concrete programming language syntax but uses a graph structure fulfilling certain structural and well-formedness properties. Moreover we have presented how to embed a simple imperative While language (without procedures) in this framework to show that the preconditions imposed on the graph structure are sensible. The formalization, including the instantiation with a While language has about 6000 lines of code and took one man-year of work.

Our next goal is to instantiate this framework with more sophisticated object oriented languages like Jinja [7] or CoreC++ [17] and to apply it to a formalization of the JVM [7] with unstructured control flow. Also we are working on expanding the framework to the intricacies of static slicing including, amongst other things, a formalization of system dependence graphs (SDGs) and an approximation of the peculiarities of dynamic binding. In the remote future we plan to extend the formalization to concurrency.

References

1. Agrawal, H., Horgan, J.R.: Dynamic program slicing. In: Proc. of PLDI 1990, pp. 246–256. ACM Press, New York (1990)
2. Amtoft, T.: Slicing for modern program structures: a theory for eliminating irrelevant loops. Information Processig Letters 106(2), 45–51 (2008)
3. Ballarin, C.: Locales and locale expressions in Isabelle/Isar. In: Berardi, S., Coppo, M., Damiani, F. (eds.) TYPES 2003. LNCS, vol. 3085, pp. 34–50. Springer, Heidelberg (2004)
4. Blech, J.O., Gesellensetter, L., Glesner, S.: Formal verification of dead code elimination in Isabelle/HOL. In: Proc. of SEFM 2005, pp. 200–209. IEEE Computer Society Press, Los Alamitos (2005)
5. Gouranton, V., Métayer, D.L.: Dynamic slicing: a generic analysis based on a natural semantics format. Journal of Logic and Computation 9(6), 835–871 (1999)
6. Jhala, R., Majumdar, R.: Path slicing. In: Proc. of PLDI 2005, pp. 38–47. ACM Press, New York (2005)
7. Klein, G., Nipkow, T.: A Machine-Checked Model for a Java-Like Language, Virtual Machine and Compiler. ACM TOPLAS 28(4), 619–695 (2006)
8. Krinke, J.: Program slicing. Handbook of Software Engineering and Knowledge Engineering 3, 307–332 (2004)
9. Lammich, P., Müller-Olm, M.: Precise fixpoint-based analysis of programs with thread-creation and procedures. In: Caires, L., Vasconcelos, V.T. (eds.) CONCUR. LNCS, vol. 4703, pp. 287–302. Springer, Heidelberg (2007)
10. Leroy, X.: Formal certification of a compiler back-end or: programming a compiler with a proof assistant. In: Proc. of POPL 2006, pp. 42–54. ACM Press, New York (2006)
11. Nielson, F., Nielson, H.R., Hankin, C.: Principles of Program Analysis. Springer, Heidelberg (1999)
12. Nipkow, T., Paulson, L.C., Wenzel, M.T.: Isabelle/HOL. LNCS, vol. 2283. Springer, Heidelberg (2002)
13. Ranganath, V.P., Amtoft, T., Banerjee, A., Hatcliff, J., Dwyer, M.B.: A new foundation for control dependence and slicing for modern program structures. ACM TOPLAS 29(5), 27 (2007)
14. Reps, T., Yang, W.: The semantics of program slicing. Technical Report CS-TR-1988-777, University of Wisconsin-Madison (1988)
15. Tip, F.: A survey of program slicing techniques. Journal of Programming Languages 3(3), 121–189 (1995)
16. Ward, M., Zedan, H.: Slicing as a program transformation. ACM TOPLAS 29(2), 1–53 (2007)
17. Wasserrab, D., Nipkow, T., Snelting, G., Tip, F.: An operational semantics and type safety proof for multiple inheritance in C++. In: Proc. of OOPSLA 2006, pp. 345–362. ACM Press, New York (2006)
18. Weiser, M.: Program slices: formal, psychological, and practical investigations of an automatic program abstraction method. PhD thesis, University of Michigan (1979)
19. Wolfe, M.J.: High Performance Compilers for Parallel Computing. Addison-Wesley, Reading (1995)

Proof Pearl: Revisiting the Mini-rubik in Coq

Laurent Théry

Marelle Project INRIA, France
Laurent.Thery@inria.fr

Abstract. The Mini-Rubik is the 2x2x2 version of the famous Rubik's cube. How many moves are required to solve the 3x3x3 cube is still unknown. The Mini-Rubik, being simpler, is always solvable in a maximum of 11 moves. This is the result that is formalised in this paper. From this formalisation, a solver is also derived inside the CoQ prover. This rather simple example illustrates how safe computation can be used to do state exploration in order to derive non-trivial properties inside a prover.

1 Introduction

A recent paper [5] has shown that 26 moves are sufficient to solve Rubik's cube. This is the best-known upper bound (the exact value is conjectured to be around 20 moves). It uses some clever approximation of the problem but relies mainly on heavy parallel computations: 8000 CPU hours are needed to get the result. In this paper, we tackle the more elementary Mini-Rubik. Instead of 26 small cubes, the Mini-Rubik is composed of 8 small cubes only. It is a well-known result that it is always solvable in a maximum of 11 moves [1]. This is the result we formalise in this paper.

In CoQ [8], there is no native data-structure. All basic types such as integer, boolean, string are tree-like structures built using the standard `Inductive` command. For example, the natural numbers use Peano representation with two constructors S and O. The natural number 3 is then internally represented as (S (S (S O))). Although this is perfectly adequate for proofs (for example, the usual inductive principle is given for free), Peano numbers are useless for computing. With a binary representation, the situation gets slightly better but is still not satisfactory. In [7], a generic mechanism is proposed for associating a dedicated data-structure for computing to the standard one for proving while preserving all the nice properties of the type system. This mechanism is applied to integer arithmetic in the following way. First, a special type `Int31` is defined that contains a single constructor with a list of 31 booleans as arguments. This is the reference data-structure. Then, this type is associated to the internal 31-bit OCAML integers[1] in a straightforward manner. So, computing within the `Int31` type directly benefits from the processor arithmetic with the corresponding speed-up. For example, the `Int31` type was used in [9] to get the primality of some large numbers using elliptic curves.

---

[1] In OCAML [6], the last bit of a word indicates if it should be interpreted as either a value or a pointer, the arithmetic has then only 32 - 1 bits.

O. Ait Mohamed, C. Muñoz, and S. Tahar (Eds.): TPHOLs 2008, LNCS 5170, pp. 310–319, 2008.

In this paper, we are going to use the `Int31` type in a slightly different manner. The Mini-Rubik has 3,674,160 possible configurations. In order to get our final result, we need to visit all these configurations while recording those which have already been encountered. It amounts to manipulating subsets of a set of size 3,674,160. In order to use as little memory as possible, we use the `Int31` type to encode subsets of small sets of size 31. We then represent a subset of the configurations with 118,522, i.e. 3,674,160/31, of these `Int31`. This capability of encoding small subsets in a single word is the key aspect that makes our formalisation work.

The paper is organised as follows. In Section 1, we present a naive formalisation that is used to get the basic properties of the problem. In Section 2, we introduce a second formalisation whose main concern is memory consumption. The main result is obtained from this second formalisation. Finally, in Section 3, we give more details about our formalisation.

2 Direct Formalisation

To represent the Mini-Rubik inside COQ, we have chosen a model that is particularly well-suited for the computation we need to perform. If we believe that one can relatively easily get convinced that we have modelled the Mini-Rubik faithfully, ultimately we should also provide a more intuitive model and formally prove the equivalence with our model.

In our model, the front-upper-left corner remains fixed. So, a configuration needs to take into consideration 7 small cubes only. Information about a small cube is split in two: its position and its orientation. Small cubes are numbered from 1 to 7. The cube on the left of Figure 1 shows which ordering has been used to label the small cubes: the fixed cube and cubes 1, 2, 3 compose the front face. Small cubes are represented by elements of the enumerate type `cube`:

```
Inductive cube: Set := C₁ | C₂ | C₃ | C₄ | C₅ | C₆ | C₇.
```

Each small cube is rigid and has 3 coloured faces only. Choosing arbitrarily the vertical direction, in any configuration, a small cube has exactly one face that belongs to either the top face or the bottom face. Knowing the colour of this face is sufficient to deduce the colours of the other faces. This means that a cube has only 3 possible orientations that can be represented by elements of the enumerate type `orientation`:

```
Inductive orientation: Set := O₁ | O₂ | O₃.
```

Note that in the following we really manipulate positions and orientations as if they were natural numbers. We have not been using directly natural numbers for efficiency reason only. With enumerate types, checking if a position is C_7 is done by one elementary pattern matching, while checking if a number is 7, i.e. `(S (S (S (S (S (S (S O)))))))`, requires a more expensive pattern matching.

A configuration of the Mini-Rubik is represented by a constructor `State` with 7 positions and 7 orientations:

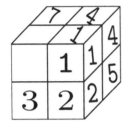 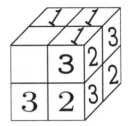

Fig. 1. Positions and Orientations of the Mini-Rubik

```
Inductive state: Set :=
  State (p₁ p₂ p₃ p₄ p₅ p₆ p₇: cube) (o₁ o₂ o₃ o₄ o₅ o₆ o₇: orientation).
```

where p_i indicates which small cube is at position i and o_i gives its orientation. We define the initial orientation in such a way that the initial state of the cube is represented as:

```
Definition init_state = State C₁ C₂ C₃ C₄ C₅ C₆ C₇ O₁ O₁ O₁ O₁ O₁ O₁ O₁.
```

The front-upper-left corner being fixed, there are only three elementary rotations of the cube: right, back and down. Following Figure 1, cubes 1-4-5-2 compose the right face, cubes 4-5-6-7 the back face and cubes 2-5-6-3 the bottom face. Note that our decision to have the front-upper-left corner fixed means that in our model a rotation of the left face (resp. front and up) is simulated by a rotation in the opposite direction of the right face (resp. back and down). Also, half-turns and anti-clockwise rotations are obtained by iterating twice, resp. thrice, the respective elementary rotation. All this is rather standard.

Orientations are the less intuitive part of our model. The cube on the right of Figure 1 tries to explain how orientations work. Orientations are numbered from 1 to 3 following the clockwise order. For each cube, we arbitrarily decide that it is the face that belongs to the top face (or the bottom face) that holds the orientation. So, from the definition of `init_state`, it follows that initially the top and bottom faces contain only 1, i.e O_1.

After a rotation, the orientation of a cube is either unaffected, modified in a clockwise manner, or modified in an anti-clockwise manner. Clockwise and anti-clockwise changes are represented by the functions `up` and `down` respectively:

```
Definition   up o = match o with O₁ ⇒ O₂ | O₂ ⇒ O₃ | O₃ ⇒ O₁ end.
Definition down o = match o with O₁ ⇒ O₃ | O₂ ⇒ O₁ | O₃ ⇒ O₂ end.
```

The three elementary rotations are modelled as functions from `State` to `State`:

```
Definition rright s := match s with
State p₁ p₂ p₃ p₄ p₅ p₆ p₇ o₁ o₂ o₃ o₄ o₅ o₆ o₇ ⇒
State p₂ p₅ p₃ p₁ p₄ p₆ p₇ (up o₂) (down o₅) o₃ (down o₁) (up o₄) o₆ o₇
end.
```

```
Definition rback s := match s with
State p₁ p₂ p₃ p₄ p₅ p₆ p₇ o₁ o₂ o₃ o₄ o₅ o₆ o₇ ⇒
State p₁ p₂ p₃ p₅ p₆ p₇ p₄ o₁ o₂ o₃ (up o₅) (down o₆) (up o₇) (down o₄)
end.
```

```
Definition rdown s := match s with
State p₁ p₂ p₃ p₄ p₅ p₆ p₇ o₁ o₂ o₃ o₄ o₅ o₆ o₇ ⇒
State p₁ p₃ p₆ p₄ p₂ p₅ p₇ o₁ o₃ o₆ o₄ o₂ o₅ o₇
end.
```

Note that our decision to use the top and bottom faces to read orientations is reflected by the fact that the down rotation does not modify any orientation.

A state is reachable if it can be reached from the initial state using the three elementary rotations. This is easily defined inductively by:

```
Inductive reachable: state → Prop :=
  reach₀: reachable init_state
| reach_r: ∀s, reachable s → reachable (rright s)
| reach_b: ∀s, reachable s → reachable (rback s)
| reach_d: ∀s, reachable s → reachable (rdown s).
```

The fact that 11 moves are sufficient to solve the Mini-Rubik is true for the half-turn metric. This means that not only elementary rotations need to be considered but also anti-clockwise rotations and half turns. This is done in the move relation:

```
Definition move (s₁ s₂: state) :=
s₂ = rright s₁ ∨ s₂ = rright (rright s₁) ∨ s₂ = rright (rright (rright s₁))
  ∨ s₂ = rback s₁ ∨ s₂ = rback(rback s₁) ∨ s₂ = rback(rback(rback s₁))
  ∨ s₂ = rdown s₁ ∨ s₂ = rdown(rdown s₁) ∨ s₂ = rdown(rdown(rdown s₁)).
```

Once moves are defined, the reachability in n moves is defined inductively as:

```
Inductive nreachable: nat → state → Prop :=
  nreach₀: nreachable 0 init_state
| nreach_S: ∀ s₁ s₂, nreachable n s₁ → move s₁ s₂ → nreachable (S n) s₂.
```

We also define the property of being reachable in *less than* n moves and the property of being reachable in *exactly* n moves:

```
Definition nlreachable n s := ∃m, m ≤ n ∧ nreachable m s.
Definition nsreachable n s :=
  nreachable n s ∧ ∀m, m < n → ¬ nreachable m s.
```

Now, the theorem we want to prove can be expressed as:

```
Lemma reach11: ∀s, reachable s → nlreachable 11 s.
```

Turning this lemma into a computational problem is quite direct. For each n, we are going to compute the states that are reachable in less than n moves and the states that are reachable in exactly n moves. We represent states by a simple list of states. On such a list, the function `in_states` checks if a state belongs to the list. We first define the list of all possible moves

```
Definition movel :=
  rright :: rright o rright :: rright o rright o rright ::
  rback  ::  rback o rback  ::  rback o  rback o rback  ::
  rdown  ::  rdown o rdown  ::  rdown o  rdown o rdown  :: nil.
```

All the states that are reachable in exactly $n+1$ states are included in the states that are within one move of states that are reachable in exactly n states. This is the basic idea of the algorithm. The function nexts does this computation for a single state:

```
Definition nexts (ps: states * states) s :=
  fold_left
    (fun (ps: states * states) f ⇒
      let (states, nstates) := ps in
      let s₁ := f s in
      if in_states s₁ states then ps else (s₁ :: states, s₁ :: nstates))
    movel ps.
```

where fold_left is the tail recursive version of the usual iterative fold function on lists. For each state s_1 that is one move from s, the nexts function checks if s_1 has already been visited. If not, it is added to the list of visited states (the first element of the pair) and to the list of the new states (the second element of the pair). Finally, to get the states that are reachable in less than n moves and the states that are reachable in exactly n moves, we just need to iterate the nexts function starting from the lists composed of the initial state only:

```
Function iters_aux n (ps: states * states) :=
  match n with
        0 ⇒ ps
  | S n₁ ⇒ let (m,p) := ps in iters_aux n1 (fold_left nexts p (m,nil))
  end.
Definition iters n := iters_aux n (init_state::nil, init_state::nil).
```

It is relatively easy to show that if the second element of the pair returned by (iters n) is the empty list, the Mini-Rubik is solvable in n-1 moves. This is formally stated by the following theorem:

```
Lemma iters_final: ∀n,
  match iters n with
    (_, nil) ⇒ ∀s, reachable s → nlreachable (pred n) s
  |       _ ⇒ True
  end.
```

It is by applying this theorem that we turn the proof of the theorem reach11 into computing (iters 12).

3 Optimising Memory Consumption

The implementation of iters is far too naive to let us prove the reach11 theorem. Computing (iters 5), which involves 12,224 states only, is already impossible inside COQ. Nevertheless, iters is useful as a reference implementation to which our optimised version is going to be proved equivalent. What iters actually does is to compute the diameter of the Cayley graph of the group generated by the three elementary rotations. As explained in [1], having a compact representation in memory of the graph is mandatory to perform this computation. If we go back to how configurations have been encoded, the values of the 14 arguments of State are strongly constrained. First of all, the seven arguments (p_1, ..., p_7) which represent positions must be a permutation of (C_1, C_2, C_3, C_4, C_5, C_6, C_7). Also, the orientation of the last cube can be guessed from the orientations of the other cubes. These constraints are captured by the predicate valid_state:

```
Definition valid_state s := match s with
State c₁ c₂ c₃ c₄ c₅ c₆ c₇ o₁ o₂ o₃ o₄ o₅ o₆ o₇ ⇒
  perm (C₁::C₂::C₃::C₄::C₅::C₆::C₇::nil) (c₁::c₂::c₃::c₄::c₅::c₆::c₇::nil)
  ∧ o₁ ⊕ o₂ ⊕ o₃ ⊕ o₄ ⊕ o₅ ⊕ o₆ ⊕ o₇ = O₁
end.
```

where perm is the permutation predicate between two lists and $\oplus$ is the projection of the addition modulo 3 to the orientation, i.e. adding O_n is done by applying the function up $n - 1$ times. The valid_state predicate is proved to hold for the initial state and to be preserved by the reachability predicate. So, we have:

```
Lemma reachable_valid: ∀s, reachable s → valid_state s.
```

Note that this theorem already indicates that there are at most $7!3^6 = 3{,}674{,}160$ configurations (7! is the contribution of the permutations, and the 3^6 corresponds to the fact that the value of the last orientation is determined by the value of the other orientations). Later, we explain how we formally prove that this is actually the exact number of configurations.

An accurate encoding of permutations of length n should take into consideration the facts that the first element of the permutation has n possible values, the second element $n - 1$ and so on. This is done with the following two functions that manipulate permutations as lists:

```
Function encode_aux l p :=
  match l with
      nil   ⇒ nil
  | m :: l₁ ⇒ (if p <_c m then down_c m else m) :: encode_aux l₁ p
  end.
Function encode l n:=
  match l, n with
    m :: l₁, (S n₁) ⇒ m :: encode (encode_aux l₁ m) n₁
  |   _   ,   _    ⇒ nil
  end.
```

where $<_c$ and down$_c$ are the projections of the comparison and the predecessor functions from the natural numbers to the enumerate type cube, i.e. $C_2 <_c C_3$ and down$_c$ C_3 = C_2. For the definition of the encode function, as the recursion is not structural, an extra argument n is required to ensure termination. It bounds the length of the resulting list. With this encoding on a permutation of length n, the ith element is ensured to be in $\{C_1, C_2, \ldots, C_{n-i+1}\}$. In particular, the last element is always C_1 and can be discarded. If n is the length of the permutation we want to encode, the extra argument of the encode function is $n - 1$. For example, if we consider the permutation of the initial configuration, its encoding is computed by (encode ($C_1::C_2::C_3::C_4::C_5::C_6::C_7::$nil) 6) and evaluates to $C_1::C_1::C_1::C_1::C_1::C_1::$nil. To sum up, the information about positions in a state can be encoded by 6 elements of type cube (p_1', p_2', p_3', p_4', p_5', p_6') with $p_i' \in \{C_1, C_2, \ldots, C_{8-i}\}$ and the information about orientations can be encoded by 6 elements of type orientation (o_1', o_2', o_3', o_4', o_5', o_6'). Furthermore, as the iters function intensively uses encoding and decoding of permutations, we actually use co-inductive types, which are evaluated lazily, to get the memoisation of these operations. This speeds up our computation by a factor of 2.

We use decision trees to represent sets of states. The 12 elements that encode a state (p_1', p_2', p_3', p_4', p_5', p_6', o_1', o_2', o_3', o_4', o_5', o_6') are used to denote a path to a boolean leaf in the decision tree. If this leaf is true, the state is in the set. In COQ, a constructor with n arguments allocates $(n + 1)$ 32-bit words. It is then better to have the elements with the largest number of arguments at the bottom of the tree structure. This is why the reordering of the path (p_6', o_1', o_2', o_3', o_4', o_5', o_6', p_5', p_4', p_3', p_2', p_1') is favoured. Furthermore, instead of boolean leaves, we can use elements of the Int31 type to encode sets of 31 elements with the usual convention that the ith element of the set is present if and only if the ith bit is set to one. In our path, p_2' has 6 possible values and p_3' has 5 possible values, this means that the pair (p_2', p_3') has 30 possible values which can be effectively represented by a single Int31 element (a single bit is then unused). The actual path that is used is then (p_6', o_1', o_2', o_3', o_4', o_5', o_6', p_5', p_4', p_1', $5(\text{index}(p_2') - 1) + \text{index}(p_3') - 1$) where index is the function that maps elements of the type cube to natural numbers, i.e. $\text{index}(C_5) = 5$. With this encoding, the last element of a path is always a natural number strictly less than 30. Two functions encode_state and decode_state are defined to relate states and paths and their composition is proved to be the identity on valid states. With this representation, a set of states requires a maximum of 295,001 32-bit words which means 2.6 bits per configuration.

It is also possible to derive a solver by slightly modifying the iters program. This follows from the observation that, given a state s that is reachable in n moves, the states which are one move from s are reachable in $n - 1$, n, or $n + 1$ moves. Out of all these states, a solver just needs to be capable to pick one state that is reachable in $n - 1$ move. Since for any n, $(n - 1) \bmod 3$, $n \bmod 3$, and $(n + 1) \bmod 3$ are always 3 distinct values, it is sufficient to be able to associate for each state s the two-bit value $n \bmod 3$ where n is the number of moves that are necessary to reach s. For this, we just need to split in two the states that are reachable in less than n moves to get the two-bit information. The modified version iters$_2$ of the function iters for the solver is the following:

```
Definition next₂s m (ps: states * states * states) s :=
  fold_left
    (fun (ps: state * states * states) f ⇒
      let (states₁, states₂, nstates) := ps in
      let s₁ := f s in
      if (in_states s₁ states₁ || in_states s₁ states₂) then ps
      else match m with
             0 ⇒ (s₁::states₁, states₂, s₁::nstates)
           | 1 ⇒ (states₁, s₁::states₂, s₁::nstates)
           | _ ⇒ (s₁::states₁, s₁::states₂, s₁::nstates)
           end)
    movel ps.
Function iter₂s_aux (n m: nat) (ps: states * states * states) :=
  match n with
        0 ⇒ ps
      | S n1 ⇒ let (ps₁,ps₂,ps₃) := ps in
      iter₂s_aux n1 ((m+1) mod 3) (fold_left (next₂s m) ps₃ (ps₁,ps₂,nil))
  end.
Definition iter₂s n :=
  iter₂s_aux n 1 (init_state::nil, nil, init_state::nil).
```

4 Running the Solver

The complete formalisation is available at

ftp://ftp-sop.inria.fr/marelle/Laurent.Thery/Rubik.zip

It is composed of 7000 lines of code: 3000 lines for the naive formalisation, 4000 for the optimised version. On a Pentium 4 with 1 Gigabyte of RAM, getting the **reach11** theorem takes 260 seconds. Most of the time is spent in computing (iters 12). Note that, once this computation has been performed, it can also be used to get another interesting result:

```
Lemma valid11: ∀s, valid_state s → reachable s.
```

This proves that the number of configurations of the Mini-Rubik is exactly 3,674,160. This is done by checking that the first element of the pair computed by (iters 12) with the optimised version has all its leaves equal to $2^{30} - 1$.

The solver returns the list of moves in the half-turn metric that leads to the initial state. We use co-inductive types and memoisation to compute only once the table that associates each state with its index of reachability modulo 3. So, the first time the solver is called, the table is actually computed:

```
Time Eval compute in solve init_state.
   = nil
Finished transaction in 384. secs (384.562537u,0.292956s)
```

The next invocations are then immediate. For example, we can try to swap two adjacent corners

```
Time Eval compute in solve (State C₂ C₁ C₃ C₄ C₅ C₆ C₇ O₁ O₁ O₁ O₁ O₁ O₁ O₁).
  = Right::Back⁻¹::Down²::Right⁻¹::Back::Right⁻¹::Back⁻¹::Right::Down²::
    Right::Back::nil
Finished transaction in 0. secs (0.00100000000009u,0.s)
```

or two opposite corners

```
Time Eval compute in solve (State C₇ C₂ C₃ C₄ C₅ C₆ O₁ O₁ O₁ O₁ O₁ O₁ O₁ 1).
  = Right::Back⁻¹::Right²::Back⁻¹::Right⁻¹::Down⁻¹::Right::Down²::Back::
    Down⁻¹::Back::nil
Finished transaction in 0. secs (0.00100000000009u,0.s)
```

5 Conclusions

Proof systems like COQ are not well-suited for dealing with state exploration. Mike Gordon has already shown in [3] how one can benefit from an external link to a BDD package to solve a solitaire game inside the HOL prover [4]. In our work, everything has been done withing the theorem prover using safe computation. The main contribution of this paper is to show that we can actually use this safe computation to effectively model problems of relatively large size like the Mini-Rubik. As in [3], what we really gain by doing this inside a prover is the formal connection between what we want to prove (the model) and what we actually compute.

The key aspect of the formalisation is its memory consumption. Most of the issues we have addressed here is not specific to theorem proving and can also be found in the model checking community. For example, in [10], the author shows how a careful design is necessary to be able to solve this problem with BDDs. Having a certified formalisation in a purely functional setting that uses 2.6 bits only per configuration is rather satisfactory. The 260 seconds to complete the exploration are less satisfactory but it is difficult to see how we could go significantly faster in a programming language without side-effects. Finally, if our decision trees are for the moment ad-hoc for the specific configurations of the Mini-Rubik, deriving a generic library that uses `Int31` to represent large finite sets could be useful for other formalisations.

A natural continuation of this work would be to tackle the full Rubik's cube. Obviously, formalising results like [5] is outside reach but getting simpler bounds like the one of 52 moves [2] seems feasible.

References

1. Cooperman, G., Finkelstein, L.: New Methods for Using Cayley Graphs in Interconnection Networks. Discrete Applied Mathematics 37(38), 95–118 (1992)
2. Frey, A.H., Singmaster, D.: Handbook of Cubik Math. Enslow Publishers (1982)
3. Gordon, M.J.C.: Reachability Programming in HOL98 using BDDs. In: Aagaard, M.D., Harrison, J. (eds.) TPHOLs 2000. LNCS, vol. 1869, pp. 179–196. Springer, Heidelberg (2000)

4. Gordon, M.J.C., Melham, T.F.: Introduction to HOL: a theorem proving environment for higher-order logic. Cambridge University Press, Cambridge (1993)
5. Kunkle, D., Cooperman, G.: Twenty-Six Moves Suffice for Rubik's Cube. In: ISSAC 2007, pp. 235–242 (2007)
6. Leroy, X.: Objective Caml (1997), `http://pauillac.inria.fr/ocaml/`
7. Spiwack, A.: Efficient Integer Computation in Type Theory, Draft paper (2007)
8. The Coq development team. The Coq Proof Assistant Reference Manual v7.2. Technical Report 255, INRIA (2002), `http://coq.inria.fr/doc`
9. Théry, L., Hanrot, G.: Primality proving with elliptic curves. In: Schneider, K., Brandt, J. (eds.) TPHOLs 2007. LNCS, vol. 4732, pp. 319–333. Springer, Heidelberg (2007)
10. Valmari, A.: What the small Rubik's cube taught me about data structures, information theory, and randomisation. International Journal for Software Tools Technology 8(3), 180–194 (2006)

Author Index

Lecture Notes in Computer Science

Sublibrary 1: Theoretical Computer Science and General Issues

For information about Vols. 1– 4854
please contact your bookseller or Springer